Lecture Notes in Mathematics 2097

For further volumes:
http://www.springer.com/series/304

Tatsuo Nishitani

Hyperbolic Systems with Analytic Coefficients

Well-posedness of the Cauchy Problem

Springer

Tatsuo Nishitani
Department of Mathematics
Graduate School of Science
Osaka University
Toyonaka, Osaka, Japan

ISBN 978-3-319-02272-7 ISBN 978-3-319-02273-4 (eBook)
DOI 10.1007/978-3-319-02273-4
Springer Cham Heidelberg New York Dordrecht London

Lecture Notes in Mathematics ISSN print edition: 0075-8434
ISSN electronic edition: 1617-9692

Library of Congress Control Number: 2013955050

Mathematics Subject Classification (2010): 35L45, 35L40, 35L55

Printed on acid-free paper

Springer is part of Springer Science+Business Media (www.springer.com)

Preface

In this monograph we discuss the C^∞ well-posedness of the Cauchy problem for hyperbolic systems. We are mainly concerned with the following two questions for differential operators of order q with smooth $m \times m$ matrix coefficients:

(A) Under which conditions on lower order terms is the Cauchy problem C^∞ well posed?
(B) When is the Cauchy problem C^∞ well posed for any lower order term?

For scalar case, that is $m = 1$, the question (B) has been answered. As for the question (A), in particular for second order scalar equations, that is $m = 1$ and $q = 2$, so many works are devoted to this question and the situation is fairly well understood. Contrary to the scalar case, for systems that is if $m \geq 2$ we have no satisfactory result.

Even for differential operators with characteristics of constant multiplicity with real analytic matrix coefficients, the question (A) has been solved very recently.

So in this monograph, assuming that the coefficients are real analytic in a neighborhood of the origin, we study these two questions. Of course this analyticity assumption is rather restrictive but which allows us to make detailed studies on the Cauchy problem. We hope that this study can throw light on the studies of the Cauchy problem for hyperbolic systems with less regular, in particular C^∞ coefficients.

The contents are organized as follows. In Chap. 1 after giving the definition of C^∞ well-posedness of the Cauchy problem we show that the Cauchy problem for symmetric hyperbolic systems is C^∞ well posed for any lower order term. Then we give an example of first order 2×2 system which is not symmetrizable but for which the Cauchy problem is C^∞ well posed for any lower order term. Actually there is a class of non-symmetrizable systems for which the Cauchy problem is C^∞ well posed for any lower order term. This is a main objection when we try to answer to the problem (B). We prove the Lax-Mizohata theorem exhibiting naive ideas which are used in Chaps. 2 and 3. For first order systems with characteristics of constant multiplicities, the necessity of the Levi condition for the C^∞ well-posedness is proved which is used in Chap. 2. In Chap. 2, we study necessary conditions about the problem (B) for $m \times m$ first order systems with real analytic coefficients. We prove rather general necessary conditions in terms of minors of the principal symbols. Contrary to the scalar case the multiplicity of characteristics is irrelevant for the problem (B) since for symmetric or symmetrizable hyperbolic systems (of first order) the Cauchy problem is always C^∞ well posed for any lower order term. Here the maximal size of the Jordan blocks, which is supposed to measure the distance from diagonal matrices, plays an important role in the problem (B).

In Chap. 3, we study two questions (A) and (B) for first order 2×2 systems with two independent variables with real analytic coefficients. For this special case we can give a necessary and sufficient condition for the questions (A) and (B), that is, in this case we have complete answer for (A) and (B). The results provide many instructive examples. For instance, we can exhibit a first order 2×2 system with analytic coefficients which is strictly hyperbolic outside the initial line for which no lower order term could be taken so that the Cauchy problem is C^∞ well posed. This cannot happen for second order hyperbolic scalar operators with two independent variables with analytic coefficients.

In Chap. 4, we introduce a new class of hyperbolic systems, that is hyperbolic systems with nondegenerate characteristics which generalizes strictly hyperbolic systems. Strictly hyperbolic systems are hyperbolic systems with nondegenerate characteristics of order one. The theory of strictly hyperbolic systems is rich, but first order strictly hyperbolic system hardly exists. We prove that the Cauchy problem for hyperbolic systems with nondegenerate characteristics is C^∞ well posed for any lower order term. We also show that nondegenerate characteristics are stable, that is any hyperbolic system which is close to a hyperbolic system with a nondegenerate characteristic of order r has a nondegenerate characteristic of the same order nearby. This shows, in particular, that near any hyperbolic system with a nondegenerate characteristic of order $r \geq 2$ there is no strictly hyperbolic system, which gives a great difference from the scalar case and shows a complexity of hyperbolic systems.

We also discuss hyperbolic systems which are perturbations of symmetric systems and prove that if the dimension of the linear space that the symbol of the symmetric system spans is large enough, then generically such hyperbolic system is similar to a symmetric system.

Osaka, Japan
March 2013

Tatsuo Nishitani

Contents

Chapter 1
Introduction

Abstract In this chapter we show that the Cauchy problem for symmetric hyperbolic systems is C^∞ well posed for any lower order term proving the existence of solutions and the bound of the supports at the same time, by steering a course somewhat close to the boundary value problems rather than the initial value problems. We give an example, which would be the simplest one, is not symmetrizable but the Cauchy problem is C^∞ well posed for any lower order term. We show the well-posedness by the classical method of characteristic curves. We also give a proof of the Lax-Mizohata theorem exhibiting naive ideas to construct an asymptotic solution to systems which will be used in Chaps. 1 and 2 in a more involved way. The Levi condition for first order systems with characteristics of constant multiplicities is discussed in a somewhat intermediate form. We use this for proving more general result in Chap. 2.

1.1 Well-Posedness of the Cauchy Problem

Let us study a differential operator of order d with $m \times m$ matrix coefficients

$$P(x, D) = \sum_{|\alpha| \le d} A_\alpha(x) D^\alpha, \quad D_j = \frac{1}{i}\frac{\partial}{\partial x_j}$$

where $A_\alpha(x)$ are $m \times m$ matrix valued smooth function defined in a neighborhood Ω of the origin of $\mathbb{R}^{n+1}$ with a system of coordinates $x = (x_0, x_1, \ldots, x_n) = (x_0, x')$. We assume that $x_0 = const.$ are non characteristic and then without restrictions we can assume that

$$A_{(d,0,\ldots,0)}(x) = I$$

T. Nishitani, *Hyperbolic Systems with Analytic Coefficients*, Lecture Notes in Mathematics 2097, DOI 10.1007/978-3-319-02273-4_1,

where I denotes the $m \times m$ identity matrix. Let $P_d(x,\xi)$ be the principal symbol of $P(x,D)$

$$P_d(x,\xi) = \sum_{|\alpha|=d} A_\alpha(x)\xi^\alpha.$$

We use the following notations.

Notations:

- $\mathscr{S}(\mathbb{R}^n)$: Schwartz space on $\mathbb{R}^n$.
- $\mathscr{S}'(\mathbb{R}^n)$: the space of tempered distributions on $\mathbb{R}^n$.
- $\mathscr{A}(W)$: the set of real analytic functions in an open set W.
- $C^\infty(W, M_m(\mathbb{C}))$: the set of $m \times m$ matrices with $C^\infty(W)$ entries.
- $\mathscr{A}(W, M_m(\mathbb{C}))$: the set of $m \times m$ matrices whose entries are real analytic in W.
- If $\omega \subset \mathbb{R}^n$ is open and $s \in \mathbb{N}$, then

$$H^s(\omega) = \{u \in L^2(\omega) \mid D^\alpha u \in L^2(\omega), \forall |\alpha| \le s\}.$$

- Let $s \in \mathbb{R}$. Then

$$H^s(\mathbb{R}^n) = \{u \in \mathscr{S}'(\mathbb{R}^n) \mid \langle\xi'\rangle^s \hat{u}(\xi') \in L^2(\mathbb{R}^n)\}, \quad \|u\|_s = \|\langle\xi'\rangle^s \hat{u}(\xi')\|_{L^2(\mathbb{R}^n)}$$

where $\langle\xi'\rangle^2 = 1 + |\xi'|^2$ and $\hat{u}(\xi')$ is the Fourier transform of u with respect to x'.

We start with

Definition 1.1. The Cauchy problem for $P(x,D)$ is said to be C^∞ well posed in the future near the origin if there exist $\epsilon > 0$ and a neighborhood ω of the origin such that; for any $|\tau| \le \epsilon$ and for any $f(x) \in C_0^\infty(\omega)$ vanishing in $x_0 < \tau$ there is a unique solution $u(x) \in H^\infty(\omega)$ to $Pu = f$ in ω vanishing in $x_0 < \tau$, where $H^\infty(\omega) = \cap_{p=0}^\infty H^p(\omega)$. If for any $|\tau| \le \epsilon$ and for any $f(x) \in C_0^\infty(\omega)$ vanishing in $x_0 > \tau$ there is a unique solution $u(x) \in H^\infty(\omega)$ to $Pu = f$ in ω vanishing in $x_0 > \tau$ we say that the Cauchy problem is C^∞ well posed in the past near the origin.

Assume that the Cauchy problem for P is C^∞ well posed in the future near the origin. If $u \in H^\infty(\omega)$, vanishing in $x_0 < \tau$ with $|\tau| < \epsilon$, satisfies $Pu = 0$ in $x_0 < t$ ($\tau < t$, $|t| < \epsilon$) then we can conclude that $u = 0$ in $x_0 < t$. To see this, note that the equation $Pw = Pu$ has a solution $w \in H^\infty(\omega)$ vanishing in $x_0 < t$. Since $w - u = 0$ in $x_0 < \tau$ and $P(w-u) = 0$, by the uniqueness we get $w = u$ and hence $u = 0$ in $x_0 < t$.

Lemma 1.1. *Assume that the Cauchy problem for P is C^∞ well posed in the future near the origin. Then for any $U \subset\subset \omega$ the following classical Cauchy problem has a unique solution $u \in H^\infty(U)$*

$$\begin{cases} Pu = f \quad in \quad U \cap \{x_0 > \tau\}, \\ D_0^j u = u_j \quad on \quad U \cap \{x_0 = \tau\}, \quad j = 0, 1, \dots, d-1 \end{cases} \tag{1.1}$$

for any given $f(x) \in C_0^\infty(\omega)$ and $u_j(x') \in C_0^\infty(\omega \cap \{x_0 = \tau\})$.

Proof. Since $x_0 = \tau$ is non characteristic we can compute $u_j(x') = D_0^j u(\tau, x')$ for $j = d, d+1, \dots$ from $u_j(x')$, $j = 0, \dots, d-1$ and the equation $Pu = f$. By a Borel's lemma we can take $\hat{u} \in C_0^\infty(\omega)$ such that $D_0^j \hat{u}(\tau, x') = u_j(x')$ on $U \cap \{x_0 = \tau\}$ for all $j \in \mathbb{N}$. Clearly we have $D_0^j(P\hat{u} - f) = 0$ on $U \cap \{x_0 = \tau\}$ for all $j \in \mathbb{N}$. The function g, defined by $g = P\hat{u} - f$ in $x_0 > \tau$ and zero in $x_0 < \tau$ is in $C_0^\infty(\omega)$. By assumption there exists $v \in H^\infty(\omega)$ such that $Pv = g$ in ω and $v = 0$ in $x_0 < \tau$. This shows that

$$\begin{cases} P(\hat{u} - v) = f \quad \text{in} \quad \omega \cap \{x_0 > \tau\}, \\ D_0^j(\hat{u} - v) = u_j(x') \quad \text{on} \quad U \cap \{x_0 = \tau\} \end{cases}$$

so that $\hat{u} - v \in H^\infty(\omega)$ is a desired solution to (1.1). □

Definition 1.2. Let $0 \in \Omega$. We say that P or $P_d(x, \xi)$ is strongly hyperbolic near the origin if the Cauchy problem for $P(x, D) + Q(x, D)$ is C^∞ well posed both in the future and the past near the origin for any differential operator $Q(x, D)$ of order at most $d - 1$ with $C^\infty(\Omega, M_m(\mathbb{C}))$ coefficients. Here it is assumed that the open neighborhood ω in Definition 1.1 can be chosen independently of Q.

Here we state some consequences of the C^∞ well-posedness.

Proposition 1.1. *Assume that the Cauchy problem for P is C^∞ well posed near the origin in the future. Then there are open neighborhood ω of the origin and $\epsilon > 0$ such that; for any compact set $K \subset\subset \omega$ and $p \in \mathbb{N}$ there exist $C > 0$, $q \in \mathbb{N}$ such that*

$$\|u\|_{H^p(K^t)} \le C \|Pu\|_{H^q(K^t)}$$

for any $u \in C_0^\infty(K_{-\epsilon})$ and any $|t| < \epsilon$ where $K^t = \{x \in K \mid x_0 \le t\}$ and similarly $K_t = \{x \in K \mid x_0 \ge t\}$.

Remark. Recall that for $u \in C_0^\infty(K_{-\epsilon}) \subset C_0^\infty(\mathbb{R}^{n+1})$ we define

$$\|u\|_{H^p(K^t)} = \inf \|U\|_{H^p(\mathbb{R}^{n+1})}$$

where the infimum is taken over all $U \in H^p(\mathbb{R}^{n+1})$ equal to u in $\{x \in \mathbb{R}^{n+1} \mid x_0 \le t\}$.

Proof. Let ω be the open set in Definition 1.1. Take an open set V so that $K \subset\subset V \subset\subset \omega$. Let us define F_M, $M = 1, 2, \dots$ by

$$F_M = \{f \in C_0^\infty(\overline{V_{-\epsilon}}) \mid \exists u \in H^p(\omega) \text{ such that}$$
$$Pu = f \text{ in } \omega, \ \|u\|_{H^p(\omega)} \le M, \ u = 0 \text{ in } x_0 \le -\epsilon\}.$$

From the C^∞ well-posedness it is clear that

$$\bigcup_{M=1}^{\infty} F_M = C_0^\infty(\overline{V_{-\epsilon}}).$$

It is also clear that F_M is symmetric, that is $-u \in F_M$ if $u \in F_M$ and convex. Let $F_M \ni f_j \to f$ in $C_0^\infty(\overline{V_{-\epsilon}})$. Then there exist u_j such that $Pu_j = f_j$ with $\|u_j\|_{H^p(\omega)} \le M$, taking a subsequence, we may assume that

$$u_j \to u \text{ locally in } H^{p-1}(\omega) \text{ and } u_j \to u \text{ weak in } H^p(\omega),\ u \in H^p(\omega).$$

It is clear that $Pu = f$ in ω and $u = 0$ in $x_0 \le -\epsilon$. This shows that F_M is closed. Since $C_0^\infty(\overline{V_{-\epsilon}})$ is a complete metric space then from the Baire's category theorem some F_M contains a neighborhood of 0 in $C_0^\infty(\overline{V_{-\epsilon}})$. That is, there is $q \ge 0$ and $\delta > 0$ such that

$$f \in C_0^\infty(\overline{V_{-\epsilon}}),\ \|f\|_{H^q(V)} \le \delta \Longrightarrow f \in F_M.$$

Thus for any $f \in C_0^\infty(\overline{V_{-\epsilon}})$ we see that $\delta f/\|f\|_{H^q(V)} \in F_M$. This shows that for any $f \in C_0^\infty(\overline{V_{-\epsilon}})$ there exists a solution to $Pu = f$ in ω vanishing in $x_0 \le -\epsilon$ which satisfies

$$\|u\|_{H^p(\omega)} \le M\delta^{-1}\|f\|_{H^q(V)}. \tag{1.2}$$

Let $u \in C_0^\infty(K_{-\epsilon})$ and take $\chi \in C_0^\infty(V)$ so that $\chi = 1$ on K. Let $g \in \mathscr{S}(\mathbb{R}^{n+1})$ be such that $g = Pu$ in $x_0 < t$. Then the solution to $Pv = \chi g$ vanishing in $x_0 \le -\epsilon$ coincides with u in $x_0 < t$ as observed after Definition 1.1 and hence

$$\|v\|_{H^p(V^t)} = \|u\|_{H^p(V^t)} \le C_0\|\chi g\|_{H^q(V)} \le C_0'\|g\|_{H^q(\mathbb{R}^{n+1})}.$$

Since this holds for any $g \in \mathscr{S}(\mathbb{R}^{n+1})$ provided $g = Pu$ in $x_0 < t$, this proves

$$\|u\|_{H^p(V^t)} \le C_0'\|Pu\|_{H^q(\{x_0<t\})} = C_0'\|Pu\|_{H^q(K^t)}$$

and hence the assertion. □

Corollary 1.1. *Assume that the Cauchy problem for P is C^∞ well posed in the future near the origin. Then there are a neighborhood ω and $\epsilon > 0$ such that for any compact set $K \subset\subset \omega$ one can find $C > 0$ and $p \in \mathbb{N}$ such that*

$$|u|_{C^0(K^t)} \le C|Pu|_{C^p(K^t)}$$

for any $u \in C_0^\infty(K_{-\epsilon})$ and any $|t| < \epsilon$ where $|u|_{C^p(K)} = \sup_{x\in K, |\alpha|\le p} |\partial_x^\alpha u(x)|$.

Proof. By the Sobolev embedding theorem. □

One can refine Corollary 1.1. Let $\sigma = (\sigma_0, \sigma_1, \ldots, \sigma_n)$, $\delta = (\delta_0, \delta_1, \ldots, \delta_n) \in \mathbb{Q}_+^{n+1}$ and we say $\sigma \geq \delta$ if $\sigma_j \geq \delta_j$ for every j where $\mathbb{Q}_+$ stands for the set of positive rational numbers. For an $m \times m$ matrix differential operator P defined near the origin with C^∞ coefficients we set with $y \in \mathbb{R}^{n+1}$

$$P_\lambda(y, x, \xi) = P(\lambda^{-\delta} y + \lambda^{-\sigma} x, \lambda^\sigma \xi),$$

where λ is a large positive parameter and $\lambda^{-\sigma} x = (\lambda^{-\sigma_0} x_0, \ldots, \lambda^{-\sigma_n} x_n)$ and $\lambda^\sigma \xi = (\lambda^{\sigma_0} \xi_0, \ldots, \lambda^{\sigma_n} \xi_n)$.

From this corollary we get a priori estimates for $P_\lambda(y, x, D)$.

Proposition 1.2. *Let $\sigma, \delta \in \mathbb{Q}_+^{n+1}$ and $\sigma \geq \delta$. Assume that the Cauchy problem for $P(x, D)$ is C^∞ well posed both in the future and the past near the origin. Then for every compact sets W, $V \subset \mathbb{R}^{n+1}$ and $T > 0$ there are positive constants C, $\bar{\lambda}$ and $p \in \mathbb{N}$ such that*

$$|u|_{C^0(W^t)} \leq C \lambda^{\bar{\sigma} p} |P_\lambda u|_{C^p(W^t)}, \quad |u|_{C^0(W_t)} \leq C \lambda^{\bar{\sigma} p} |P_\lambda u|_{C^p(W_t)}$$

with $\bar{\sigma} = \max_j \sigma_j$ for any $u \in (C_0^\infty(W))^m$, $\lambda \geq \bar{\lambda}$, $y \in V$, $|t| < T$.

Proof. Let Φ be the map : $z \to \lambda^\sigma z + \lambda^{\sigma - \delta} y$. Then it is clear that there is a compact set $K \subset \Omega$ such that $\Phi^* u(z) = u(\lambda^\sigma z - \lambda^{\sigma-\delta} y) \in (C_0^\infty(K))^m$ for $u(x) \in (C_0^\infty(W))^m$, $y \in V$ when λ is large. Assume that the Cauchy problem for P is C^∞ well posed in the future near the origin. Then from Corollary 1.1 we have

$$|\Phi^* u|_{C^0(K^t)} \leq C |\Phi^*(P_\lambda u)|_{C^p(K^t)}, \ |t| < \tau.$$

On the other hand we have $|\lambda^{-\sigma_0} s + \lambda^{-\delta_0} y_0| < \tau$, if $|s| < T$, $y \in V$ when λ is large. Thus we get

$$|u|_{C^0(W^s)} \leq C \lambda^{\bar{\sigma} p} |P_\lambda u|_{C^p(W^s)}$$

which is the desired first assertion. To prove the second assertion it is enough to repeat the same arguments. □

1.2 Symmetric Hyperbolic Systems

We show that first order symmetric hyperbolic systems, which are the most important hyperbolic systems, are strongly hyperbolic (see [14, 15]). Let us consider

$$Pu = \sum_{j=0}^{n} A_j(x) D_j u + B(x) u = f$$

where it is assumed that $A_j(x)$ are symmetric and $A_0(x)$ is definite. Such systems are called symmetric hyperbolic system with respect to the hyperplane $x_0 = const.$ We study the forward Cauchy problem: for any given f which vanishes in $x_0 < a$ we look for u verifying the following

$$\begin{cases} Pu = f, \quad f = 0 \ \text{in} \ x_0 < a, \\ u = 0 \quad \text{in} \ x_0 < a. \end{cases}$$

In what follows, for simplicity, we assume that $A_j(x)$, $B(x)$ are $C^\infty(\mathbb{R}^{n+1})$ with bounded derivatives of all orders in $\mathbb{R}^{n+1}$. Replacing P by $-P$ we can assume that $A_0(x)$ is positive definite. Moreover considering $A_0(x)^{-1/2}PA_0(x)^{-1/2}$ one can assume that $A_0(x) = I$ since the unique solvability for P is equivalent to that of $A_0(x)^{-1/2}PA_0(x)^{-1/2}$.

We introduce the following function spaces [21, Appendix B.1].

Definition 1.3. For any $m, s \in \mathbb{R}$ we set

$$H_{(m,s)}(\mathbb{R}^{n+1}) = \{u \in \mathscr{S}'(\mathbb{R}^{n+1}) \mid \int |\hat{u}(\xi)|^2 \langle\xi\rangle^{2m} \langle\xi'\rangle^{2s} d\xi < +\infty\}$$

where $\langle\xi\rangle^2 = 1 + |\xi|^2 = 1 + \xi_0^2 + |\xi'|^2 = \xi_0^2 + \langle\xi'\rangle^2$ and $\hat{u}(\xi)$ denotes the Fourier transform of u with respect to x

$$\hat{u}(\xi) = (2\pi)^{-(n+1)/2} \int e^{-ix\xi} u(x) dx.$$

The norm $\|u\|_{(m,s)}$ is given by

$$\|u\|_{(m,s)} = \|\langle\xi\rangle^m \langle\xi'\rangle^s \hat{u}(\xi)\|_{L^2(\mathbb{R}^{n+1})}.$$

We note [21, Appendix B.1]

$$H_{(m_1,s_1)} \subset H_{(m_2,s_2)} \Longleftrightarrow m_2 \le m_1 \text{ and } m_2 + s_2 \le m_1 + s_1. \tag{1.3}$$

Lemma 1.2. *Let $u \in H_{(1,l)}(\mathbb{R}^{n+1})$. Then we have*

$$\big|\|u(t_1,\cdot)\|_l^2 - \|u(t_2,\cdot)\|_l^2\big| \le \int_{t_1}^{t_2} \|D_0 u(s,\cdot)\|_l^2 ds + \int_{t_1}^{t_2} \|u(s,\cdot)\|_l^2 ds.$$

Proof. Let $u \in \mathscr{S}$ and put $\langle D'\rangle^l u = w$ where

$$\langle D'\rangle^l u = (2\pi)^{-(n+1)/2} \int e^{ix\xi} \hat{u}(\xi) \langle\xi'\rangle^l d\xi.$$

Note $w \in \mathscr{S}$ and

$$\begin{aligned}
\|w(t_1,\cdot)\|^2 - \|w(t_2,\cdot)\|^2 &= \int_{t_1}^{t_2} \frac{d}{dx_0}\|w(s,\cdot)\|^2 ds \\
&= \int_{t_1}^{t_2}\int \frac{d}{dx_0}w(s,x')\overline{w(s,x')} + w(s,x')\overline{\frac{d}{dx_0}w(s,x')}dsdx' \\
&\le 2\int_{t_1}^{t_2}\int |\frac{d}{dx_0}w(s,x')||w(s,x')|dsdx' \\
&\le \int_{t_1}^{t_2} \|D_0u(s,\cdot)\|_l^2 ds + \int_{t_1}^{t_2} \|u(s,\cdot)\|_l^2 ds
\end{aligned}$$

which proves the assertion for $u \in \mathscr{S}$. Let $u \in H_{(1,l)}(\mathbb{R}^{n+1})$. Since $\mathscr{S}$ is dense in $H_{(m,s)}$ we take $u_\epsilon \in \mathscr{S}$ so that $u_\epsilon \to u$ in $H_{(1,l)}$. Then it is clear that

$$\int_{t_1}^{t_2} \|D_0u_\epsilon(s,\cdot)\|_l^2 ds \to \int_{t_1}^{t_2} \|D_0u(s,\cdot)\|_l^2 ds$$

as $\epsilon \to 0$ because

$$\|u\|_{(1,l)}^2 = \int \|D_0u(s,\cdot)\|_l^2 ds + \int \|u(s,\cdot)\|_{l+1}^2 ds. \tag{1.4}$$

On the other hand the mapping

$$H_{(1,l)} \ni u \mapsto u(t,\cdot) \in H^{l+1/2}(\mathbb{R}^n)$$

is continuous [21, Appendix B.1] then letting $\epsilon \downarrow 0$ we get the desired assertion for $u \in H_{(1,l)}$. □

Lemma 1.3. *Let $u \in H_{(1,l)}(\mathbb{R}^{n+1})$. Then we have*

$$\lim_{t\to\pm\infty} \|u(t,\cdot)\|_l^2 = 0.$$

Proof. Note that the limits exist by Lemma 1.2. Since

$$\int \|u(s,\cdot)\|_l^2 ds < +\infty$$

the assertion is clear. □

Remark. Note that the assertion holds even for $l < 0$.

Suppose that f with $\operatorname{supp} f \subset \{x_0 > a\}$ is given. We look for u satisfying

$$Pu = f, \quad \operatorname{supp} u \subset \{x_0 > a\}$$

in the form $e^{\gamma x_0}u$. Since

$$e^{-\gamma x_0}Pe^{\gamma x_0} = P(x, D_0 - i\gamma, D')$$

then we have

$$P(x, D_0 - i\gamma, D')u = e^{-\gamma x_0} f \Longrightarrow P(e^{\gamma x_0}u) = f$$

where $D' = (D_1, \dots, D_n)$. Let us set

$$P_\gamma(x, D) = P(x, D_0 - i\gamma, D').$$

Lemma 1.4. *For any $l \in \mathbb{R}$ there is γ_l such that*

$$\int_{-\infty}^{t} \|u(s,\cdot)\|_l^2 ds \le \int_{-\infty}^{t} \|P_\gamma u(s,\cdot)\|_l^2 ds$$

holds for any $u \in H_{(1,l)}$ and $\gamma \ge \gamma_l$.

Proof. Let $u, v \in \mathscr{S}(\mathbb{R}^{n+1})$ and denote by (u, v) the inner product in $L^2(\mathbb{R}^n)$:

$$(u, v) = \int u(x_0, x')\overline{v(x_0, x')}dx'.$$

Note that

$$\begin{aligned}((D_0 - i\gamma)u, u) &= (P_\gamma u, u) - \sum_{j=1}^{n}(A_j(x)D_j u, u) - (B(x)u, u)\\ &= (P_\gamma u, u) - (u, \sum_{j=1}^{n} A_j(x)D_j u) - (B(x)u, u) - (u, Z(x)u)\end{aligned}$$

where $Z(x) = \sum_{j=1}^{n} D_j A_j(x)$ and we have used $A_j(x) = A_j^*(x)$. We insert

$$-\sum_{j=1}^{n} A_j(x)D_j u = (D_0 - i\gamma)u - P_\gamma u + B(x)u \tag{1.5}$$

into the above identity to obtain

$$\begin{aligned}((D_0 - i\gamma)u, u) - (u, (D_0 - i\gamma)u) &= (P_\gamma u, u) - (u, P_\gamma u)\\ &\quad +(u, B(x)u) - (B(x)u, u) - (u, Z(x)u).\end{aligned}$$

Taking the imaginary part we get

$$\frac{d}{dx_0}\|u(x_0,\cdot)\|^2 + 2\gamma\|u(x_0,\cdot)\|^2$$
$$= 2\mathrm{Im}(u, P_\gamma u) + 2\mathrm{Im}(Bu, u) + \mathrm{Im}(u, Zu). \tag{1.6}$$

From the Cauchy–Schwarz inequality we have

$$|\mathrm{Im}\,(Bu, u)| + |\mathrm{Im}\,(u, Zu)| \le C\,\|u(x_0,\cdot)\|^2$$

and hence the right-hand side of (1.6) is bounded by

$$2\|u(x_0)\|\,\|P_\gamma u(x_0)\| + C\,\|u(x_0)\|^2.$$

Thus we get

$$\frac{d}{dx_0}\|u(x_0)\|^2 + (2\gamma - C - 1)\|u(x_0)\|^2 \le \|P_\gamma u(x_0)\|^2.$$

Choose γ such that $2\gamma - C - 1 \ge 1$ and integrate this from T to t in x_0 which gives

$$\int_T^t \|u(x_0)\|^2 dx_0 \le \|u(T)\|^2 + \int_T^t \|P_\gamma u(x_0)\|^2 dx_0.$$

Letting $T \to -\infty$ we get the assertion for $u \in \mathscr{S}$ and $l = 0$.

We next consider $\langle D'\rangle^l P_\gamma = P_\gamma \langle D'\rangle^l - [P_\gamma, \langle D'\rangle^l]$. Put $[P_\gamma, \langle D'\rangle^l]\langle D'\rangle^{-l} = R_l$ then

$$\langle D'\rangle^l P_\gamma = (P_\gamma + R_l)\langle D'\rangle^l.$$

Here we recall (see for example [21, Theorem 18.1.8])

Lemma 1.5. *Let $u \in \mathscr{S}(\mathbb{R}^n)$. Then we have*

$$|(R_l(x_0)u, u)| \le C_l\,\|u\|^2.$$

Since $\langle D'\rangle^l u \in \mathscr{S}$ repeating the same arguments we conclude that the assertion holds for any l and any $u \in \mathscr{S}$.

Let $u \in H_{(1,l)}$ and put

$$u_\epsilon(x) = (2\pi)^{-(n+1)/2} \int e^{ix\xi - \epsilon|\xi|^2}\hat{u}(\xi)d\xi \in \mathscr{S}(\mathbb{R}^{n+1}).$$

As $\epsilon \downarrow 0$ it is clear that $u_\epsilon \to u$ in $H_{(1,l)}$. Note that

$$\int_{-\infty}^t \|u_\epsilon(s)\|_l^2 ds \to \int_{-\infty}^t \|u(s)\|_l^2 ds, \quad \int_{-\infty}^t \|P_\gamma u_\epsilon(s)\|_l^2 ds \to \int_{-\infty}^t \|P_\gamma u(s)\|_l^2 ds$$

which proves the assertion. □

Proposition 1.3. *Assume that* $Pw = f$ *and*

$$e^{-\gamma x_0} w \in H_{(0,l+1)}, \quad e^{-\gamma x_0} f \in H_{(0,l)}$$

for $\gamma \geq \gamma_l$. *Then we have*

$$\int_{-\infty}^{t} \|e^{-\gamma s} w(s,\cdot)\|_l^2 ds \leq \int_{-\infty}^{t} \|e^{-\gamma s} f(s,\cdot)\|_l^2 ds.$$

In particular

$$f = 0 \text{ in } x_0 < a \Longrightarrow w = 0 \text{ in } x_0 < a.$$

Proof. Take γ_l in Lemma 1.4. Set $u = e^{-\gamma x_0} w$, $g = e^{-\gamma x_0} f$ so that $u \in H_{(0,l+1)}$, $g \in H_{(0,l)}$ and $P_\gamma u = g$. From (1.5) and (1.4) it follows that $u \in H_{(1,l)}$. From Lemma 1.4 we have

$$\int_{-\infty}^{t} \|u(s,\cdot)\|_l^2 ds \leq \int_{-\infty}^{t} \|g(s,\cdot)\|_l^2 ds$$

which proves the assertion. The second assertion follows from this inequality immediately. □

We turn to the proof of the existence of solutions. Denote by $\langle u, v\rangle$ the inner product in $L^2(\mathbb{R}^{n+1})$. Note that the adjoint of P is given by

$$P^* = D_0 + \sum_{j=1}^{n} D_j A_j(x) + B^*(x).$$

Since $P_\gamma = e^{-\gamma x_0} P e^{\gamma x_0}$ we see $\langle P_\gamma u, v\rangle = \langle u, e^{\gamma x_0} P^* e^{-\gamma x_0}\rangle$. Thus it is clear that

$$(P_\gamma)^* = P^*_{-\gamma} = P^*(x, D_0 + i\gamma, D').$$

Lemma 1.6. *For any* $l \in \mathbb{R}$ *there exists* γ_l^* *such that*

$$\int_{t}^{\infty} \|u(s,\cdot)\|_l^2 ds \leq \int_{t}^{\infty} \|P_\gamma^* u(s,\cdot)\|_l^2 ds$$

holds for any $u \in H_{(1,l)}$ *and* $\gamma \geq \gamma_l^*$.

Proof. Repeating the same arguments for P_γ^* we have for $u \in \mathscr{S}(\mathbb{R}^{n+1})$

$$-\frac{d}{dx_0}\|u(x_0)\|^2 + (2\gamma - C - 1)\|u(x_0)\|^2 \leq \|P^*_{-\gamma} u(x_0)\|^2.$$

The rest of the proof is just a repetition. □

Note that the bilinear form

$$\langle u, v\rangle = \int u(x)\overline{v(x)}dx, \quad u, v \in \mathscr{S}(\mathbb{R}^{n+1})$$

extends to the bilinear form on $H_{(m,s)} \times H_{(-m,-s)}$ by the continuity. Since $H_{(m,s)}(\mathbb{R}^{n+1})$ is isometric isomorphic to $L^2(\mathbb{R}_\xi^{n+1}; \langle\xi\rangle^{2m}\langle\xi'\rangle^{2s}d\xi)$ then $H_{(m,s)}$ and $H_{(-m,-s)}$ are mutually dual with respect to $\langle\cdot,\cdot\rangle$. Let us consider the following linear space

$$E = \{P_{-\gamma}^* u \mid u \in C_0^\infty(\mathbb{R}^{n+1})\}.$$

We define a linear map T on E by

$$T : E \ni P_{-\gamma}^* u \mapsto \langle u, g\rangle$$

where $g \in H_{(0,l)}$ is given beforehand. From Lemma 1.6 we have $\|u\|_{(0,-l)} \leq \|P_{-\gamma}^* u\|_{(0,-l)}$ and hence

$$|\langle u, g\rangle| \leq \|g\|_{(0,l)}\|u\|_{(0,-l)} \leq \|g\|_{(0,l)}\|P_{-\gamma}^* u\|_{(0,-l)}.$$

This shows that T is well-defined on $E \subset H_{(0,-l)}$, that is $P_{-\gamma}^* u = 0$ implies $\langle u, g\rangle = 0$. Therefore from the Hahn-Banach theorem T extends to $H_{(0,-l)}$ with the same upper bound. Since $H_{(0,l)}$ is dual to $H_{(0,-l)}$ with respect to $\langle\cdot,\cdot\rangle$ there exists $w \in H_{(0,l)}$ with $\|w\|_{(0,l)} \leq \|g\|_{(0,l)}$ satisfying

$$T(\varphi) = \langle\varphi, w\rangle, \quad \forall\varphi \in H_{(0,-l)}.$$

From the definition

$$T(P_{-\gamma}^* u) = \langle P_{-\gamma}^* u, w\rangle = \langle u, g\rangle, \quad u \in C_0^\infty(\mathbb{R}^{n+1})$$

that is $P_\gamma w = g$. On the other hand since $g \in H_{(0,l)}$ from (1.5) and (1.4) we have $w \in H_{(1,l-1)}$.

Proposition 1.4. *Assume $e^{-\gamma x_0} f \in H_{(0,l)}$ for some $\gamma \geq \gamma_l$. Then there exists u such that $e^{-\gamma x_0} u \in H_{(1,l-1)}$ satisfying*

$$Pu = f$$

where $u = 0$ in $x_0 < a$ if $f = 0$ in $x_0 < a$.

Proof. Set $g = e^{-\gamma x_0} f \in H_{(0,l)}$. As we just have seen above there exists w verifying $P_\gamma w = g$. Since

$$e^{-\gamma x_0} P(e^{\gamma x_0} w) = e^{-\gamma x_0} f$$

then $u = e^{\gamma x_0} w$ is the desired one. The second assertion is nothing but Proposition 1.3. □

1.3 Systems Which Are Not Symmetrizable

Here we give a simple example which is not symmetrizable hyperbolic system which is strongly hyperbolic. We start with giving the definition of symmetrizable systems.

Definition 1.4 ([15]). We say that a system of first order differential operators

$$Lu = \sum_{j=0}^{n} A_j(x) D_j u$$

is symmetrizable if there exists a positive definite Hermitian (or symmetric) matrix $S(x,\xi)$ such that $S(x,\xi)L(x,\xi)$ becomes to be Hermitian (symmetric). Here $L(x,\xi) = \sum_{j=0}^{n} A_j(x)\xi_j$.

Remark. If L is symmetrizable then there is a non singular matrix $T(x,\xi)$ such that $T(x,\xi)L(x,\xi)T(x,\xi)^{-1}$ is Hermitian (symmetric). In fact it is enough to set $T(x,\xi) = S(x,\xi)^{1/2}$. The converse is also true. Indeed if T is given then it suffices to define S by $S = T^*T$.

Let us consider the following 2×2 system in $\mathbb{R}^2$

$$Lu = \partial_t u + \begin{bmatrix} 0 & 1 \\ t^2 & 0 \end{bmatrix} \partial_x u + B(t,x)u$$

where $(t,x) \in \mathbb{R}^2$, $u = {}^t(u_1, u_2)$. This system is not symmetrizable indeed if this were symmetrizable then there is a non singular T such that

$$T \begin{bmatrix} 0 & 1 \\ 0 & 0 \end{bmatrix} T^{-1}$$

is symmetric and hence should be the zero matrix which is a contradiction.

We prove that L is strongly hyperbolic near the origin. That is for any $B(t,x)$ the following Cauchy problem

$$\begin{cases} Lu = f \\ u(0,x) = u_0(x) \end{cases} \tag{1.7}$$

has C^∞ solution in a neighborhood of the origin for any $f \in C^\infty(U)$ and any $u_0(x) \in C^\infty(U \cap \{t = 0\})$ where U is some neighborhood of the origin. Without restrictions we can assume that $u_0 = 0$ and $\partial_t^j f(0,x) = 0$, $j = 0, 1, \ldots, M-1$ for M given beforehand. Indeed differentiating $Lu = f$ with respect to t one can find $\partial_t^j u(0,x) = u_j(x)$ successively and using these $u_j(x)$ we put

$$u_M(t,x) = \sum_{j=0}^{M} \frac{1}{j!} u_j(x) t^j$$

then $w = u - u_M$ satisfies

$$\begin{cases} Lw = f_M \\ w(0,x) = 0. \end{cases} \tag{1.8}$$

Note that $\partial_t^j f_M(0,x) = 0$, $j = 0, 1, \dots, M-1$ and $\partial_t^j w(0,x) = 0$, $j = 0, 1, \dots, M$. If we get a solution w to (1.8) then $u = w + u_M$ is a solution to (1.7).

To prove the existence of solution we consider

$$L_\epsilon = \partial_t + \begin{bmatrix} \epsilon & 1 \\ t^2 & -\epsilon \end{bmatrix} \partial_x + B(t,x)$$

instead of L where $\epsilon > 0$ is a small positive parameter. We first show that the Cauchy problem

$$\begin{cases} L_\epsilon u^{(\epsilon)} = f \\ u^{(\epsilon)}(0,x) = 0 \end{cases} \tag{1.9}$$

has a solution $u^{(\epsilon)}$ and then we prove that $u^{(\epsilon)}$ converges as $\epsilon \downarrow 0$, taking a subsequence if necessary, to some u which is the desired solution.

Note that the eigenvalues of

$$\begin{bmatrix} \epsilon & 1 \\ t^2 & -\epsilon \end{bmatrix} \tag{1.10}$$

are $\pm\sqrt{t^2+\epsilon^2}$ which are real distinct when $\epsilon > 0$. That is the system is strictly hyperbolic whenever $\epsilon > 0$ and hence the existence of solution is well known. Since our case is one space dimensional then we can show the existence of solution applying the Picard's iteration method. Let $\epsilon > 0$ then with a smooth non singular matrix $T(t,\epsilon)$ one can diagonalize (1.10)

$$T(t,\epsilon)^{-1} \begin{bmatrix} \epsilon & 1 \\ t^2 & -\epsilon \end{bmatrix} T(t,\epsilon) = \begin{bmatrix} \sqrt{t^2+\epsilon^2} & 0 \\ 0 & -\sqrt{t^2+\epsilon^2} \end{bmatrix}.$$

Considering $T(t,\epsilon)^{-1} L_\epsilon T(t,\epsilon)$ it is enough to study

$$L_\epsilon u = \partial_t u + \begin{bmatrix} \sqrt{t^2+\epsilon^2} & 0 \\ 0 & -\sqrt{t^2+\epsilon^2} \end{bmatrix} \partial_x u - C(t,x,\epsilon) u = f. \tag{1.11}$$

Let us solve this Cauchy problem by the method of characteristic curves. For any (t, x), $t \geq 0$ we set

$$x_1(s;t,x) = x + \int_t^s \sqrt{\tau^2 + \epsilon^2}\, d\tau, \quad x_2(s;t,x) = x - \int_t^s \sqrt{\tau^2 + \epsilon^2}\, d\tau.$$

Assume that u verifies (1.11) then we have

$$\frac{d}{ds} u_i(s, x_i(s;t,x)) = \sum_{j=1}^{2} c_{ij}(s, x_i(s;t,x), \epsilon) u_j(s, x_i(s;t,x)) + f_i(s, x_i(s;t,x)) \tag{1.12}$$

where $i = 1, 2$ and $C = (c_{ij}(t, x, \epsilon))$, $f = {}^t(f_1, f_2)$. Conversely if $u(t, x) \in C^1$ verifies (1.12) then $u(t, x)$ is a solution to (1.11). We rewrite (1.12) as an integral equation

$$u_i(t,x) = \int_0^t \{\sum_{j=1}^{2} c_{ij}(s, x_i(s;t,x), \epsilon) u_j(s, x_i(s;t,x)) + f_i(s, x_i(s;t,x))\} ds.$$

We solve this integral equation by the iteration method. Define $u_i^{(0)}(t, x) = 0$ and $u_i^{(n+1)}(t, x)$, $i = 1, 2$ by

$$u_i^{(n+1)}(t,x) = \int_0^t \{\sum_{j=1}^{2} c_{ij}(s, x_i(s;t,x), \epsilon) u_j^{(n)}(s, x_i(s;t,x)) + f_i(s, x_i(s;t,x))\} ds.$$

Let $(\bar{t}, \bar{x})$ be fixed and denote by D the region surrounded by two characteristic curves $x_1(s; \bar{t}, \bar{x})$, $x_2(s; \bar{t}, \bar{x})$ $(s \leq \bar{t})$ and x axis. Then there exist C, B such that for any $(t, x) \in D$ we have

$$|u_i^{(n+1)}(t,x) - u_i^{(n)}(t,x)| \leq C B^n t^n / n!$$

for $i = 1, 2$, $n = 0, 1, 2, \ldots$. Thus $u_i^{(n)}(t, x)$ $(i = 1, 2)$ converges to $u_i(t, x)$ uniformly in D. Thanks to (1.12) the uniqueness of the solution follows from the standard arguments. Repeating the same arguments proving the differentiability with respect to parameters we can prove that the solution is smooth.

We turn to the next step. We derive estimates for the solution $u^{(\epsilon)}$ to (1.9) independent of ϵ. Set

$$M_\epsilon = \partial_t - \begin{bmatrix} \epsilon & 1 \\ t^2 & -\epsilon \end{bmatrix} \partial_x$$

then we have

$$M_\epsilon L_\epsilon = \partial_t^2 - (t^2+\epsilon^2)\partial_x^2 + P(t,x,\epsilon)\partial_x + B(t,x)\partial_t + Q(t,x,\epsilon) \tag{1.13}$$

where

$$P(t,x,\epsilon) = \begin{bmatrix} 0 & 0 \\ 2t & 0 \end{bmatrix} - \begin{bmatrix} \epsilon & 1 \\ t^2 & -\epsilon \end{bmatrix} B(t,x),$$

$$Q(t,x,\epsilon) = \partial_t B(t,x) - \begin{bmatrix} \epsilon & 1 \\ t^2 & -\epsilon \end{bmatrix} \partial_x B(t,x).$$

Let us put

$$h = \partial_t^2 - (t^2+\epsilon^2)\partial_x^2$$

then we see

$$hu \cdot \overline{\partial_t u} + \overline{hu} \cdot \partial_t u = \partial_t G_1(u) + \partial_x G_2(u) - R(u) \tag{1.14}$$

where

$$G_1(u) = |\partial_t u|^2 + (t^2+\epsilon^2)|\partial_x u|^2,$$
$$G_2(u) = -(t^2+\epsilon^2)(\partial_t u \cdot \overline{\partial_x u} + \overline{\partial_t u} \cdot \partial_x u), \quad R(u) = 2t|\partial_x u|^2.$$

Let $T > 0$ be fixed and choose $c > 0$ so that

$$cG_1(u) \pm G_2(u) \geq 0 \tag{1.15}$$

holds for $0 \leq t \leq T$ and $0 < \epsilon \leq 1$. Multiply $e^{-\theta t} t^{-N}$ to (1.14) and integrate over $S = \{\delta \leq t \leq T \mid |x| \leq c(T-t)\}$. Since we have $e^{-\theta t} t^{-N} \partial_t = \partial_t(e^{-\theta t} t^{-N}) + \theta e^{-\theta t} t^{-N} + N e^{-\theta t} t^{-N-1}$ then

$$\begin{aligned} &\int_{\partial S} e^{-\theta t} t^{-N} (G_1(u)dx - G_2(u)dt) + 2\int_S e^{-\theta t} t^{-N} |hu||\partial_t u| dxdt \\ &\geq N \int_S e^{-\theta t} t^{-N-1} G_1(u) dxdt \\ &\quad + \theta \int_S e^{-\theta t} t^{-N} G_1(u) dxdt - \int_S e^{-\theta t} t^{-N} R(u) dxdt. \end{aligned}$$

By the Cauchy–Schwarz inequality we have

$$2t^{-N}|hu||\partial_t u| \leq 3N^{-1} t^{-N+1} |hu|^2 + 3^{-1} N t^{-N-1} |\partial_t u|^2$$

and noting (1.15) we get

$$3N^{-1}\int_S e^{-\theta t}t^{-N+1}|hu|^2dxdt + \int_{|x|\le c(T-\delta)} e^{-\theta\delta}\delta^{-N}G_1(u(\delta,x))dx$$
$$\ge (2N/3)\int_S e^{-\theta t}t^{-N-1}(|\partial_t u|^2 + (t^2+\epsilon^2)|\partial_x u|^2)dxdt \quad (1.16)$$
$$+\int_S e^{-\theta t}t^{-N-1}(3^{-1}N(t^2+\epsilon^2)-2t^2)|\partial_x u|^2dxdt$$
$$+\theta\int_S e^{-\theta t}t^{-N}G_1(u)dxdt.$$

We multiply $e^{-\theta t}t^{-N-2}$ to

$$\partial_t u\cdot\overline{u} + \overline{\partial_t u}\cdot u = \partial_t|u|^2$$

and integrate over S to get

$$2\int_S e^{-\theta t}t^{-N-2}|\partial_t u||u|dxdt \ge \int_{\partial S} e^{-\theta t}t^{-N-2}(-|u|^2dx)$$
$$+(N+2)\int_S e^{-\theta t}t^{-N-3}|u|^2dxdt$$
$$+\theta\int_S e^{-\theta t}t^{-N-2}|u|^2dxdt.$$

From the Cauchy–Schwarz inequality it follows

$$2t^{-N-2}|\partial_t u||u| \le 3N^{-1}t^{-N-1}|\partial_t u|^2 + 3^{-1}Nt^{-N-2}|u|^2.$$

Multiply N^2 to (1.16) and we get

$$3N\int_S e^{-\theta t}t^{-N-1}|\partial_t u|^2dxdt + N^2\int_{|x|\le c(T-\delta)} e^{-\theta\delta}\delta^{-N-2}|u(\delta,x)|^2dx$$
$$\ge (2N^3/3)\int_S e^{-\theta t}t^{-N-3}|u|^2dxdt + N^2\theta\int_S e^{-\theta t}t^{-N-2}|u|^2dxdt. \quad (1.17)$$

Now set

$$E(u) = |\partial_t u|^2 + t^2|\partial_x u|^2 + N^2t^{-2}|u|^2,$$
$$E_\epsilon(u) = |\partial_t u|^2 + (t^2+\epsilon^2)|\partial_x u|^2 + N^2t^{-2}|u|^2 \ge E(u)$$

and choosing N so that $N/3 \ge 2$ then from (1.16) and (1.17) we have

Lemma 1.7. *We have*

$$3\int_S e^{-\theta t}t^{-N+1}|hu|^2dxdt + Ne^{-\theta\delta}\delta^{-N}\int_{|x|\le c(T-\delta)} E_\epsilon(u(\delta,x))dx$$
$$\ge (N^2/3)\int_S e^{-\theta t}t^{-N-1}E(u)dxdt + \theta N\int_S e^{-\theta t}t^{-N}E(u)dxdt.$$

Let $P(t,x,\epsilon)$ be a smooth 2×2 matrix in (t,x,ϵ), $0\le\epsilon\le 1$ it is easy to see that there is C independent of ϵ such that

$$\int_S e^{-\theta t}t^{-N+1}|P\partial_x u|^2dxdt \le C\int_S e^{-\theta t}t^{-N-1}E(u)dxdt,$$
$$\int_S e^{-\theta t}t^{-N+1}|P\partial_t u|^2dxdt \le C\int_S e^{-\theta t}t^{-N+1}E(u)dxdt.$$

We apply Lemma 1.7 to get

Proposition 1.5. *There exists N such that*

$$\theta\int_S e^{-\theta t}t^{-N}E(u)dxdt + \int_S e^{-\theta t}t^{-N-1}E(u)dxdt$$
$$\le C\int_S e^{-\theta t}t^{-N+1}|M_\epsilon L_\epsilon u|^2dxdt + Ce^{-\theta\delta}\delta^{-N}\int_{|x|\le c(T-\delta)} E_\epsilon(u(\delta,x))dx$$

where C is independent of $0\le\epsilon\le 1$.

Let $u^{(\epsilon)}$ be a solution to

$$\begin{cases} L_\epsilon u^{(\epsilon)} = f \\ u^{(\epsilon)}(0,x) = 0. \end{cases} \tag{1.18}$$

Since one can assume $\partial_t^j f(0,x) = 0$, $j = 0,1,\dots,M$, $2M-N-1\ge 0$ we get $M_\epsilon L_\epsilon u^{(\epsilon)} = M_\epsilon f = O(t^M)$. Differentiating (1.18) in t and put $t=0$ we get $\partial_t^j u^{(\epsilon)}(0,x) = 0$, $j=0,1,\dots,M$. Thus $u^{(\epsilon)} = O(t^{M+1})$ and hence

$$E_\epsilon(u^{(\epsilon)}(\delta,\cdot)) = O(\delta^{2M})$$

then we get

$$\lim_{\delta\to 0}\delta^{-N}\int_{|x|\le c(T-\delta)} E_\epsilon(u^{(\epsilon)}(\delta,x))dx = 0.$$

Therefore one can make $\delta\downarrow 0$ in the right-hand side of (1.9) from which we obtain estimates of $u^{(\epsilon)}$ independent of ϵ. In particular there is C independent of ϵ such that

$$\int_{S_0} (|\partial_t u^{(\epsilon)}(t,x)|^2 + |\partial_x u^{(\epsilon)}(t,x)|^2 + |u^{(\epsilon)}(t,x)|^2) dt dx \le C$$

where

$$S_0 = \{0 \le t \le T, |x| \le c(T-t)\}.$$

From this inequality it follows that $u^{(\epsilon)}$ is bounded in S_0 uniformly in ϵ. Considering the equation which is obtained by differentiating (1.9) with respect to x, which is the equation for $\partial_x u^{(\epsilon)}$ and repeating the same arguments we conclude that $\partial_x u^{(\epsilon)}$ is bounded in S_0 uniformly in ϵ. Similarly we can show that $\partial_t^k \partial_x^j u^{(\epsilon)}$, $k+j \le p$ are bounded in S_0 uniformly in ϵ. Take p so that $p \ge 2$ then from Ascoli-Arzela's theorem one can pick a uniformly convergent subsequence of $\{u^{(\epsilon)}\}$ such that their derivatives also uniformly convergent. Let us denote by w the limit function. Then it is clear that

$$\begin{cases} Lw = f \\ w(0,x) = 0. \end{cases}$$

Thus we conclude the proof.

1.4 Lax-Mizohata Theorem

We first recall the Lax-Mizohata theorem.

Theorem 1.1 ([33, 43]). *Let $h(x,\xi) = \det P_d(x,\xi)$. If the Cauchy problem for $P(x,D)$ is C^∞ well posed in the future (or in the past) near the origin then we have*

$$h(x, \xi + \tau(1,0,\dots,0)) = 0 \Longrightarrow \tau \in \mathbb{R}, \quad \forall \xi \in \mathbb{R}^{n+1}, \quad \forall x \ \textit{close to}\ 0.$$

Proof. Assume that with some $\bar{\xi}' \in \mathbb{R}^n$

$$h(0, \xi_0, \bar{\xi}') = 0$$

has a root $\bar{\xi}_0$ with $\operatorname{Im} \bar{\xi}_0 \neq 0$. Since $h(x,\xi)$ is homogeneous of degree md, we may assume that

$$\operatorname{Im} \bar{\xi}_0 = -1$$

taking $a\bar{\xi}'$, $a \in \mathbb{R}$ in place of $\bar{\xi}'$. It is clear that $\bar{\xi}'_\mu \neq 0$ with some μ and then with a new system of local coordinates

$$y_0 = x_0, \ \ y_1 = \langle x', \bar{\xi}' \rangle + x_0 \operatorname{Re} \bar{\xi}_0, \ \ y_j = x_j \ (j \neq 1, \mu), \ \ y_\mu = x_1$$

we can assume that

$$\bar{\xi}' = (1,0,\dots,0), \quad \bar{\xi}_0 = -i.$$

Hence we can write with some $q \in \mathbb{N}$

$$h(0,\xi_0,\xi_1,0,\dots,0) = (\xi_0 + i\xi_1)^q g(\xi_0,\xi_1), \quad g(-i,1) \neq 0. \tag{1.19}$$

We denote by $\tilde{M}(\xi_0,\xi_1)$ the cofactor matrix of $\tilde{P}_d(\xi_0,\xi_1) = P_d(0,\xi_0,\xi_1,0,\dots,0)$ so that $\tilde{P}_d(\xi_0,\xi_1)\tilde{M}(\xi_0,\xi_1) = (\xi_0 + i\xi_1)^q g(\xi_0,\xi_1)$. We factor out $\xi_0 + i\xi_1$ from $\tilde{M}(\xi_0,\xi_1)$ so that

$$\tilde{M}(\xi_0,\xi_1) = (\xi_0 + i\xi_1)^{q_1} H(\xi_0,\xi_1), \quad H(-i,1) \neq O$$

where O denotes the zero matrix. Here we note that $0 \le q_1 \le q-1$ because if $q_1 \ge q$ then we would have

$$(\xi_0 + i\xi_1)^{q_1}\tilde{P}_d(\xi_0,\xi_1)H(\xi_0,\xi_1) = (\xi_0 + i\xi_1)^q g(\xi_0,\xi_1)$$

and hence

$$\det H(\xi_0,\xi_1) = (\xi_0 + i\xi_1)^{-m(q_1-q)-q} g(\xi_0,\xi_1)^{m-1}.$$

Since $g(-i,1) \neq 0$ and $\det H(\xi_0,\xi_1)$ is a polynomial this would give a contradiction. Thus one can write

$$\tilde{P}_d(\xi_0,\xi_1)H(\xi_0,\xi_1) = (\xi_0 + i\xi_1)^r g(\xi_0,\xi_1)$$

with some $1 \le r \in \mathbb{N}$. Let us set

$$Q(x,D) = P(x,D)H(D_0,D_1) = \sum_{|\alpha|\le\bar{q}} A_\alpha(x)D^\alpha$$

with $\bar{q} = md - q_1$. For a large $N_1 \in \mathbb{N}$ write

$$A_\alpha(x) = \sum_{|\beta|\le N_1-1} \frac{1}{\beta!}\partial_x^\beta A_\alpha(0)x^\beta + \tilde{A}_\alpha(x) = A_{\alpha,N_1}(x) + \tilde{A}_\alpha(x)$$

and define

$$Q(x,D) = Q_{N_1}(x,D) + R_{N_1}(x,D), \quad Q_{N_1}(x,D) = \sum A_{\alpha,N_1}(x)D^\alpha.$$

Take $\nu(> r)$ sufficiently large and take a new system of local coordinates depending on a large parameter λ

$$y_j = \lambda^{2\nu} x_j \ (j = 0, 1), \ \ y_j = \lambda^{\nu} x_j \ (j \geq 2)$$

and consider

$$\begin{aligned} Q_\lambda(y, D) &= Q(\lambda^{-\tilde{\nu}} y, \lambda^{\tilde{\nu}} D) = Q_{N_1}(\lambda^{-\tilde{\nu}} y, \lambda^{\tilde{\nu}} D) + R_{N_1}(\lambda^{-\tilde{\nu}} y, \lambda^{\tilde{\nu}} D) \\ &= Q_{N_1,\lambda}(y, D) + R_{N_1,\lambda}(y, D) \end{aligned}$$

where $\tilde{\nu} = \nu(2, 2, 1, \ldots, 1)$. Note that the coefficients of $R_{N_1,\lambda}(y, D)$ are $O(\lambda^{-\nu N_1})$. In what follows we denote by $O(\lambda^{-k})$ any matrix whose entries are differential operators of order at most $\bar{q}$ with coefficients of order $O(\lambda^{-k})$. Then it is clear that

$$\lambda^{-2\bar{q}\nu}(Q_{N_1,\lambda}(y, D) - \tilde{h}(\lambda^{\tilde{\nu}} D)) = O(\lambda^{-\nu}), \ \ R_{N_1,\lambda}(y, D) = O(\lambda^{-\nu N_1})$$

with $\tilde{h}(\xi) = \tilde{P}_d(\xi_0, \xi_1) H(\xi_0, \xi_1)$. Since

$$\tilde{h}(\lambda^{\tilde{\nu}} D) = \lambda^{2\bar{q}\nu}(D_0 + iD_1)^r g(D_0, D_1)$$

this shows that

$$\lambda^{-2\bar{q}\nu} Q_{N_1,\lambda}(y, D) = (D_0 + iD_1)^r g(D_0, D_1) + O(\lambda^{-\nu}).$$

Here we remark that

$$D_0 + iD_1 = \frac{1}{i}\left(\frac{\partial}{\partial y_0} + i\frac{\partial}{\partial y_1}\right)$$

is the Cauchy–Riemann operator in $z = y_0 + iy_1$ plane. Take

$$\psi = -i(y_0 + iy_1) - i(y_0 + iy_1)^2$$

which is a polynomial in z verifying

$$\psi(0) = 0, \ \ \operatorname{Im} \psi = -y_0 - y_0^2 + y_1^2 \geq 0$$

for $y_0 \leq 0$ and $|z|$ small. We now set

$$\varphi(y) = \psi(y_0 + iy_1) + i(y_2^2 + \cdots + y_n^2)$$

then we have $(D_0 + iD_1)\varphi = 0$ and

$$\operatorname{Im} \varphi(y) \geq c|y|^2 \quad \text{when} \ \ y_0 \leq 0, \ \ |y| \leq \delta \tag{1.20}$$

with some $c > 0$ and $\delta > 0$.

In what follows we look for an asymptotic solution to $P_\lambda u = 0$ of the form

$$u = \sum_{j=0}^{\infty} e^{i\lambda\varphi(y)} v_j(y)\lambda^{-j}. \tag{1.21}$$

Note that

$$\lambda^{-2\bar{q}\nu} e^{-\lambda\varphi} R_{N_1,\lambda} e^{i\lambda\varphi} = O(\lambda^{-N_1\nu+\bar{q}-2\bar{q}\nu}),$$
$$\lambda^{-2\bar{q}\nu} e^{-i\lambda\varphi} Q_{N_1,\lambda} e^{i\lambda\varphi} = \lambda^{\bar{q}-r} a(D_0 + iD_1)^r + O(\lambda^{\bar{q}-r-1})$$

where $a = g(\partial\varphi/\partial y_0, \partial\varphi/\partial y_1)$ because

$$e^{-i\lambda\varphi}(D_0 + iD_1)e^{i\lambda\varphi} = D_0 + iD_1,$$
$$e^{-i\lambda\varphi} g(D_0, D_1) e^{i\lambda\varphi} = \lambda^{\bar{q}-r} g(\partial\varphi/\partial y_0, \partial\varphi/\partial y_1) + O(\lambda^{\bar{q}-r-1}).$$

Therefore we get

$$\lambda^{-2\bar{q}\nu} e^{-i\lambda\varphi} P_{N_1,\lambda} e^{i\lambda\varphi} \sum_{j=0} v_j \lambda^{-j}$$
$$= \lambda^{\bar{q}-r} \sum_{j=0} \{a(D_0 + iD_1)^r v_j - F_j(y, v_0, \ldots, v_{j-1})\}\lambda^{-j}$$

where $F_0 = 0$ and $F_j(y, v_0, \ldots, v_{j-1})$ are polynomials in x and derivatives of $v_0, \ldots, v_{j-1}$. Thus we are led to the equations

$$a(D_0 + iD_1)^r v_0 = 0, \quad a(D_0 + iD_1)^r v_j = F_j(y, v_0, \ldots, v_{j-1}), \quad j \geq 1. \tag{1.22}$$

Here we choose constant vector v_0 so that

$$H(-i, 1)v_0 \neq 0$$

which is possible because $H(-i, 1) \neq O$. We note that $a \neq 0$ near the origin because $g(\partial\varphi/\partial y_0, \partial\varphi/\partial y_1) = g(-i, 1)$ at the origin. Thanks to the Cauchy–Kowalevsky theorem we can find smooth $v_j(y)$ which verifies (1.22) near the origin. For any N we choose N_1 and N_2 so that

$$u_\lambda^{(N)} = \sum_{j=0}^{N_2} e^{i\lambda\varphi} v_j(y)\lambda^{-j}$$

satisfies

$$e^{-i\lambda\varphi} Q_\lambda u_\lambda^{(N)} = O(\lambda^{-N}). \tag{1.23}$$

Here we recall that

$$Q_\lambda(y,D) = P_\lambda(y,D)H_\lambda(D) = \lambda^{2\nu(\bar{q}-d)} P_\lambda H(D)$$

and set

$$U_\lambda = H(D)\sum_{j=0}^{N_2} e^{i\lambda\varphi} v_j(y)\lambda^{-j} = H(D)u_\lambda^{(N)}.$$

Take $\chi \in C_0^\infty(\mathbb{R}^{n+1})$ which is 1 near the origin so that (1.20) and (1.23) hold in a neighborhood of the support of χ. Note that

$$P_\lambda \chi U_\lambda = \chi P_\lambda U_\lambda + [P_\lambda, \chi]U_\lambda.$$

Since $-\mathrm{Im}\ \varphi \le -\epsilon$ if $[P_\lambda, \chi]U_\lambda \ne 0$ and $y_0 \le 0$ this proves that

$$|P_\lambda \chi U_\lambda|_{C^p(K^0)} \le C_N \lambda^{-N+p}$$

where $K = \mathrm{supp}\chi$. On the other hand we have

$$|\chi U_\lambda|_{C^0(K^0)} \ge c\lambda^{\bar{q}-d}$$

with some $c > 0$ since $\varphi(0) = 0$ and

$$H(D)e^{i\lambda\varphi} = \lambda^{\bar{q}-d} e^{i\lambda\varphi}\{H(\partial\varphi/\partial y_0, \partial\varphi/\partial y_1) + O(\lambda^{-1})\}$$

with $H(-i,1)v_0 \ne 0$. Taking N large enough we get a contradiction to the a priori estimate in Proposition 1.2. □

1.5 Levi Condition

Let us consider

$$Pu = D_0 - \sum_{j=1}^{n} A_j(x)D_j u + B(x)u = L(x,D)u + B(x)u = f$$

where we assume that $A_j(x)$, $B(x) \in \mathscr{A}(\Omega)$. In this section we prove the necessity of the Levi condition for the well-posedness for systems with characteristics of constant multiplicity.

Proposition 1.6 ([27]). *Assume that L is strongly hyperbolic near the origin and there exist open sets $0 \in W \subset \mathbb{R}^{n+1}$, $U \subset \mathbb{R}^n \setminus \{0\}$ and $\omega(x,\xi') \in \mathscr{A}(W \times U)$ such that*

$$\det L(x,\xi) = (\xi_0 - \omega(x,\xi'))^r e(x,\xi), \quad e(x,\omega(x,\xi'),\xi') \ne 0$$

holds in $W \times U$ where $r \in \mathbb{N}$ is independent of (x, ξ'). Then we have

$$\operatorname{rank} L(x, \omega(x, \xi'), \xi') = m - r$$

for $(x, \xi') \in W \times U$. Equivalently we have $\dim \operatorname{Ker} L(x, \omega(x, \xi'), \xi') = r$ *for $(x, \xi') \in W \times U$.*

Before proving the proposition we improve Corollary 1.1 in another direction.

Lemma 1.8. *Assume that $A_j(x)$, $B(x) \in \mathscr{A}(\Omega)$ and the Cauchy problem for P is C^∞ well posed in the future near the origin. Then there exist a convex cone $\Gamma = \{x \in \mathbb{R}^{n+1} \mid -x_0 \geq c|x'|\}$ with $c > 0$, a neighborhood V of the origin and $\delta > 0$ such that one can find $C > 0$ and $p \in \mathbb{N}$ such that*

$$|u(y)| \leq C\,|Pu|_{C^p(y+\Gamma)}$$

for any $u \in C_0^\infty(\overline{V})$ and any $|y| < \delta$ where $y + \Gamma = \{y + x \mid x \in \Gamma\}$ and Γ is independent of $B(x)$.

Proof. Let $V \subset\subset \omega$. Then for any $f \in C_0^\infty(\overline{V}_{-\epsilon})$ there is a unique $u \in H^\infty(\omega)$ such that $u = 0$ in $x_0 \leq -\epsilon$. Since $A_j(x)$, $B(x)$ are real analytic it follows from the Holmgren uniqueness theorem (see [44], Theorem 4.8) that $u(y) = 0$ if $f = 0$ in $y + \Gamma$. Consider the distribution $T : f \mapsto u(y)$. Then from Corollary 1.1 and the Whitney's extension theorem (see [20, Theorem 2.3.6]) we get the assertion. □

Proof of Proposition 1.6. Let us write $A(x, \xi') = \sum_{j=1}^n A_j(x)\xi_j$. From the assumption there is an open set $W_1 \times U_1 \subset W \times U$ and $S(x, \xi') \in \mathscr{A}(W_1 \times U_1, M_m(\mathbb{C}))$ such that

$$S^{-1}(x, \xi')A(x, \xi')S(x, \xi') = A_1(x, \xi') \oplus A_2(x, \xi')$$

where

$$\det(\xi_0 I_r - A_1) = (\xi_0 - \omega)^r.$$

For a proof we refer to [71]. We can assume that W_1 is contained in V in Lemma 1.8. Note that one can find $W_2 \times U_2 \subset W_1 \times U_1$and $S_1(x, \xi') \in \mathscr{A}(W_2 \times U_2, M_r(\mathbb{C}))$ such that

$$S_1^{-1}(x, \xi')A_1 S_1(x, \xi') = \oplus_{i=1}^s \Lambda_i$$

where

$$\Lambda_i = \begin{bmatrix} \omega & 1 & & & \\ & \omega & 1 & & \\ & & \ddots & \ddots & \\ & & & \ddots & 1 \\ & & & & \omega \end{bmatrix} \in \mathscr{A}(W_2 \times U_2, M_{p_i}(\mathbb{C})), \quad \sum_{i=1}^s p_i = r.$$

(see [27,72]). Thus with

$$T(x,\xi') = (S_1 \oplus I_{m-r})S(x,\xi') \in \mathscr{A}(W_2 \times U_2, M_m(\mathbb{C}))$$

we have

$$T^{-1}LT = (\xi_0 I_r - \oplus_{i=1}^{s}\Lambda_i) \oplus (\xi_0 I_{m-r} - A_2).$$

We now show that $p_1 = \cdots = p_s = 1$ by contradiction. Without restrictions we suppose $p_1 = k \geq 2$. Let $(\hat{x},\hat{\xi}') \in W_2 \times U_2$ and let $\varphi(x)$ be a solution to

$$\partial_{x_0}\varphi = \omega(x,\varphi_{x'}), \quad \varphi(\hat{x}_0, x') = \langle x', \hat{\xi}'\rangle.$$

Let us set $N(x) = T(x,\varphi_{x'})$ so that

$$N^{-1}(x)A(x,\varphi_{x'})N(x) = \Lambda \oplus D$$

where

$$\Lambda = \begin{bmatrix} \omega(x,\varphi_{x'}) & 1 & & \\ & \omega(x,\varphi_{x'}) & \ddots & \\ & & \ddots & 1 \\ & & & \omega(x,\varphi_{x'}) \end{bmatrix}$$

is a $k \times k$ matrix. Now we have

$$\begin{aligned} &N^{-1}(x)(D_0 - A + B(x))N(x) \\ &\quad = D_0 - A_N + N^{-1}\{(D_0 - A)N\} + N^{-1}BN \end{aligned}$$

where

$$A_N(x,D) = \sum_{j=1}^{n} N^{-1}(x)A_j(x)N(x)D_j.$$

We define B by

$$N^{-1}\{(D_0 - A)N\} + N^{-1}BN = \begin{bmatrix} 0 & & \\ & \ddots & \\ i & & 0 \end{bmatrix} \oplus O = C$$

where O is the zero matrix of order $m - k$. It is clear that

$$N^{-1}(x)(D_0 - A + B)N(x) = D_0 - A_N + C.$$

Let us set

$$\psi = \sum_{j=0}(\varphi(x) + \tau_j(x_0 - \hat{x}_0))\mu^{-j/k}, \quad v(\mu) = {}^t(w(\mu), 0), \ 0 = (0, \ldots, 0) \in \mathbb{R}^{m-k}$$

where $w(\mu) \in \mathbb{R}^k$ and $\tau_j \in \mathbb{C}$ will be determined later. It is easy to see

$$\begin{aligned} e^{-i\mu\psi} A_N e^{i\mu\psi} v(\mu) &= \mu \sum_{j=0}\sum_{l=1}^{n} N^{-1} A_l N \varphi_{x_l} \mu^{-j/k} v(\mu) \\ &= \mu \sum_{j=0} \mu^{-j/k} N^{-1} A(x, \varphi_{x'}) N v(\mu) \\ &= \mu \sum_{j=0} \mu^{-j/k} \begin{bmatrix} \Lambda & O \\ 0 & D \end{bmatrix} v(\mu) \\ &= \begin{bmatrix} \mu \sum_{j=0} \mu^{-j/k} \Lambda w(\mu) \\ 0 \end{bmatrix}. \end{aligned}$$

Thus we are led to solve

$$\Big\{\mu(\sum_{j=0}(\varphi_{x_0} + \tau_j)\mu^{-j/k} - \sum_{j=0} \mu^{-j/k}\Lambda) + \begin{bmatrix} 0 & & \\ & \ddots & \\ i & & 0 \end{bmatrix}\Big\} w(\mu) = 0.$$

Since $\partial_{x_0}\varphi - \omega(x, \varphi_{x'}) = 0$ we have

$$\Big\{\sum_{j=0} \tau_j \mu^{-j/k} - \begin{bmatrix} 0 & 1 & & \\ & \ddots & \ddots & \\ & & \ddots & 1 \\ & & & 0 \end{bmatrix} \sum_{j=0} \mu^{-j/k} + \begin{bmatrix} 0 & & & \\ & \ddots & \ddots & \\ & & & \ddots \\ i/\mu & & & 0 \end{bmatrix}\Big\} w(\mu) = 0.$$

With

$$\tau(\mu) = \sum_{j=0} \tau_j \mu^{-j/k}, \quad e(\mu) = \sum_{j=0} \mu^{-j/k}$$

one can write

$$\begin{bmatrix} \tau(\mu) & -e(\mu) & & \\ & \ddots & \ddots & \\ & & \ddots & -e(\mu) \\ i/\mu & & & \tau(\mu) \end{bmatrix} w(\mu) = 0. \tag{1.24}$$

We choose $\tau(\mu)$ so that (1.24) has a non trivial solution $w(\mu)$, that is

$$\tau(\mu)^k + \frac{i}{\mu} e(\mu)^{k-1} = 0. \tag{1.25}$$

We take $\tau_0 = 0$ so that $\tau_1^k + i = 0$. We choose τ_1 such that $\mathrm{Im}\tau_1 > 0$. Successively τ_j, $j \geq 2$ will be determined from (1.25). Let us denote $w(\mu) = {}^t(w_1(\mu), \ldots, w_k(\mu))$ and put

$$w_j(\mu) = \left(\frac{e(\mu)}{\tau(\mu)}\right)^{k-j} w_k(\mu), \quad j = 1, \ldots, k-1$$

then it is clear that (1.24) is satisfied. With $w_k(\mu) = \tau(\mu)^{k-1}$ we have $w_j(\mu) = e(\mu)^{k-j}\tau(\mu)^{j-1}$. It is clear that one can write

$$w(\mu) = \sum_{j=0} W_j \mu^{-j/k}, \quad W_j \in \mathbb{C}^k.$$

Thus we conclude

$$N^{-1}(x)e^{-i\mu\psi}(L + B(x))e^{i\mu\psi}N(x)v(\mu) \sim 0.$$

Let $\chi(x) \in C_0^\infty(W_2)$ be 1 near $\hat{x}$ and set $v_\mu(x) = \chi(x)e^{i\mu\psi}N(x)v(\mu)$. Note that

$$(L + B(x))v_\mu(x) = O(\mu^{-m})e^{i\mu\psi} + \tilde{\chi}(x)e^{i\mu\psi}$$

for any $m \in \mathbb{N}$ where $\tilde{\chi}(x)$ vanishes on $\{\chi(x) = 1\}$ and hence we have $-\mu\mathrm{Im}\,\psi \leq -c\mu^{(k-1)/k}$ with some $c > 0$ on $(\hat{x} + \Gamma) \cap \mathrm{supp}\tilde{\chi}$. Therefore for any large $m \in \mathbb{N}$ we have

$$\sup_{\hat{x}+\Gamma, |\beta| \leq p} |D^\beta(L + B(x))v_\mu| = O(\mu^{-m})$$

which contradicts with the a priori estimate in Lemma 1.8 because $|v_\mu(\hat{x})| = |N(\hat{x})v(\mu)|$. Then we conclude that $p_1 = \cdots = p_s = 1$. Since $(\hat{x}, \hat{\xi}') \in W_2 \times U_2$ is arbitrary this implies that $\mathrm{rank}\, L(x, \omega(x, \xi'), \xi') = m - r$ for $(x, \xi') \in W_2 \times U_2$. Let $\ell > m - r$ and let $d(x, \xi')$ be any ℓ-th minor of $\omega(x, \xi')I - A(x, \xi')$. Then we have $d(x, \xi') = 0$ on $W_2 \times U_2$. Since $d(x, \xi')$ is real analytic in $W \times U$ and hence $d(x, \xi') = 0$ on $W \times U$. This proves the assertion. □

1.6 A Lemma on Hyperbolic Polynomials

We start with

Definition 1.5. Let $P(\zeta)$ be a (monic) polynomial in ζ. Then we say that $P(\zeta)$ is a hyperbolic polynomial if and only if all zeros of $P(\zeta)$ are real.

Let us study

$$f(t,s) = t^r + f_1(s)t^{r-1} + \cdots + f_r(s) \tag{1.26}$$

where $f_i(s) \in C^\infty(J)$ and J is an open interval containing the origin. We also assume that

$$f_i(0) = 0,\ i = 1, 2, \ldots, r \tag{1.27}$$

that is $t = 0$ is a r folded root of $f(t, 0)$.

Lemma 1.9 ([23]). *Assume (1.27). Then we have*

$$f_i(s) = O(s^i) \text{ as } s \to 0, \quad i = 1, 2, \ldots, r$$

and one can write

$$f(t,s) = f_{(0,0)}(t,s) + O(|t| + |s|)^{r+1}$$

where $f_{(0,0)}(t, s)$ is of homogeneous of degree r and hyperbolic with respect to t for all $s \in \mathbb{R}$.

Remark. Note that $f_{(0,0)}(t, s)$ is given by

$$f(\mu t, \mu s) = \mu^r\{f_{(0,0)}(t,s) + O(\mu)\}, \quad \mu \to 0.$$

Proof of Proposition 1.6. Take $\sigma_j \in \mathbb{N}$ such that $f_j(s) = O(s^{\sigma_j})$ (if $f_j(s) = O(s^k)$ for any k then we take σ_j sufficiently large). Put

$$\min_{1\le j\le r} \frac{\sigma_j}{j} = \lambda = \frac{q}{p} > 0$$

where p, q are relatively prime. We first prove $f_i(s) = O(s^i)$. It is enough to prove $\lambda \ge 1$. We suppose $0 < \lambda < 1$ and derive a contradiction. Plug $t = w|s|^\lambda$ into $f(t, s) = 0$ which yields

$$0 = \sum_{j=0}^{r} w^j |s|^{\lambda j} f_{r-j}(s), \ f_0(s) = 1.$$

Multiplying $|s|^{-\lambda r}$ we get

$$0 = \sum_{j=0}^{r} w^j f_{r-j}(s)|s|^{-\lambda(r-j)}.$$

Let $s \to \pm 0$ then we have

$$0 = \sum_{j=0}^{r} w^j f_{r-j}^{\pm} = 0, \quad f_{r-j}^{\pm} = \lim_{s \to \pm 0} |s|^{-\lambda(r-j)} f_{r-j}(s). \tag{1.28}$$

By the assumption there is at least one $0 \le j \le r-1$ such that $f_{r-j}^{\pm} \neq 0$. We first note that (1.28) has r real roots. Otherwise since $f_0^{\pm} = f_0(s) = 1$, by Rouché's theorem, $f(t,s) = 0$ would have a non real root for small s which contradicts the assumption. We first treat the case $q > 2$. If $f_j^{\pm} \neq 0$ then $\sigma_j q = pj$ and hence $j = nq$ with some $n \in \mathbb{N}$. Then (1.28) with $+$ sign is reduced to

$$w^r + a_1 w^{r-q} + \cdots + a_l w^{r-lq} = 0$$

and (1.28) with $-$ sign is reduced to a similar equation. One can express

$$w^r \left(1 + a_1 (\frac{1}{w})^q + \cdots + a_l (\frac{1}{w})^{lq}\right) = 0, \ (a_l \neq 0).$$

With $W = (1/w)^q$ this turns out to be

$$a_l W^l + \cdots + a_1 W + 1 = 0. \tag{1.29}$$

Noting that (1.29) has a non zero root W, we get a non real root w from $w^q = 1/W$ because $q > 2$ and hence a contradiction. We turn to the case $q = 2$ and hence $p = 1$. From the same arguments (1.28) reduced to

$$w^r + a_1^{\pm} w^{r-2} + \cdots + a_l^{\pm} w^{r-2l} = 0.$$

Since $f_{2k}(s) = s^k(a_{2k} + O(s))$, $s \to 0$ we see that $a_k^+ = a_k^-$ if k is even and $a_k^+ = -a_k^-$ if k is odd. As before we are led to

$$w^r \left(1 + a_1^{\pm} (\frac{1}{w})^2 + \cdots + a_l^{\pm} (\frac{1}{w})^{2l}\right) = 0.$$

With $W = (1/w)^2$ we have

$$a_l^{\pm} W^l + \cdots + a_1^{\pm} W + 1 = 0. \tag{1.30}$$

As observed above, W and $-W$ are the root of (1.30) at the same time and hence from $w^2 = 1/W$ we get a non real root and a contradiction. Thus we have proved that $\lambda \ge 1$ and hence the result.

We turn to the second assertion. Set $t = ws$ and insert this into $f(t,s) = 0$. Then we have

$$\begin{aligned} s^{-r} f(t,s) &= w^r + a_1 w^{r-1} + \cdots + a_r + s g(w,s) \\ &= f_{(0,0)}(w,1) + s g(w,s). \end{aligned}$$

From this we see that $f_{(0,0)}(w, 1) = 0$ has only real roots. Since

$$f_{(0,0)}(t, s) = s^r f_{(0,0)}(\frac{t}{s}, 1)$$

we get the desired assertion. □

Let $h(x, \xi) = \det P_d(x, \xi)$. Taking the Lax-Mizohata theorem into account we assume that h is hyperbolic with respect to $(1, 0, \dots, 0)$.

Definition 1.6. We say that $\rho = (\bar{x}, \bar{\xi}) \in \Omega \times (\mathbb{R}^{n+1} \setminus \{0\})$ is a characteristic of order r for $h(x, \xi)$ (or for P_d) if

$$\partial_x^\alpha \partial_\xi^\beta h(\rho) = 0, \ \ \forall|\alpha + \beta| < r, \quad \partial_x^\alpha \partial_\xi^\beta h(\rho) \neq 0, \ \ \exists|\alpha + \beta| = r.$$

We define the localization of $h(x, \xi)$ at ρ by

$$\sum_{|\alpha+\beta|=r} \frac{1}{\alpha!\beta!} \partial_x^\alpha \partial_\xi^\beta h(\rho) x^\alpha \xi^\beta$$

which is a homogeneous polynomial of degree r in $X = (x, \xi) \in \mathbb{R}^{n+1} \times \mathbb{R}^{n+1}$ and denoted by $h_\rho(X)$.

Here we remark

Lemma 1.10. *Let $\rho = (0, e_n)$, $e_n = (0, \dots, 0, 1)$ be a characteristic of order r for $h(x, \xi)$. Then the localization $h_\rho(x, \xi)$ is independent of ξ_n, that is h_ρ is a homogeneous polynomial in $(x, \tilde{\xi})$, $\tilde{\xi} = (\xi_0, \xi_1, \dots, \xi_{n-1})$.*

Proof of Proposition 1.6. It suffices to show that $\partial_x^\alpha \partial_\xi^\beta h(0, e_n) = 0$ for $|\alpha + \beta| = r$ if $\beta_n \geq 1$. Let us set $g(t) = \partial_x^\alpha \partial_\xi^{\beta - e_n} h(0, te_n)$ which is homogeneous in t of degree $m - r + 1$ and $g(1) = 0$. From the Euler's identity it follows that

$$\partial_x^\alpha \partial_\xi^\beta h(0, e_n) = \frac{d}{dt} g(1) = (m - r + 1) g(1) = 0$$

which proves the assertion. □

From Lemma 1.9 we have

Lemma 1.11. *$h_\rho(X)$ is a hyperbolic polynomial with respect to $\theta = (0, \dots, 0, 1, 0, \dots, 0)$, that is*

$$h_\rho((x, \xi) + \tau\theta) = 0 \Longrightarrow \tau \in \mathbb{R}, \ \ \forall (x, \xi) \in \mathbb{R}^{n+1} \times \mathbb{R}^{n+1}.$$

Chapter 2
Necessary Conditions for Strong Hyperbolicity

Abstract In this chapter we study the Cauchy problem for a first order differential operator defined near the origin of $\mathbb{R}^{n+1}$. We give necessary conditions on L for the Cauchy problem for $P = L + B$ to be C^∞ well posed for any lower order term B, that is necessary conditions for L to be strongly hyperbolic. Denoting by h and $M = (m_{ij})$ the determinant and the cofactor matrix of $L(x, \xi)$ respectively, this necessary condition for strong hyperbolicity is roughly stated that if L is strongly hyperbolic then the Cauchy problem for scalar operators $h + m_{ij}$ is C^∞ well posed for all m_{ij}. In particular, from this condition we see that if L is strongly hyperbolic then at a multiple characteristic point (x, ξ) the maximal size of Jordan blocks in the Jordan canonical form of $L(x, \xi)$ is at most two, which corresponds to a well known Ivrii–Petkov necessary condition for scalar strongly hyperbolic operators. We also see that if the multiple characteristic point (x, ξ) is involutive then $L(x, \xi)$ is diagonalizable for L to be strongly hyperbolic which recovers the necessary condition when L is a system with characteristics of constant multiplicity.

2.1 Necessary Conditions for Strong Hyperbolicity

Let us consider the Cauchy problem for a first order system

$$P(x, D) = \sum_{j=0}^{n} A_j(x)D_j + B(x) = L(x, D) + B(x), \quad A_0(x) = I.$$

We will give necessary conditions in terms of $(m - 1)$-th minors of the principal symbol $L(x, \xi)$ for L to be strongly hyperbolic near the origin. Denoting by $h(x, \xi)$ the determinant of L we state the main result in this chapter.

Theorem 2.1. *Assume that $A_j(x)$ are real analytic in Ω and $0 \in \Omega$. Let $\rho = (0, \bar{\xi})$, $\bar{\xi} \in \mathbb{R}^{n+1} \setminus \{0\}$ be a characteristic of order r for $h(x, \xi)$. Then if L is strongly*

T. Nishitani, *Hyperbolic Systems with Analytic Coefficients*, Lecture Notes in Mathematics 2097, DOI 10.1007/978-3-319-02273-4_2,

hyperbolic near the origin, it follows that every $(m-1)$-th minor of $L(x,\xi)$ vanishes of order $r-2$ at ρ, that is for any $(m-1)$-th minor $q(x,\xi)$ of $L(x,\xi)$ we have

$$\partial_x^\beta \partial_\xi^\alpha q(\rho) = 0, \quad \forall |\alpha+\beta| < r-2.$$

Moreover for any $(m-1)$-th minor $q(x,\xi)$ of $L(x,\xi)$

$$q_\rho(x,\xi) = \sum_{|\alpha+\beta|=r-2} \frac{1}{\alpha!\beta!} \partial_x^\alpha \partial_\xi^\beta q(\rho) x^\alpha \xi^\beta$$

is divisible by $\prod g_j(x,\xi)^{r_j-1}$ where $\prod g_j(x,\xi)^{r_j}$ is the irreducible factorization of $h_\rho(x,\xi)$.

Corollary 2.1. *Assume that $A_j(x)$ are real analytic in Ω and $0 \in \Omega$. Let $\rho = (0,\bar{\xi})$, $\bar{\xi} \in \mathbb{R}^{n+1} \setminus \{0\}$ be a multiple characteristic for $h(x,\xi)$ and V_0 be the generalized eigenspace for $L(\rho)$ associated to the zero eigenvalue. Then if L is strongly hyperbolic near the origin we have*

$$(L(\rho)|V_0)^2 = O$$

where $L(\rho)|V_0$ is the restriction of $L(\rho)$ to V_0.

Proof. We first note that the strong hyperbolicity is invariant under a change of coordinates preserving the x_0 coordinate and also invariant under a change of basis for $\mathbb{C}^m$. Then one can assume that $\rho = (0, e_n)$, $e_n = (0,\ldots,0,1) \in \mathbb{R}^{n+1}$ so that $L(\rho) = A_n(0)$ and $A_n(0) = J \oplus C$ where $C \in M_{m-r}(\mathbb{C})$, $\det C \neq 0$, $J = \oplus_1^s J(r_j) \in M_{r_j}(\mathbb{C})$, $r_1 \geq \cdots \geq r_s$, r is the multiplicity of ρ and

$$J(p) = \begin{bmatrix} 0 & 1 & & \\ & \ddots & \ddots & \\ & & \ddots & 1 \\ & & & 0 \end{bmatrix} \in M_p(\mathbb{C}).$$

We now consider

$$L(0,\xi_0,0,\ldots,0,1) = \xi_0 I_m + (J \oplus C).$$

Then it is clear that the cofactor of $(r_1,1)$-th entry in $L(0,\xi_0,0,\ldots,0,1)$ is

$$(-1)^{r_1+1}\xi_0{}^{r-r_1}\det(\xi_0 I_{m-r} + C) = (-1)^{r_1+1}\xi_0{}^{r-r_1}\det C + O(\xi_0^{r-r_1+1})$$

as $\xi_0 \to 0$. From Theorem 2.1 this must vanish of order $r-2$ and hence $r_1 \leq 2$ which proves the assertion. □

Corollary 2.2. *Assume that* $A_j(x)$ *are real analytic in* $\Omega \ni 0$ *and* L *is strongly hyperbolic near the origin. Then for any* x *close to the origin and for any* ξ'*, the maximal size of Jordan blocks in the Jordan canonical form of* $A(x,\xi') = \sum_{j=1}^{n} A_j(x)\xi_j$ *is at most two.*

Proof. Let $(\bar{x},\bar{\xi}')$ be fixed. We may assume $\bar{x} = 0$ without restrictions. Let $\lambda_1,\ldots,\lambda_s$ be different eigenvalues of $A(0,\bar{\xi}')$ of multiplicities $m_1,\ldots,m_s$. This implies that $(0,-\lambda_j,\bar{\xi}')$ is a characteristic of order m_j for $h(x,\xi)$ and the generalized eigenspace for $A(0,\bar{\xi}')$ associated to λ_j is the generalized eigenspace for $L(0,-\lambda_j,\bar{\xi}')$ associated to zero eigenvalue. Since the assertion is invariant under changes of basis for $\mathbb{C}^m$ we may assume that $A(0,\bar{\xi}')$ has the Jordan canonical form. Then it is clear that $(L(0,-\lambda_j,\bar{\xi}')|V_0)^2 = O$ implies that the size of Jordan blocks associated to the eigenvalue λ_j is at most two. □

Corollary 2.2 clearly corresponds to the well known result of Ivrii and Petkov [23]; if a scalar differential operator P is strongly hyperbolic then every multiple characteristic of the principal symbol of P is at most double.

2.2 Key Propositions

In this section we give a key proposition to prove Theorem 2.1. Let us take $\sigma = (\sigma_0,\sigma_1,\ldots,\sigma_n)$, $\delta = (\delta_0,\delta_1,\ldots,\delta_n) \in \mathbb{Q}_+^{n+1}$. For a differential operator K with C^∞ $m \times m$ matrix coefficients we set for some $y \in \mathbb{R}^{n+1}$

$$K_\lambda(y,x,\xi) = K(\lambda^{-\delta}y + \lambda^{-\sigma}x, \lambda^{\sigma}\xi),$$

as in Sect. 1.1. Note that $K_\lambda(y,x,\xi)$ is a formal meromorphic function in $\lambda^{-\epsilon}$;

$$K_\lambda(y,x,\xi) = \sum_{j=s}^{\infty} K_j(y,x,\xi)\lambda^{-\epsilon j}$$

with some $s \in \mathbb{Z}$ and some $\epsilon \in \mathbb{Q}_+$ where every entry of $K_j(y,x,\xi)$ is a polynomial in (y,x,ξ).

Here we list up several notations which are frequently used in the following.

Notations: Let $\epsilon \in \mathbb{Q}_+$ and let W be an open set in $\mathbb{R}^N$ with a system of local coordinates z.

- $\mathscr{A}(W)\{\{\lambda^{-\epsilon}\}\}$: the set of formal meromorphic functions in $\lambda^{-\epsilon}$ with coefficients in $\mathscr{A}(W, M_m(\mathbb{C}))$, that is the set of all

$$A(z,\lambda) = \sum_{j=t}^{\infty} A_j(z)\lambda^{-\epsilon j}, \quad A_j(z) \in \mathscr{A}(W, M_m(\mathbb{C})), \quad A_t(z) \neq 0 \qquad (2.1)$$

with some $t \in \mathbb{Z}$.

– $\sigma_p(A)(z) = A_t(z)$: the leading coefficient of $A(z, \lambda)$,
– $\mathrm{Ord}\, A = -t\epsilon$: the leading exponent of λ in $A(z, \lambda)$,
– $\sigma_0(A)(z) = A_0(z)$: the coefficient of λ^0 in $A(z, \lambda)$,
– $A(z, \lambda) = O(\lambda^k)$, $k \in \mathbb{Q}$ if $\mathrm{Ord}\, \lambda^{-k} A(z, \lambda) \le 0$,
– $A(z, \lambda) = o(\lambda^k)$ if $\mathrm{Ord}\, \lambda^{-k} A(z, \lambda) \le 0$ and $\sigma_0(\lambda^{-k} A) = 0$.

- $\mathscr{A}(W)\{\{\lambda^{-\epsilon}\}\} \supset \mathscr{A}(W)\{\lambda^{-\epsilon}\}$: the set of convergent meromorphic functions in $\lambda^{-\epsilon}$ with coefficients in $\mathscr{A}(W, M_m(\mathbb{C}))$, that is (2.1) converges and the sum is analytic in W for large λ.
- $\mathscr{A}(W)\{\{\lambda^{-\epsilon}\}\} \supset \mathscr{A}(W)[[\lambda^{-\epsilon}]]$: the set of formal power series in $\lambda^{-\epsilon}$ with coefficients in $\mathscr{A}(W, M_m(\mathbb{C}))$.
- $\mathscr{A}(W)\{\lambda^{-\epsilon}\} \supset \mathscr{A}(W)[\lambda^{-\epsilon}]$: the set of convergent power series in $\lambda^{-\epsilon}$ with coefficients in $\mathscr{A}(W, M_m(\mathbb{C}))$.
- $\mathscr{A}(W)\{\{\tilde{\lambda}\}\} = \cup_{\epsilon \in \mathbb{Q}_+} \mathscr{A}(W)\{\{\lambda^{-\epsilon}\}\}$.
- $\mathscr{A}[[\tilde{\lambda}]] = \cup_{\epsilon \in \mathbb{Q}_+} \mathscr{A}(W)[[\lambda^{-\epsilon}]]$.

Definition 2.1. We say that $A(z, \lambda) \in \mathscr{A}(W)\{\{\tilde{\lambda}\}\}$ is invertible if there exist $B_i(z, \lambda) \in \mathscr{A}(W)\{\{\tilde{\lambda}\}\}$ such that $B_1(z, \lambda)A(z, \lambda) = A(z, \lambda)B_2(z, \lambda) = I$. It is clear that $B_1 = B_2$ in this case and hence we write $A(z, \lambda)^{-1}$ instead of B_i.

We now state a key proposition for the proof of Theorem 2.1. In the following proposition we only assume $A_j(x)$, $B(x) \in C^\infty(\Omega, M_m(\mathbb{C}))$ so that the analyticity of $A_j(x)$, $B(x)$ are not required.

Proposition 2.1. *Assume that $\sigma \ge \delta$, $0 \in \Omega$ and there are a differential operator N with $C^\infty(\Omega)$ $m \times m$ matrix coefficients and a real scalar $\varphi(y, x, \lambda) \in \mathscr{A}(W)[[\tilde{\lambda}]]$ with some open set W in $\mathbb{R}^{2(n+1)}$ satisfying the followings*

$$L(x, \xi)N(x, \xi) = G(x, \xi), \quad G_\lambda(y, x, \varphi_x(y, x, \lambda)) = O,$$
$$G_\lambda^{(\alpha)}(y, x, \varphi_x(y, x, \lambda)) = c_\alpha(y, x, \lambda)K(y, x, \lambda), \ \forall |\alpha| = 1,$$

with scalar $c_\alpha(y, x, \lambda) \in \mathscr{A}(W)[[\tilde{\lambda}]]$ and an invertible $K(y, x, \lambda) \in \mathscr{A}(W)\{\{\tilde{\lambda}\}\}$ where $\sigma_0(c_{(1,0,\ldots,0)})(y, x, \lambda) = c_{(1,0,\ldots,0),0}(y, x) \neq 0$. Then if L is strongly hyperbolic near the origin we have

$$N_\lambda(y, x, \varphi_x(y, x, \lambda))K(y, x, \lambda)^{-1} = O(1).$$

Here $\varphi_x(y, x, \lambda) = (\partial\varphi/\partial x_0, \ldots, \partial\varphi/\partial x_n)$ and $G_\lambda^{(\alpha)}(y, x, \xi) = (\partial/\partial\xi)^\alpha G_\lambda(y, x, \xi)$.

To interpret the result we recall the formulation of the Levi condition for scalar operators with characteristic of constant multiplicities (see Flaschka and Strang [12]). Let P be a scalar differential operator of order m with principal symbol p. Assume that p has the form $p = eq^r$ near $(\bar{x}, \bar{\xi})$ where $e(\bar{x}, \bar{\xi}) \neq 0$, $q(\bar{x}, \bar{\xi}) = 0$, $q_\xi(\bar{x}, \bar{\xi}) \neq 0$ with some $r \in \mathbb{N}$, $r \ge 2$. Then in order that the Cauchy problem for P is C^∞ well posed it is necessary that P verifies the

LEVI CONDITION: *for any* $\varphi \in C^\infty$ *defined near* $\bar{x}$ *such that* $q(x, \varphi_x(x)) = 0$ *with* $\varphi_x(\bar{x}) = \bar{\xi}$ *and for any* $f \in C^\infty$ *one has*

$$P(e^{i\lambda\varphi} f) = O(\lambda^{m-r}), \quad \lambda \to \infty$$

near $\bar{x}$.

Since $q(x, \varphi_x(x)) = 0$ implies $p(x, D)(e^{i\lambda\varphi} f) = O(\lambda^{m-r})$, roughly speaking the Levi condition means

$$p(e^{i\lambda\varphi} f) = O(\lambda^{m-r}) \Longrightarrow P(e^{i\lambda\varphi} f) = O(\lambda^{m-r}). \tag{2.2}$$

Let us now consider $L_\lambda(y, x, D) N_\lambda(y, x, D) e^{i\varphi(y,x,\lambda)} f$. Let $K = O(\lambda^g)$ then the assumption of Proposition 2.1 means that

$$\big(L_\lambda N_\lambda\big)(e^{i\varphi} f) = O(\lambda^g).$$

If we assume that a counter part of (2.2)

$$\big((L_\lambda + B)N_\lambda\big)(e^{i\varphi} f) = O(\lambda^g)$$

would hold for any B then it is quite natural to conclude that $N_\lambda(e^{i\varphi} f) = O(\lambda^g)$ which is nothing but the assertion of the proposition.

Taking $N(x, \xi) = M(x, \xi)$, the cofactor matrix of $L(x, \xi)$, we have the following corollary.

Corollary 2.3. *Assume* $\sigma \geq 8$. *Let* W *be an open set in* $\mathbb{R}^{2(n+1)}$ *and* $\varphi(y, x, \lambda) \in \mathscr{A}(W)[[\lambda]]$ *be a real scalar function satisfying*

$$\begin{aligned} &h_\lambda(y, x, \varphi_x(y, x, \lambda)) = 0, \\ &h_\lambda^{(\alpha)}(y, x, \varphi_x(y, x, \lambda)) - \lambda^\mu a_\alpha(y, x) = o(\lambda^\mu), \quad \forall |\alpha| = 1 \end{aligned} \tag{2.3}$$

with some $\mu \in \mathbb{Q}$ *and* $a_{(1,0,\ldots,0)}(y, x) \neq 0$. *Assume* $0 \in \Omega$ *and* L *is strongly hyperbolic near the origin. Then we have*

$$M_\lambda(y, x, \varphi_x(y, x, \lambda)) = O(\lambda^\mu).$$

Proof. With $N(x, \xi) = M(x, \xi)$, $G(x, \xi) = h(x, \xi)I$ and $K(y, x, \lambda) = \lambda^\mu I$ the hypothesis of Proposition 2.1 is verified. Then we have the assertion by Proposition 2.1. □

In the rest of this section we make detailed looks at the "*principal part*" of $h_\lambda(y, x, \xi)$. Let $h(x, \xi)$ be a polynomial in ξ_0 of degree m with coefficients in $\mathscr{A}(U_1)$ where $U_1 = \Omega \times \Gamma$ is a conic neighborhood of $(0, e_n')$, $e_n' = (0, \ldots, 0, 1) \in \mathbb{R}^n$. Let U be any neighborhood of $(0, 0, e_n') \in \mathbb{R}^{n+1} \times \mathbb{R}^{n+1} \times (\mathbb{R}^n \setminus \{0\})$ which is conic in ξ'. Then we note that $h_\lambda(y, x, \xi)$ is defined in U if λ is large. Let

$\sigma, \delta \in \mathbb{Q}_+^{n+1}$ with $\sigma = (\sigma_0, \ldots, \sigma_{n-1}, 1) = (\tilde{\sigma}, 1) \geq \delta$, $1 > \sigma_j$, $j < n$ and develop $\lambda^{-\sigma_0 r-(m-r)} h_\lambda(y, x, \xi)$ in ascending power of $\lambda^{-\epsilon}$ with $\epsilon \in \mathbb{Q}_+$ such that $\delta/\epsilon, \sigma/\epsilon \in \mathbb{N}_+^{n+1}$

$$\lambda^{-\gamma} h_\lambda(y, x, \xi) = \lambda^{-\gamma} h(\lambda^{-\delta} y + \lambda^{-\sigma} x, \lambda^\sigma \xi) = \sum_{j=p} h_j(y, x, \xi)\lambda^{-\epsilon j}, \quad p \in \mathbb{Z}$$

where $\gamma = \gamma(\sigma_0, r) = \sigma_0 r + (m - r)$.

Definition 2.2. We define $I(h, \rho)$ as the set of $(\sigma, \delta) \in \mathbb{Q}_+^{n+1} \times \mathbb{Q}_+^{n+1}$ such that $\sigma = (\sigma_0, \ldots, \sigma_{n-1}, 1) = (\tilde{\sigma}, 1) \geq \delta$, $1 > \sigma_j$, $j < n$ and

$$h_j = 0 \text{ for } j < 0 \ \ (\text{equivalently } \mathrm{Ord}(\lambda^{-\gamma} h_\lambda) \leq 0).$$

Let $(\sigma, \delta) \in I(h, \rho)$ then we define $\bar{h}_\rho^{\sigma,\delta}(y, x, \xi)$ and $h_\rho^{\sigma,\delta}(y, x, \tilde{\xi})$ by

$$\bar{h}_\rho^{\sigma,\delta}(y, x, \xi) = \lim_{\lambda\to\infty} \lambda^{-\gamma} h_\lambda(y, x, \xi) = \sigma_0(\lambda^{-\gamma} h_\lambda)(y, x, \xi),$$
$$h_\rho^{\sigma,\delta}(y, x, \tilde{\xi}) = \bar{h}_\rho^{\sigma,\delta}(y, x, \tilde{\xi}, 1) = \sigma_0(\lambda^{-\gamma} h_\lambda)(y, x, \tilde{\xi}, 1)$$

where $\tilde{\xi} = (\xi_0, \ldots, \xi_{n-1})$. It is clear that $\bar{h}_\rho^{\sigma,\delta}(y, x, \xi) = \xi_n^m h_\rho^{\sigma,\delta}(y, x, \tilde{\xi}/\xi_n)$.

We state some properties of $h_\rho^{\sigma,\delta}$.

Lemma 2.1. *Let* $(\sigma, \delta) \in I(h, \rho)$ *then* $h_\rho^{\sigma,\delta}(y, x, \tilde{\xi})$ *is a polynomial in* $(y, x, \tilde{\xi})$ *of degree* r *in* ξ_0 *which is a hyperbolic polynomial with respect to* ξ_0.

Proof. By the Taylor expansion of h at $(0, e_n)$ it is obvious that $h_\rho^{\sigma,\delta}(y, x, \tilde{\xi})$ is a polynomial in $(y, x, \tilde{\xi})$. We first note that $h(x, \xi)$ can be factorized as follows;

$$h(x, \xi) = \tilde{h}(x, \xi)e(x, \xi)$$

where $\tilde{h}(x, \xi)$ is a monic polynomial in ξ_0 of degree r for which ρ is a characteristic of order r and $e(\rho) \neq 0$. It is clear that $\lim_{\lambda\to\infty} \lambda^{-(m-r)} e_\lambda(y, x, \tilde{\xi}, 1) = e(0, e_n)$ because $1 > \sigma_j$. From this it follows that

$$h_\rho^{\sigma,\delta}(y, x, \tilde{\xi}) = e(0, e_n) \lim_{\lambda\to\infty} \lambda^{-\sigma_0 r} \tilde{h}_\lambda(y, x, \tilde{\xi}, 1)$$

and hence $h_\rho^{\sigma,\delta}$ is a polynomial in ξ_0 of degree r. This proof also shows that $(\sigma, \delta) \in I(\tilde{h}, \rho)$ if $(\sigma, \delta) \in I(h, \rho)$. Since

$$\lambda^{-\gamma} h_\lambda(y, x, \xi_0, \xi') = \{\bar{h}_\rho^{\sigma,\delta}(y, x, \xi_0, \xi') + \lambda^{-\epsilon} h_1(y, x, \xi_0, \xi', \lambda^{-\epsilon})\}$$

the hyperbolicity of $h_\rho^{\sigma,\delta}$ follows from that of $h_\lambda(x, \xi)$ and Rouché's theorem. □

In what follows to simplify notations we denote

$$\partial_{\tilde{\xi}}^{\tilde{\alpha}} \partial_x^{\beta} h(\rho) = h_{(\beta)}^{(\tilde{\alpha})}(\rho).$$

Lemma 2.2. *Let* $\sigma = (\bar{\sigma}, \ldots, \bar{\sigma}, 1)$, $\delta = (\bar{\delta}, \ldots, \bar{\delta})$, $\bar{\delta} = 1 - \bar{\sigma}$. *If* $\bar{\sigma} > \bar{\delta} > 0$ *then* $(\sigma, \delta) \in I(h, \rho)$ *and*

$$h_{\rho}^{\sigma,\delta}(y, x, \tilde{\xi}) = h_{\rho}(y, \tilde{\xi}) = \sum_{|\tilde{\alpha}+\beta|=r} h_{(\beta)}^{(\tilde{\alpha})}(0, e_n) y^{\beta} \tilde{\xi}^{\tilde{\alpha}} / (\tilde{\alpha}! \beta!)$$

where $\tilde{\alpha} = (\alpha_0, \ldots, \alpha_{n-1})$ *and hence with these* σ, δ, *the polynomial* $h_{\rho}^{\sigma,\delta}$ *is invariantly defined. If* $\bar{\sigma} = \bar{\delta} = 1/2$ *and* $y = (0, \ldots, 0, y_n)$ *then we have* $(\sigma, \delta) \in I(h, \rho)$ *and*

$$h_{\rho}^{\sigma,\delta}(y, x, \tilde{\xi}) = h_{\rho}(\tilde{x}, y_n, \tilde{\xi})$$

with $\tilde{x} = (x_0, \ldots, x_{n-1})$.

Proof. Noticing $h_{(\beta)}^{(\tilde{\alpha})}(0, e_n) = 0$, $|\tilde{\alpha}+\beta| < r$ the assertion follows from Lemma 1.10 and the Taylor expansion of h at $(0, e_n)$. □

2.3 Proof of Theorem 2.1 (Simplest Case)

In this section we give a proof of Theorem 2.1 in the simplest case. To do so we first discuss the zeros of h_λ when λ is large. Recall

$$\gamma = \sigma_0 r + (m - r).$$

Proposition 2.2. *Let* $\rho = (0, e_n)$ *be a characteristic of order* m *for* h *and* $(\sigma, \delta) \in I(h, \rho)$ *so that* $\mathrm{Ord}(\lambda^{-\sigma_0 m} h_\lambda) \leq 0$. *Then for every open conic set* $V \subset U$ *there are an open conic set* $W \subset V$ *and* $\epsilon \in \mathbb{Q}_+$ *such that* $\lambda^{-\sigma_0 m} h_\lambda(y, x, \xi) = 0$ *admits* m *roots in* $\mathscr{A}(W)[\lambda^{-\epsilon}]$, *that is one can write*

$$\lambda^{-\sigma_0 m} h_\lambda(y, x, \xi) = \prod_{j=1}^{m} (\xi_0 - \omega^j(y, x, \xi', \lambda))$$

where

$$\omega^j(y, x, \xi', \lambda) = \sum_{p=0}^{\infty} \omega_p^j(y, x, \xi') \lambda^{-\epsilon p}, \quad \omega_p^j(y, x, \xi') \in \mathscr{A}(W)$$

and $\omega_0^j(y, x, \xi')$ *are the roots of* $\bar{h}_{\rho}^{\sigma,\delta}(y, x, \xi) = 0$.

We will postpone the proof until the end of the proof of Theorem 2.1.

Remark. Although we need another proof this proposition is also valid in C^∞ category, that is we get ω^j not in $\mathscr{A}(W)[\lambda^{-\epsilon}]$ but in $\mathscr{A}(W)[[\lambda^{-\epsilon}]]$ assuming that the coefficients of h is in $C^\infty(U_1)$. Note that $\omega^j(y,x,\xi',\lambda)$ are real which follow from the hyperbolicity of h.

Remark. Let $h(x,\xi)$ be a monic polynomial in ξ_0 with coefficients in $\mathscr{A}(U)$. Then for any open set $V \subset U$ there exist an open set $W \subset V$ and $\omega^j(x,\xi') \in \mathscr{A}(W)$ such that

$$h(x,\xi) = \prod_{j=1}^{m}(\xi_0 - \omega^j(x,\xi')).$$

Thus the main part of the assertion of the proposition is that, as far as the neighborhood of λ is concerned, one can always choose a neighborhood of $\lambda = \infty$.

Corollary 2.4. *Let $\rho = (0, e_n)$ be a characteristic of order r for h and $(\sigma, \delta) \in I(h,\rho)$ so that $\mathrm{Ord}(\lambda^{-\gamma}h_\lambda) \le 0$. Then for every open conic set $V \subset U$ there are an open conic set $W \subset V$ and $\epsilon \in \mathbb{Q}_+$ such that $h_\lambda(y,x,\xi) = 0$ admits r roots $\omega^j(y,x,\xi',\lambda) \in \mathscr{A}(W)[\lambda^{-\epsilon}]$*

$$h_\lambda(y,x,\omega^j(y,x,\xi',\lambda),\xi') = 0.$$

Proof. Recall that $h(x,\xi)$ can be factorized as $h(x,\xi) = \tilde{h}(x,\xi)e(x,\xi)$ where $\tilde{h}(x,\xi)$ is a polynomial in ξ_0 of degree r for which ρ is a characteristic of order r and $e(\rho) \neq 0$. Noting

$$e_\lambda(y,x,\xi) = e(\lambda^{-\delta}y + \lambda^{-\sigma}x, \lambda^\sigma\xi) = \lambda^{m-r}\big(e(0,\xi_n e_n) + O(\lambda^{-\tilde{\epsilon}})\big)$$

and $\mathrm{Ord}(\lambda^{-\sigma_0 r}\tilde{h}_\lambda) \le 0$ the assertion follows from Proposition 2.2. □

We turn to the proof of Theorem 2.1. Let $\rho = (0, e_n)$ be a characteristic of order r for h. In Corollary 2.4 shrinking W and taking $\bar{\lambda}$ large, if necessary, we may assume that for $\lambda \ge \bar{\lambda}$ we have either

$$|\omega^i(y,x,\xi',\lambda) - \omega^j(y,x,\xi',\lambda)| \ge c\lambda^{a_{ij}}$$

with some $c > 0$ and $a_{ij} \in \mathbb{Q}$ or

$$\omega^i(y,x,\xi',\lambda) = \omega^j(y,x,\xi',\lambda)$$

in W. To simplify notations we write $z = (y,x)$. Thus one can write

$$\lambda^{-\gamma}h_\lambda = \prod_{j=1}^{t}(\xi_0 - \omega^j(z,\xi',\lambda))^{r_j}\lambda^{-(m-r)}e_\lambda, \quad \omega^j \in \mathscr{A}(U)[[\lambda^{-\epsilon}]] \tag{2.4}$$

where ω^j admits the following development

$$\omega^j(z,\xi',\lambda) = \sum_{p=0}^{\infty} \omega^j_p(z,\xi')\lambda^{-\epsilon p}, \quad \omega^j_p \in \mathscr{A}(U)$$

where U is an open set in $\mathbb{R}^{2(n+1)} \times \mathbb{R}^n$.

Definition 2.3. Let $\omega(z,\xi',\lambda) \in \mathscr{A}(U)[[\tilde{\lambda}]]$ be a root of $h_\lambda(z,\xi_0,\xi')$ in ξ_0. We say that $\varphi(z,\lambda) \in \mathscr{A}(W)[[\tilde{\lambda}]]$ is a characteristic function to ω at $\hat{z} = (\hat{y},\hat{x})$ if φ satisfies the following

$$\partial_{x_0}\varphi = \omega(z,\varphi_{x'},\lambda), \; (\varphi_0)_x(\hat{z}) \neq 0.$$

Now let $\varphi(z,\lambda) = \sum_{p=0}\varphi_p(z)\lambda^{-\epsilon p} \in \mathscr{A}(U)[[\lambda^{-\epsilon}]]$ be a characteristic function to ω^j. Let us define $a_j \in \mathbb{Q}$ and $b_j \in \mathbb{Q}$ by

$$\begin{cases} \partial_{\xi_0}^{r_j} h_\lambda(z,\varphi_x) = \lambda^{-a_j}(c_j(z)+o(1)), \quad c_j(z) \not\equiv 0, \\ \displaystyle\prod_{k=1,k\neq j}^{t} (\omega^j(z,\varphi_x,\lambda)-\omega^k(z,\varphi_x,\lambda))^{r_k} = \lambda^{-b_j}(c'_j(z)+o(1)), \quad c'_j(z)\not\equiv 0. \end{cases} \tag{2.5}$$

Then it is clear that a_j and b_j are independent of $\varphi(z,\lambda)$ and $a_j = b_j - \gamma$.

In this section, as the simplest case, we assume that ω^j are mutually distinct so that $t = r$ and $r_j = 1$. Let φ be a characteristic function to ω^j. From (2.5) we see that $h_\lambda^{(\alpha)}(z,\varphi_x) = O(\lambda^{-a_j})$ for $|\alpha| = 1$. Let $q(x,\xi)$ be any $(m-1)$-th minor of $L(x,\xi)$. From Corollary 2.3 it follows that

$$q_\lambda(z,\varphi_x) = O(\lambda^{-a_j}). \tag{2.6}$$

From this we can derive information about the vanishing order of q at $(0,e_n)$. To do so we prepare

Lemma 2.3. *Let* $\sigma = (\bar{\sigma},\ldots,\bar{\sigma},1)$, $\delta = (\bar{\delta},\ldots,\bar{\delta})$, $\bar{\delta} = 1-\bar{\sigma}$, $3\bar{\delta} > 1$. *Assume that one of the following conditions holds*

(a) $\mathrm{Ord}(\lambda^{-\gamma}q_\lambda) \le 0$,
(b) *the degree of* $\sigma_p(\lambda^{-\gamma}q_\lambda)$ *with respect to* ξ_0 *is at least* $r-2$.

Then we have

$$\partial_x^\beta\partial_\xi^\alpha q(0,e_n) = 0, \quad \forall |\alpha+\beta| < r-2.$$

Proof. Let us write $x = (\tilde{x},x_n) = (x_0,\ldots,x_{n-1},x_n)$ and the same notation for ξ. We also write $\alpha = (\tilde{\alpha},\alpha_n) = (\alpha_0,\ldots,\alpha_{n-1},\alpha_n)$. Suppose that there were $s \le r-3$

such that $\partial_x^\beta \partial_\xi^\alpha q(0, e_n) = 0$ for all $|\alpha + \beta| < s$ and $\partial_x^\beta \partial_\xi^\alpha q(0, e_n) \neq 0$ for some $|\alpha + \beta| = s$. From the Euler's identity (see the proof of Lemma 1.10) we have

$$\partial_x^\beta \partial_\xi^\alpha q(0, e_n) = q_{(\beta)}^{(\alpha)}(0, e_n) = 0$$

for any $|\alpha + \beta| \leq s$ with $\alpha_n \neq 0$. Thus we would have $q_{(\beta)}^{(\tilde{\alpha})}(0, e_n) \neq 0$ with some $|\tilde{\alpha} + \beta| = s \leq r - 3$. Let us set

$$S = \{(\tilde{\alpha}, \beta) \mid |\tilde{\alpha} + \beta| = s, q_{(\beta)}^{(\tilde{\alpha})}(0, e_n) \neq 0\}, \quad \kappa = \max_{(\tilde{\alpha},\beta)\in S} F(\tilde{\alpha}, \beta) \quad (2.7)$$

$$F(\tilde{\alpha}, \beta) = -\gamma - \bar{\delta}|\beta| + \bar{\sigma}|\tilde{\alpha}| + m - |\tilde{\alpha}| = \bar{\delta}(r - |\tilde{\alpha} + \beta|).$$

Note that one can write

$$\lambda^{-\gamma} q_\lambda(y, x, \xi) = \lambda^{-\gamma} q(\lambda^{-\delta} y + \lambda^{-\sigma} x, \lambda^\sigma \xi) = \sum I_j$$

where

$$I_j = \lambda^{-\gamma} \sum_{|\tilde{\alpha}+\beta|=j} q_{(\beta)}^{(\tilde{\alpha})}(0, \lambda \xi_n e_n)(\lambda^{-\delta} y + \lambda^{-\sigma} x)^\beta (\lambda^\sigma \tilde{\xi})^{\tilde{\alpha}} / \tilde{\alpha}! \beta!.$$

Thus I_j is a sum of such terms

$$q_{(\beta)}^{(\tilde{\alpha})}(0, e_n)(y_n + \lambda^{\bar{\delta}-1} x_n)^{\beta_n} (\tilde{y} + \lambda^{\bar{\delta}-\bar{\sigma}} \tilde{x})^{\tilde{\beta}} \tilde{\xi}^{\tilde{\alpha}} \xi_n^{m-1-|\tilde{\alpha}|} \lambda^{F(\tilde{\alpha},\beta)-1} / \tilde{\alpha}! \beta$$

with $|\tilde{\alpha} + \beta| = j$. Note that

$$I_s = \lambda^{\kappa-1}(J(z, \xi) + o(1))$$

where $J(z, \xi)$ is a polynomial in (z, ξ) of degree (with respect to ξ_0) at most s. Since $S \neq \emptyset$ then we have $J \neq 0$. For $j \geq s + 1$ we have

$$\max_{|\tilde{\alpha}+\beta|=j} F(\tilde{\alpha}, \beta) < \max_{|\tilde{\alpha}+\beta|=s} F(\tilde{\alpha}, \beta)$$

so that

$$\sigma_p(\lambda^{-\gamma} q_\lambda) = J(z, \xi).$$

This proves that the condition (b) does not hold. For $|\tilde{\alpha}+\beta| = s$ we have $r-|\tilde{\alpha}+\beta| \geq 3$ since $s \leq r - 3$ and hence $\bar{\delta}(r - |\tilde{\alpha} + \beta|) \geq 3\bar{\delta} > 1$. This implies $\kappa > 1$ so that $\mathrm{Ord}(\lambda^{-\gamma} q_\lambda) > 0$ and hence (a) does not hold. Thus neither (a) nor (b) holds. This contradiction proves the assertion. □

Lemma 2.4. *Assume that there is an open set $V \subset U$ such that*

$$q_\lambda(z, \varphi_x) = O(\lambda^{-a_j})$$

holds for any characteristic function $\varphi(z, \lambda)$ to ω^j with $(z, \varphi_{x'}(z, \lambda)) \in V$, $j = 1, \dots, r$. We further assume $\mathrm{Ord}(\lambda^{-\gamma} q_\lambda) > 0$. Then $\sigma_p(\lambda^{-\gamma} q_\lambda)$ is a polynomial of degree at least r in ξ_0.

Proof. We note that $(\sigma, \delta) \in I(h, \rho)$ and hence $\bar{h}_\rho^{\delta,\sigma}$ is a polynomial in ξ_0 of degree r. From the assumption there exists $\tau \in \mathbb{Q}_+$ such that

$$q_\lambda(z, \xi) = \lambda^\gamma \sum_{j=0} \tilde{q}_j(z, \xi)\lambda^{-\epsilon j + \tau}, \quad \tilde{q}_0(z, \xi) \neq 0.$$

We will show that

$$\tilde{q}_0(z, \xi) = \sigma_p(\lambda^{-\gamma} q_\lambda)(z, \xi)$$

is divisible by $\bar{h}_\rho^{\sigma,\delta}(z, \xi)$ as a polynomial in ξ_0 when (z, ξ') is in some open set. In particular the degree of $\sigma_p(\lambda^{-\gamma} q_\lambda)$ as a polynomial in ξ_0 is greater or equal to that of $\bar{h}_\rho^{\sigma,\delta}$ which is the desired assertion.

Let $g = \lambda^{-\tau-\gamma} q_\lambda$ then $g = \sum_{j=0} \tilde{q}_j(z, \xi)\lambda^{-\epsilon j}$ and $g_0(z, \xi) = \tilde{q}_0(z, \xi)$. Let $\varphi(z, \lambda)$ be a characteristic function to ω^j. Then from the assumption we have

$$g(z, \varphi_x, \lambda) = O(\lambda^{-\tau-\gamma-a_j}) = o(\lambda^{-b_j}).$$

Note that for any $(z, \xi') \in V$ there exists a characteristic function $\varphi(z, \lambda)$ to ω^j with $\varphi_{x'}(z, \lambda) = \xi'$ then we can assume that

$$g(z, \omega^j(z, \xi', \lambda), \xi', \lambda) = o(\lambda^{-b_j}) \tag{2.8}$$

holds for any $(z, \xi') \in V$. Repeating the same arguments we can assume that

$$\prod_{k=1, k\neq j}^{r} (\omega^j(z, \xi', \lambda) - \omega^k(z, \xi', \lambda)) = \lambda^{-b_j}(c_j'(z, \xi') + o(1)) \tag{2.9}$$

holds for any $(z, \xi') \in V$ where $c_j'(z, \xi') \neq 0$. Set

$$f(z, \xi, \lambda) = \prod_{j=1}^{r} (\xi_0 - \omega^j(z, \xi', \lambda))$$

and we divide g by f as polynomials in ξ_0

$$g(z, \xi, \lambda) = k(z, \xi, \lambda) f(z, \xi, \lambda) + \ell(z, \xi, \lambda) \tag{2.10}$$

where the degree of ℓ with respect to ξ_0 is at most $r-1$. From (2.8) it follows that

$$\ell(z,\omega^j,\lambda)=o(\lambda^{-b_j}). \tag{2.11}$$

Since one can write

$$\frac{\ell(z,\xi,\lambda)}{f(z,\xi,\lambda)}=\sum_{j=1}^{r}\frac{C_j(z,\xi',\lambda)}{\xi_0-\omega^j(z,\xi',\lambda)}$$

which gives

$$\ell(z,\xi,\lambda)=\sum_{j=1}^{r}C_j(z,\xi',\lambda)\prod_{p=1,p\neq j}^{r}(\xi_0-\omega^p(z,\xi',\lambda)).$$

From (2.9) and (2.11) we obtain $C_j(z,\xi',\lambda)=o(1)$ and hence

$$\lim_{\lambda\to\infty}\ell(z,\xi,\lambda)=0.$$

Now from (2.10) we get the assertion. □

Proof of Theorem 2.1 (simplest case). Since we have (2.6) then the assertion follows immediately from Lemmas 2.3 and 2.4. □

Proof of Proposition 2.2. Since a real analytic function in an open set $V\subset\mathbb{R}^N$ can be continued to a holomorphic function in a complex neighborhood $\tilde{V}$ of V then Proposition 2.2 is a consequence of the following proposition. □

Proposition 2.3. *Let $h(y,x,s)$ be*

$$h(y,x,s)=y^m+f_1(x,s)y^{m-1}+\cdots+f_m(x,s),$$

$$f_j(x,s)\in\mathscr{O}(\Delta_N^{\delta}(\hat{x})\times\Delta_1^{r}(0))$$

where $\Delta_N^{\delta}(\hat{x})=\{x\in\mathbb{C}^N\mid|x-\hat{x}|<\delta\}$, $\hat{x}\in\mathbb{R}^N$ and $\mathscr{O}(W)$ denotes the set of all holomorphic functions in W. Then there are $\Delta_N^{\delta'}(a)\subset\Delta_N^{\delta}(\hat{x})$, $a\in\mathbb{R}^N$ and $r'>0$ such that $h(y,x,s)$ has m roots which are Puiseux series in s, $|s|<r'$ with coefficients in $\mathscr{O}(\Delta_N^{\delta'}(a))$

$$h(y,x,s)=\prod_{j=1}^{m}(y-\varphi^j(x,s))$$

where with some $p\in\mathbb{N}$ we have

$$\varphi^j(x,s)=\sum_{k=0}^{\infty}\varphi_k^j(x)s^{k/p},\quad \varphi_k^j(x)\in\mathscr{O}(\Delta_N^{\delta'}(a)).$$

Proof. Denote by $\mathscr{O}_{(\hat{x},0)}$ the ring of holomorphic germs at $(\hat{x},0) \in \mathbb{C}^N \times \mathbb{C}$. Since $\mathscr{O}_{(\hat{x},0)}[y]$ (polynomial ring over $\mathscr{O}_{(\hat{x},0)}$) is a unique factorization domain we can write

$$h(y,x,s) = \prod_{j=1}^{k} p_j(y,x,s)^{r_j}$$

where $p_j(y,x,s) \in \mathscr{O}_{(\hat{x},0)}[y]$ are irreducible. Let

$$p_j(y,x,s) = f_{j0}(x,s)y^{m_j} + \cdots + f_{jm_j}(x,s)$$

where one can assume that $f_{j0}(x,s) = 1$. Note that it is enough to prove the assertion for each p_j. We first show the following lemma.

Lemma 2.5. *Let $\omega_j(x,s)$ be the discriminant of $p_j(y,x,s)$. Then there are $\Delta_N^{\delta'}(a) \subset \Delta_N^{\delta}(\hat{x})$, $a \in \mathbb{R}^N$ and $r' > 0$ such that*

$$\omega_j(x,s) \neq 0, \quad \forall x \in \Delta_N^{\delta'}(a), \ \forall s \in \Delta_1^{r'}(0)^*$$

where $\Delta_1^{r'}(0)^* = \Delta_1^{r'}(0) \setminus \{0\}$.

Proof. To simplify notations we set $\omega(x,s) = \omega_1(x,s)$, $p(y,x,s) = p_1(y,x,s)$. If $\omega(x,s) \equiv 0$ then $p(y,x,s)$ is reducible in $\mathscr{O}_{(\hat{x},0)}[y]$ which is a contradiction. Thus one can assume $\omega(x,s) \neq 0$ in $\mathscr{O}_{(\hat{x},0)}$. Here we note that

$$\begin{aligned} &\omega(x,s) = 0, \ \forall x \in \Delta_N^{\delta}(\hat{x}) \cap \mathbb{R}^N, \ \forall |s| < r \\ &\Longrightarrow \omega(x,s) = 0, \ \forall x \in \Delta_N^{\delta}(\hat{x}), \ \forall |s| < r. \end{aligned} \tag{2.12}$$

If there is $a \in \Delta_N^{\delta}(\hat{x}) \cap \mathbb{R}^N$ with $\omega(a,0) \neq 0$, taking δ', r' small we get the desired assertion. Otherwise we have $\omega(x,0) = 0$, $x \in \Delta_N^{\delta}(\hat{x}) \cap \mathbb{R}^N$. Since $\omega(x,s) \not\equiv 0$ from (2.12) there are $x' \in \Delta_N^{\delta}(\hat{x}) \cap \mathbb{R}^N$, $|s'| < r$ with $\omega(x',s') \neq 0$. Hence we can take $b \in \Delta_N^{\delta}(\hat{x}) \cap \mathbb{R}^N$ so that

$$\omega(b,s) \not\equiv 0, \quad \omega(b,0) = 0.$$

From the Weierstrass preparation theorem it follows that

$$\omega(x,s) = e(x,s)\{s^n + g_1(x)s^{n-1} + \cdots + g_n(x)\}$$

near $(b,0)$ with $g_j(b) = 0$, $e(b,0) \neq 0$. Let i be the integer satisfying $g_{i+1} = \cdots = g_n = 0$ and $g_i \neq 0$ in $\mathscr{O}_b$. Since $b \in \Delta_N^{\delta}(\hat{x}) \cap \mathbb{R}^N$ there is $a \in \Delta_N^{\delta}(\hat{x}) \cap \mathbb{R}^N$ such that

$$g_{i+1} = \cdots = g_n = 0 \text{ in } \mathscr{O}_a, \quad g_i(a) \neq 0, \ e(a,0) \neq 0.$$

Therefore one can find δ', r' with $\Delta_N^{\delta'}(a) \subset \Delta_N^{\delta}(\hat{x})$ such that

$$\omega(x,s) = e(x,s)s^i\{s^{n-i} + \cdots + g_i(x)\}, \quad x \in \Delta_N^{\delta'}(a), \quad |s| < r'.$$

Since

$$|\omega(x,s)| \geq |e(x,s)||s|^i\{|g_i(x)| - C|s|\}$$

with some constant C, taking r', δ' sufficiently small, it follows that $\omega(x,s) \neq 0$, $x \in \Delta_N^{\delta'}(a)$, $s \in \Delta_1^{r'}(0)^*$. Now by induction we get the desired result. □

We turn to the proof of the proposition. Let $p(y,x,s) = p_j(y,x,s)$, $\omega(x,s) = \omega_j(x,s)$, $m = m_j$. Since $\omega(a,r'/2) \neq 0$ by Lemma 2.5 there are m different holomorphic functions $\varphi_j(x,s)$ defined near $(a,r'/2)$ which solve $p(y,x,s) = 0$. Here we note that any closed curve in $\Delta_N^{\delta'}(a) \times \Delta_1^{r'}(0)^*$ can be deformed continuously in $\Delta_N^{\delta'}(a) \times \Delta_1^{r'}(0)^*$ to a curve $C^h = C \cdots C$ (h times) with some $h \in \mathbb{N}$ where C is the curve given by $C : (x(t), s(t)) = (a, r'e^{it}/2)$, $0 \leq t \leq 2\pi$. Therefore the arguments proving the assertion in the case x is fixed can be applied without modifications (see, for example [1]). This argument shows that there are a partition of the set $I = \{1,2,\ldots,m\}$; $I = \cup_{j=1}^{l} I_j$ and the functions $F_j(x,s)$ which are holomorphic in $\Delta_N^{\delta'}(a) \times \Delta_1^{r'}(0)$ satisfying

$$F_j(x,s) = \tilde{\varphi}_i(\pi_j(x,s)), \quad i \in I_j$$

where π_j is the map: $(x,s) \to (x, s^{h_j})$ with $h_j = |I_j|$, the cardinal number of I_j, and $\tilde{\varphi}_i(x,s)$ is the analytic continuation of $\varphi_i(x,s)$. Since we can write

$$F_j(x,s) = \sum_{k=0}^{\infty} f_{jk}(x)s^k, \quad f_{jk}(x) \in \mathscr{O}(\Delta_N^{\delta'}(a))$$

it is clear that we have a development of the form

$$\varphi_i(x,s) = \sum_{k=0}^{\infty} \varphi_{jik}(x)s^{k/h_j}, \quad \varphi_{jik}(x) \in \mathscr{O}(\Delta_N^{\delta'}(a)), \quad i \in I_j.$$

This is the desired conclusion. □

2.4 Proof of Theorem 2.1 (General Case)

When $h_\lambda = 0$ has a multiple root, that is some $r_j \geq 2$ in (2.4), the choice of $N = M$ does not work and we need another choice of N. A naive idea for finding

N is found in the proof of Theorem 1.1, that is we factor out the multiple factor from the cofactor matrix and we take the resulting (factored out) matrix as N. We begin with the following proposition.

Proposition 2.4. *Assume that L is strongly hyperbolic near the origin. Let $\omega^j \in \mathscr{A}(U)[[\tilde{\lambda}]]$ be a root of h_λ with multiplicity $r_j \geq 2$. Then there is a positive R such that*

$$M_\lambda^{(\alpha)}(z, \omega^j(z, \xi', \lambda), \xi') = O$$

for $(z, \xi') \in U$, $\forall |\alpha| \leq r_j - 2$ and $\lambda > R$.

From (2.5) we have

$$\partial_{\xi_0}^{r_j} h_\lambda(z, \omega^j(z, \xi', \lambda), \xi') = \lambda^{-a_j}(c_j(z, \xi') + o(1)) \tag{2.13}$$

with $c_j(z, \xi') \not\equiv 0$. Let us set

$$N^j(x, \xi) = \partial_{\xi_0}^{r_j - 1} M(x, \xi).$$

Then we have

Proposition 2.5. *Notations being as above. Assume that L is strongly hyperbolic near the origin. Let*

$$L(x, \xi) N^j(x, \xi) = G^j(x, \xi).$$

Then for every open set $V \subset U$ there is an open set $W \subset V$ such that $G^j(x, \xi)$ satisfies the followings

$$G_\lambda^j(z, \omega^j(z, \xi', \lambda), \xi') = O,$$
$$G_\lambda^{j(\alpha)}(z, \omega^j(z, \xi', \lambda), \xi') = c_\alpha^j(z, \xi', \lambda) K^j(z, \xi', \lambda), \quad \forall |\alpha| = 1$$

with scalar $c_\alpha^j(z, \xi', \lambda) \in \mathscr{A}(W)[[\tilde{\lambda}]]$ where $c_{(1,0,\ldots,0)}^j(z, \xi', \lambda) = 1$ and invertible $K^j(z, \xi', \lambda) \in \mathscr{A}(W)\{\{\tilde{\lambda}\}\}$ which verifies

$$r_j(\partial_{\xi_0}^{r_j} h_\lambda(z, \omega^j, \xi'))^{-1} \partial_{\xi_0}^{r_j - 1} M_\lambda(z, \omega^j, \xi') = N_\lambda^j(z, \omega^j, \xi') K^j(z, \xi', \lambda)^{-1}.$$

In particular

$$\mathrm{Ord}(N_\lambda^j(z, \omega^j, \xi') K^j(z, \xi', \lambda)^{-1}) = \mathrm{Ord}(\partial_{\xi_0}^{r_j - 1} M_\lambda(z, \omega^j, \xi')) + a_j.$$

We will give the proofs of these two propositions in the next section. Admitting these propositions we give a proof of Theorems 2.1 (general case).

Take σ, δ as in Lemma 2.3 with $3\bar{\delta} > 1$ and let $\varphi(z,\lambda)$ be a characteristic function to $\omega^j(z,\xi',\lambda)$. By virtue of Proposition 2.5 we can apply Proposition 2.1 with $N = N^j$, $G = G^j$ to get

$$\partial_{\xi_0}^{r_j-1} q_\lambda(z, \varphi_x(z,\lambda)) = O(\lambda^{-a_j}) \tag{2.14}$$

for any $(m-1)$-th minor q of L.

Lemma 2.6. *Assume* $\mathrm{Ord}(\lambda^{-\gamma} q_\lambda) > 0$. *We further assume that there exists an open set* $V \subset U$ *such that we have*

$$q_\lambda^{(\alpha)}(z,\varphi_x) = 0,\ |\alpha| \le r_j - 2,\ \partial_{\xi_0}^{r_j-1} q_\lambda(z,\varphi_x) = O(\lambda^{-a_j})$$

for any characteristic function $\varphi(z,\lambda)$ *to* ω^j *with* $(z,\varphi_{x'}(z,\lambda)) \in V$, $j = 1,\dots,t$. *Then* $\sigma_p(\lambda^{-\gamma} q_\lambda)(z,\xi)$ *is divisible by* $\bar{h}_\rho^{\sigma,\delta}(z,\xi)$ *as polynomials in* ξ_0 *when* (z,ξ') *in some open set. In particular the degree of* $\sigma_p(\lambda^{-\gamma} q_\lambda)$ *as a polynomial in* ξ_0 *is at least* r.

Proof. From the assumption there is $\tau \in \mathbb{Q}_+$ such that

$$q_\lambda(z,\xi) = \lambda^\gamma \sum_{j=0} \tilde{q}_j(z,\xi)\lambda^{-\epsilon j+\tau}, \quad \tilde{q}_0(z,\xi) \ne 0.$$

Let $g = \lambda^{-\tau-\gamma} q_\lambda$ so that $g = \sum_{j=0} \tilde{q}_j(z,\xi)\lambda^{-\epsilon j}$ where $g_0(z,\xi) = \tilde{q}_0(z,\xi)$. From the assumption we have

$$g^{(\alpha)}(z,\varphi_x,\lambda) = 0,\ |\alpha| \le r_j - 2,\ \partial_{\xi_0}^{r_j-1} g(z,\varphi_x,\lambda) = O(\lambda^{-\tau-\gamma-a_j}) = O(\lambda^{-\tau-b_j}).$$

Note that for any $(z,\xi') \in V$ there is a characteristic function $\varphi(z,\lambda)$ (to ω^j) with $\varphi_{x'}(z,\lambda) = \xi'$ and hence one can assume

$$g^{(\alpha)}(z,\omega^j(z,\xi',\lambda),\xi',\lambda) = 0, \quad |\alpha| \le r_j - 2,$$
$$\partial_{\xi_0}^{r_j-1} g(z,\omega^j(z,\xi',\lambda),\xi',\lambda) = O(\lambda^{-\tau-b_j})$$

for any $(z,\xi') \in V$. Similarly we can assume that

$$\prod_{k=1,k\ne j}^{t} (\omega^j(z,\xi',\lambda) - \omega^k(z,\xi',\lambda))^{r_k} = \lambda^{-b_j}(c_j'(z,\xi') + o(1))$$

for any $(z,\xi') \in V$ where $c_j'(z,\xi') \ne 0$. We adapt simpler notations. Let $\omega^j(x,s)$, $1 \le j \le t$ be formal power series in s which are mutually different

$$\omega^j(x,s) = \sum_{p=0} \omega_p^j(x)s^p, \quad \omega_p^j(x) \in \mathscr{A}(U)$$

where U is some open set in $\mathbb{R}^{n^*}$. Let $f(x, y, s)$ be a formal power series in s with coefficients in $\mathscr{A}(U)[y]$

$$f(x, y, s) = \sum_{p=0} f_p(x, y)s^p$$

where $f_p(x, y) \in \mathscr{A}(U)[y]$ of which degree in y is at most m_0.

Lemma 2.7. *Assume that*

$$\prod_{k=1,k\neq j}^{t} (\omega^j(x, s) - \omega^k(x, s))^{r_k} = s^{b_j}(c_j(x) + o(s))$$

as $s \to 0$ where $c_j(x) \neq 0$ and

$$f^{(k)}(x, \omega^j(x, s), s) = 0,\ k \leq r_j - 2,\ \ f^{(r_j-1)}(x, \omega^j(x, s), s) = o(s^{b_j})$$

for $0 \leq j \leq t$. Here $f^{(k)}(x, y, s) = \partial^k f(x, y, s)/\partial y^k$. Then $f_0(x, y)$ is divisible by $\Omega(x, y) = \prod_{j=1}^{t}(y - \omega_0^j(x))^{r_j}$.

Completion of the proof of Lemma 2.6. It is enough to apply Lemma 2.7 with $s = \lambda^{-\epsilon}$, $x = (z, \xi')$, $y = \xi_0$, $f = g(z, \xi, \lambda)$. Since $g_0(z, \xi) = \tilde{q}_0(z, \xi)$ the assertion is clear. □

Proof of Lemma 2.7. It suffices to prove the assertion for every fixed x. Thus it is enough to prove that $f(0, y)$ is divisible by $\prod_{j=1}^{t}(y - \omega^j(0))^{r_j}$ assuming that for every $j = 1, \ldots, t$ there are $b_j \in \mathbb{N}$, $c_j \neq 0$ such that

$$\prod_{k=1,k\neq j}^{t} (\omega^j(s) - \omega^k(s))^{r_k} = s^{b_j}(c_j + o(1)) \quad (s \to 0) \tag{2.15}$$

where $r_j \in \mathbb{N}\setminus\{0\}$, $\omega^j(s) \in \mathbb{C}[[s]]$ $(j = 1, \ldots, t)$ and $f(s, y) \in (\mathbb{C}[[s]])[y]$ verifies

$$(\partial_y^k f)(s, \omega^j(s)) = o(s^{b_j(r_j-k)}) \quad (s \to 0) \quad (0 \leq k \leq r_j - 1)$$

for $j = 1, \ldots, t$. We proceed to the proof. One can assume that $\omega^j \in \mathbb{C}[s]$ dropping off enough high powers in s in the expansion of ω^j. Set

$$F(s, y) = \prod_{j=1}^{t}(y - \omega^j(s))^{r_j}, \quad \deg_y F = \sum_{j=1}^{t} r_j = r$$

then one can write

$$f(s, y) = q(s, y)F(s, y) + R(s, y), \quad q, R \in (\mathbb{C}[[s]])[y]$$

where $\deg_y R \le r - 1$. From the assumption we have

$$(\partial_y^k R)(s, \omega^j(s)) = (\partial_y^k f)(s, \omega^j(s)) = o(s^{b_j(r_j-k)}) \tag{2.16}$$

as $s \to 0$ $(0 \le k \le r_j - 1)$. We show $R(0, y) \equiv 0$. Choosing $\epsilon > 0$ enough small one can assume that $\{\omega^j(s) \mid j = 1, \dots, t\}$ are different from each other in $0 < |s| < \epsilon$. Fixing such s one can write

$$\frac{R(s,y)}{F(s,y)} = \sum_{j=1}^{t} \frac{U_j(s,y)}{(y - \omega^j(s))^{r_j}}, \quad \deg_y U_j \le r_j - 1$$

where $U_j(s, y) \in \mathbb{C}[y]$ are uniquely determined. This gives

$$R(s,y) = \sum_{j=1}^{t} U_j(s,y) \prod_{p=1, p\neq j}^{t} (y - \omega^p(s))^{r_p} \tag{2.17}$$

and hence we have

$$(\partial_y^k U_j)(s, \omega^j(s)) = o(s^{b_j(r_j-k-1)}) \quad (s \to 0,\ 0 \le k \le r_j - 1)$$

for any $j = 1, \dots, t$ by induction on k. Indeed since

$$R(s, \omega^j(s)) = U_j(s, \omega_j(s)) \prod_{p=1, p\neq j}^{t} (\omega^j(s) - \omega^p(s))^{r_p}$$

the assertion for $k = 0$ follows from (2.15) and (2.16) with $k = 0$. Assume that the assertion holds for $k - 1$, $1 \le k \le r_j - 1$. From (2.17) it follows that

$$\begin{aligned}
(\partial_y^k R)(s, \omega^j(s)) &= (\partial_y^k U_j)(s, \omega^j(s)) \prod_{p=1, p\neq j}^{t} (\omega^j(s) - \omega^p(s))^{r_p} \\
&\quad + \sum_{l=1}^{k} \binom{k}{l} (\partial_y^{k-l} U_j)(s, \omega^j(s)) (\partial_y^l \prod_{p=1, p\neq j}^{t} (y - \omega^p(s))^{r_p})|_{y=\omega^j(s)} \\
&= (\partial_y^k U_j)(s, \omega^j(s)) \prod_{p=1, p\neq j}^{t} (\omega^j(s) - \omega^p(s))^{r_p} + O(s^{b_j(r_j-k)})
\end{aligned}$$

then from (2.15) and (2.16) we have $(\partial_y^k U_j)(s, \omega^j(s)) = O(s^{b_j(r_j-k-1)})$. Therefore one sees

$$U_j(s,y) = \sum_{j=0}^{r_j-1} \frac{1}{k!} (\partial_y^k U_j)(s, \omega^j(s))(y - \omega^j(s))^k = o(1)$$

as $s \to 0$ so that $U_j(0, y) \equiv 0$ $(1 \le j \le t)$. Thus we have $R(0, y) \equiv 0$ which is the desired assertion. □

Proof of Theorem 2.1 (general case). Proposition 2.4 and (2.14) imply that the hypothesis of Lemma 2.6 is verified. Hence $\sigma_p(\lambda^{-\gamma} q_\lambda)(z, \xi)$ is divisible by $\bar{h}_\rho^{\sigma,\delta}(z, \xi)$ when (z, ξ') is in an open set and hence the degree of $\sigma_p(\lambda^{-\gamma} q_\lambda)$ with respect to ξ_0 is greater than or equal to r if $\mathrm{Ord}(\lambda^{-\gamma} q_\lambda) > 0$. Thus Lemma 2.3 proves that

$$q_{(\beta)}^{(\alpha)}(0, e_n) = 0, \quad |\alpha + \beta| < r - 2$$

which is the desired assertion. □

To prove the last assertion of Theorem 2.1 we need to improve Lemma 2.6. Let $q(x, \xi)$ be a homogeneous polynomial in ξ of degree $m-1$ with $C^\infty(\Omega)$ coefficients. We set

$$q_\lambda(z, \xi) = \lambda^\gamma \sum_j q_j(z, \xi) \lambda^{-\epsilon j}.$$

For the proofs of the following lemmas we refer to [50].

Lemma 2.8. *Assume (2.5) and* $\mathrm{Ord}(\lambda^{-\gamma} q_\lambda) > 0$. *Let* $I \subset \{1, 2, \ldots, t\}$. *We assume that there is an open set* $V \subset U$ *such that we have*

$$q_\lambda^{(\alpha)}(z, \varphi_x, \lambda) = 0, \ |\alpha| \le r_j - 2, \ \ \partial_{\xi_0}^{r_j - 1} q_\lambda(z, \varphi_x, \lambda) = O(\lambda^{-a_j}) \tag{2.18}$$

for every characteristic function $\varphi(z, \lambda)$ *to* ω^j, $j \in I$ *with* $(z, \varphi_{x'}(z, \lambda)) \in V$. *Then* $\sigma_p(\lambda^{-\gamma} q_\lambda)(z, \xi)$ *is divisible by*

$$\prod_{j \in I} (\xi_0 - \omega_0^j(z, \xi'))^{r_j}.$$

Lemma 2.9. *Assume that* $\mathrm{Ord}(\lambda^{-\gamma} q_\lambda) = 0$ *and there is an open set* $V \subset U$ *such that (2.18) holds for any characteristic function* φ *to* ω^j *with* $(z, \varphi_{x'}(z, \lambda)) \in V$, $j = 1, \ldots, t$. *Then* $\sigma_0(\lambda^{-\gamma} q_\lambda)(z, \xi)$ *is divisible by* $\prod g_j(z, \xi)^{k_j - 1}$ *where* $\prod g_j(z, \xi)^{k_j}$ *is the irreducible factorization of* $\bar{h}_\rho^{\sigma,\delta}(z, \xi)$.

We take $\bar{\sigma} = \bar{\delta} = 1/2$ and $y = (0, \ldots, 0, y_n)$. Thanks to Lemmas 2.2 and 2.9 we get the second assertion of Theorem 2.1 since

Lemma 2.10. *Let* $\sigma = (\bar{\sigma}, \ldots, \bar{\sigma}, 1)$, $\delta = (\bar{\delta}, \ldots, \bar{\delta})$ *and* $\bar{\sigma} = \bar{\delta} = 1/2$ *and* $y = (0, \ldots, 0, y_n)$. *Assume that* $q_{(\beta)}^{(\tilde{\alpha})}(0, e_n) = 0$, $|\tilde{\alpha} + \beta| < r - 2$ *and* $q_{(\beta)}^{(\tilde{\alpha})}(0, e_n) \ne 0$ *for some* $|\tilde{\alpha} + \beta| = r - 2$. *Then* $\mathrm{Ord}(\lambda^{-\gamma} q_\lambda) = 0$ *and*

$$\sigma_0(\lambda^{-\gamma} q_\lambda)(y, x, \xi) = \sum_{|\tilde{\alpha} + \beta| = r - 2} q_{(\beta)}^{(\tilde{\alpha})}(0, e_n)(\tilde{x}\xi_n)^{\tilde{\beta}} (y_n \xi_n)^{\beta_n} \tilde{\xi}^{\tilde{\alpha}} / \tilde{\alpha}! \beta!.$$

2.5 Proofs of Propositions 2.4 and 2.5

In this section we prove Propositions 2.4 and 2.5. Let $\omega^j(z,\xi',\lambda) \in \mathscr{A}(U)[\lambda^{-\epsilon}]$ be the roots of h_λ with multiplicity r_j so that

$$\lambda^{-\gamma} h_\lambda = \prod_{j=1}^{t} (\xi_0 - \omega^j)^{r_j} \lambda^{m-r} e_\lambda$$

where $\gamma = \sigma_0 r + (m-r)$. Let $(\hat{y}, \hat{x}, \hat{\xi}') \in U$ and let Ω' and Ω'' be an open neighborhood of $\hat{x}$ and a conic open neighborhood of $\hat{\xi}'$ respectively such that $\{\hat{y}\} \times \Omega' \times \Omega'' \subset U$. Here the main observation is that for any large "*fixed*" λ it is clear that $L_\lambda(\hat{y}, x, D)$ is a differential operator on Ω' and that

$$\omega^j(\hat{y}, x, \xi') \in \mathscr{A}(U)$$

and moreover ω^j is a characteristic root of $h_\lambda(\hat{y}, x, \xi)$ with multiplicity r_j which is constant on $\Omega' \times \Omega''$. Clearly strong hyperbolicity of L implies that of $L_\lambda(\hat{y}, x, D)$. Thus one can apply Proposition 1.6 for $L_\lambda(\hat{y}, x, D)$ for large each fixed λ to get

Proposition 2.6. *Under the same assumptions as in Proposition 2.4 there is a positive constant R such that*

$$\mathrm{rank}(\omega^j(z,\xi',\lambda) I - \lambda^{-\sigma_0} A_\lambda(z,\xi')) = m - r_j$$

for $(z,\xi') \in U$ *and* $\lambda > R$, $j = 1, \ldots, t$.

Proof of Proposition 2.4. Let fix a large λ. As noted in the proof of Proposition 1.6 there exists $U_\lambda \subset U$ and $T(z,\xi') \in \mathscr{A}(U_\lambda, M_m(\mathbb{C}))$ such that

$$T^{-1}(z,\xi')\lambda^{-\sigma_0} A_\lambda T(z,\xi') = \omega^j I_{r_j} \oplus A_{2\lambda}$$

where

$$\det(\omega^j(z,\xi') I_{r_j} - A_{2\lambda}(z,\xi')) \neq 0, \quad (z,\xi') \in U_\lambda.$$

Set $Q = (\xi_0 - \omega^j) I_{r_j} \oplus (\xi_0 - A_{2\lambda})$. Then it follows that

$$T^{-1}\lambda^{-\sigma_0(m-1)} M_\lambda T = {}^{co}Q(z,\xi,\lambda)$$

in $U_\lambda \times \mathbb{C}_{\xi_0}$. It is clear that every $(m-1)$-th minor of Q contains the factor $(\xi_0 - \omega^j)^{r_j - 1}$ and hence

$${}^{co}Q^{(\alpha)}(z,\omega^j,\xi',\lambda) = 0$$

in U_λ for $|\alpha| \le r_j - 2$. This shows that

$$M_\lambda^{(\alpha)}(z, \omega^j, \xi', \lambda) = 0 \tag{2.19}$$

in U_λ for $|\alpha| \le r_j - 2$. Since $\omega^j(z, \xi', \lambda) \in \mathscr{A}(U)[\lambda^{-\epsilon}]$ and $M_\lambda^{(\alpha)}(z, \xi_0, \xi')$ is analytic in $U \times \mathbb{C}_{\xi_0}$ so that (2.19) holds in U which is the desired conclusion. □

We turn to the proof of Proposition 2.5. We first discuss triangulations in $\mathscr{A}(U)\{\{\lambda^{-\epsilon}\}\}$ where U is an open set in $\mathbb{R}^N$ with a system of local coordinates x.

Lemma 2.11. *Assume that $A(x,\lambda), B(x,\lambda) \in \mathscr{A}(U)\{\{\lambda^{-\epsilon}\}\}$. Then we have*

$$\det(A(x,\lambda)B(x,\lambda)) = \det A(x,\lambda) \cdot \det B(x,\lambda).$$

Proof. It is enough to show that

$$\det(AB) - \det A \cdot \det B = O(\lambda^{-q})$$

for any $q \in \mathbb{Q}$ and then the proof reduced to the finite case, that is A, B are polynomials in $\lambda^{-\epsilon}$. □

Lemma 2.12. *Let $B(x,\lambda) \in \mathscr{A}(U)\{\{\lambda^{-\epsilon}\}\}$. Then for every open set $V \subset U$ there are an open set $W \subset V$ and $P(x,\lambda)$, $Q(x,\lambda) \in \mathscr{A}(W)[[\lambda^{-\epsilon}]]$ with $\det \sigma_0(P)(x) \neq 0$, $\det \sigma_0(Q)(x) \neq 0$ such that*

$$P(x,\lambda)B(x,\lambda)Q(x,\lambda) = \operatorname{diag}(a^1(x,\lambda), \dots, a^m(x,\lambda))$$

where the right-hand side is the diagonal matrix and $a^k(x,\lambda) \in \mathscr{A}(W)\{\{\lambda^{-\epsilon}\}\}$.

Proof. We first note that if $P_i \in \mathscr{A}(V)[[\lambda^{-\epsilon}]]$, $\det \sigma_0(P_i) \neq 0$ then $P = P_1 \cdots P_r \in \mathscr{A}(V)[[\lambda^{-\epsilon}]]$ and $\sigma_0(P) = \sigma_0(P_1) \cdots \sigma_0(P_r)$ so that $\det \sigma_0(P) \neq 0$. It is also clear that for a scalar $b(x,\lambda) \in \mathscr{A}(V)\{\{\lambda^{-\epsilon}\}\}$ which is not zero in $\mathscr{A}(V)\{\{\lambda^{-\epsilon}\}\}$ we can choose an open set $W \subset V$ on which $\sigma_p(b)$ never vanish. Hence there is $b^{-1}(x,\lambda) \in \mathscr{A}(W)\{\{\lambda^{-\epsilon}\}\}$ with $\operatorname{Ord} b^{-1} = -\operatorname{Ord} b$. In virtue of these facts the proof is reduced to the constant matrix case. □

Lemma 2.13. *Let $A(x,\lambda) \in \mathscr{A}(U)\{\{\lambda^{-\epsilon}\}\}$ and assume that*

$$\det(\tau - A(x,\lambda)) = \prod_{j=1}^m (\tau - \omega^j(x,\lambda)), \quad \omega^j(x,\lambda) \in \mathscr{A}(U)\{\{\lambda^{-\epsilon}\}\}.$$

Then for every open set $V \subset U$ there are an open set $W \subset V$ and $T(x,\lambda) \in \mathscr{A}(W)[[\lambda^{-\epsilon}]]$ with $\det \sigma_0(T) \neq 0$ such that

$$T^{-1}(x,\lambda)A(x,\lambda)T(x,\lambda)$$

is upper triangular with diagonal entries $\omega^1, \omega^2, \dots, \omega^m$.

Proof. From Lemma 2.12 there are an open set $V_1 \subset V$ and $P, Q \in \mathscr{A}(V_1)[[\lambda^{-\epsilon}]]$ with $\det \sigma_0(P)$, $\det \sigma_0(Q) \neq 0$ such that $P(\omega^1 - A)Q = \oplus_{i=1}^m a^i$. Since $\det(\omega^1 - A) = 0$ we may assume that $a^1 = 0$. Take $\mathscr{V}^1 = Q\mathscr{W}$ with $\mathscr{W} = {}^t(1, 0, \ldots, 0)$ so that $\mathscr{V}^1 \in \mathscr{A}(V_1)[[\lambda^{-\epsilon}]]$ and $A(x,\lambda)\mathscr{V}^1 = \omega^1\mathscr{V}^1$. Since $\sigma_0(\mathscr{V}^1) \neq 0$ one can find $m-1$ constant vectors $\mathscr{V}^2, \ldots, \mathscr{V}^m \in \mathbb{C}^m$ so that $\det(\sigma_0(\mathscr{V}^1), \mathscr{V}^2, \ldots, \mathscr{V}^m) \neq 0$ shrinking V_1 if necessary. Then setting $T = (\mathscr{V}^1, \mathscr{V}^2, \ldots, \mathscr{V}^m)$ we see that the first column of $T^{-1}AT$ is ${}^t(\omega^1, 0, \ldots, 0)$. Hence by induction we get the result. □

Now we turn to study M_λ.

Lemma 2.14. *Let* $\omega^j(z, \xi', \lambda) \in \mathscr{A}(U)[[\lambda^{-\epsilon}]]$ *are the roots of* h_λ *with multiplicity* r_j, $\sum_{j=1}^t r_j = r$. *Assume that L is strongly hyperbolic near the origin. Then for every open set* $V \subset U$ *there are an open set* $W \subset V$ *and* $T(z, \xi', \lambda) \in \mathscr{A}(W)[[\lambda^{-\epsilon}]]$ *with* $\det \sigma_0(T) \neq 0$ *such that*

$$T^{-1}\lambda^{-\sigma_0}A_\lambda T = A'_{1\lambda} \oplus A'_{2\lambda}$$

where $A'_{1\lambda}(z, \xi')$ *is upper block triangular with diagonal blocks*

$$\omega^1 I_{r_1}, \omega^2 I_{r_2}, \ldots, \omega^t I_{r_t}$$

and $A'_{2\lambda}(z, \xi')$ *is a* $(m-r) \times (m-r)$ *matrix satisfying*

$$\det(\xi_0 - A'_{2\lambda}(z, \xi')) = \lambda^{(1-\sigma_0)(m-r)}(c + o(1))$$

with a constant $c \neq 0$.

Proof. Since $\rho = (0, e_n)$ is a characteristic of order r for $h(x, \xi)$ then there exist an open conic neighborhood $\Omega' \times \Gamma$ of $(0, e_n')$ and $S(x, \xi') \in \mathscr{A}(\Omega' \times \Gamma, M_m(\mathbb{C}))$ such that

$$S^{-1}(x, \xi')A(x, \xi')S(x, \xi') = A_1(x, \xi') \oplus A_2(x, \xi')$$

where $\det(\xi_0 - A_1) = \tilde{h}$ and $\det(\xi_0 - A_2) = e$. Note that $S_\lambda(z, \xi') = S(\lambda^{-\delta}y + \lambda^{-\sigma}x, \lambda^{\tilde{\sigma}}\xi') \in \mathscr{A}(U)[\lambda^{-\epsilon}]$ and

$$\sigma_0(S_\lambda) = S(0, e_n').$$

In particular $\det \sigma_0(S_\lambda) \neq 0$. Thus we have

$$S_\lambda^{-1}(z, \xi')\lambda^{-\sigma_0}A_\lambda(z, \xi')S_\lambda(z, \xi') = \lambda^{-\sigma_0}A_{1\lambda} \oplus \lambda^{-\sigma_0}A_{2\lambda}.$$

From Lemma 2.13 one can find an open set $W \subset V$ and $K(z, \xi', \lambda) \in \mathscr{A}(W)[[\lambda^{-\epsilon}]]$ with $\det \sigma_0(K) \neq 0$ such that $K^{-1}\lambda^{-\sigma_0}A_{1\lambda}K$ is upper triangular with diagonal entries $\omega^1, \ldots, \omega^1, \omega^2, \ldots, \omega^t, \ldots, \omega^t$. Setting $T = S_\lambda(K \oplus I_{m-r}) \in \mathscr{A}(W)[[\lambda^{-\epsilon}]]$ with $\det \sigma_0(T) \neq 0$ it follows that

$$T^{-1}\lambda^{-\sigma_0} A_\lambda T = \{K^{-1}\lambda^{-\sigma_0} A_{1\lambda} K\} \oplus \lambda^{-\sigma_0} A_{2\lambda}.$$

We now apply Proposition 2.6 to get

$$\mathrm{rank}(\omega^j I_r - K^{-1}\lambda^{-\sigma_0} A_{1\lambda} K) = r - r_j$$

in W. This shows that $K^{-1}\lambda^{-\sigma_0} A_{1\lambda} K$ must be block upper triangular and hence the assertion. □

From this lemma for any given open set $V \subset U$ and $1 \le k \le t$ one can find an open set $W \subset V$ and $T^k \in \mathscr{A}(W)[[\lambda^{-\epsilon}]]$ with $\sigma_0(T^k) \neq 0$ such that

$$(T^k)^{-1}\lambda^{-\sigma_0} A_\lambda T^k = \begin{bmatrix} \omega^k I_{r_k} & C^k \\ O & B^k \end{bmatrix}.$$

Setting $\mu_k = (\xi_0 - \omega^k)$ and $\Lambda_k = (\xi_0 - B^k)$ it is clear that

$$(T^k)^{-1}\lambda^{-\sigma_0(m-1)} M_\lambda T^k = \begin{bmatrix} \mu_k^{r_k - 1} \det \Lambda_k I_{r_k} & D^k \\ O & \mu_k^{r_k}\, {}^{co}\Lambda_k \end{bmatrix}. \tag{2.20}$$

In what follows we simply write $T(\omega^k)$ for $T(z, \omega^k, \xi', \lambda)$ when $T(z, \xi, \lambda) \in \mathscr{A}(U \times \mathbb{C}_{\xi_0})\{\{\tilde{\lambda}\}\}$. Here we note that

$$\det\Lambda_k = \lambda^{-\sigma_0(m-r)} e_\lambda \prod_{j \neq k} (\xi_0 - \omega^j)^{r_j},\ \partial_{\xi_0}^{r_k} \lambda^{-\sigma_0 m} h_\lambda(\omega^k) = r_k! \det \Lambda_k(\omega^k).$$

This shows that

$$\det \Lambda_k(\omega^k) = (r_k!)^{-1} \partial_{\xi_0}^{r_k} \lambda^{-\sigma_0 m} h_\lambda(\omega^k). \tag{2.21}$$

Then it is clear from (2.20) and (2.21) that

$$(T^k)^{-1}\lambda^{\sigma_0} \partial_{\xi_0}^{r_k - 1} M_\lambda(\omega^k) T^k = \begin{bmatrix} r_k^{-1} \partial_{\xi_0}^{r_k} h_\lambda(\omega^k) I_{r_k} & D'^k \\ O & O \end{bmatrix}. \tag{2.22}$$

Now we give a proof of Proposition 2.5.

Proof of Proposition 2.5. Since it is enough to prove the assertion for each k, $1 \le k \le t$ we write $r_k = r$, $\omega^k = \omega$, $a_k = a$, $N^k = N$, $K^k = K$ and $G^k = G$ to simplify notations. We first recall that $LN = \partial_{\xi_0}^{r-1} h - (r-1)\partial_{\xi_0}^{r-2} M$ because $LM = hI_m$. From this it follows that

$$L_\lambda \lambda^{-\sigma_0(r-1)} \partial_{\xi_0}^{r-1} M_\lambda = \lambda^{-\sigma_0(r-1)} \partial_{\xi_0}^{r-1} h_\lambda - (r-1)\lambda^{-\sigma_0(r-2)} \partial_{\xi_0}^{r-2} M_\lambda.$$

Set $g = \partial_{\xi_0}^{r-1} h_\lambda$, $f = \partial_{\xi_0}^{r} h_\lambda$, $H = \lambda^{\sigma_0}\partial_{\xi_0}^{r-2} M_\lambda$, $F = \partial_{\xi_0}^{r-1} M_\lambda$, $S = gI - (r-1)H$ so that

$$L_\lambda F = gI - (r-1)H = S.$$

Since $h_\lambda = h'(\xi_0 - \omega)^r$ it is clear that $g^{(\alpha)}(\omega) = f(\omega)c_\alpha$ with $c_\alpha(z, \xi', \lambda) = (\xi_0 - \omega)^{(\alpha)}$ for $|\alpha| = 1$. On the other hand from Proposition 2.4 it follows that

$$M_\lambda(z, \xi) = M'(z, \xi, \lambda)(\xi_0 - \omega(z, \lambda))^{r-1}$$

and hence $H^{(\alpha)}(\omega) = \lambda^{\sigma_0} F(\omega)c_\alpha$. These imply that

$$S^{(\alpha)}(\omega) = (f(\omega)I - (r-1)\lambda^{\sigma_0} F(\omega))c_\alpha = Q(\omega)c_\alpha.$$

From (2.22) there is $T \in \mathscr{A}(W)[[\lambda^{-\epsilon}]]$ with $\det \sigma_0(T) \neq 0$ such that

$$T^{-1}\lambda^{\sigma_0} F(\omega)T = \begin{bmatrix} r^{-1} f(\omega)I_r & C' \\ O & O \end{bmatrix}$$

and hence

$$T^{-1} Q(\omega)T = \begin{bmatrix} r^{-1} f(\omega)I_r & -(r-1)C' \\ O & f(\omega)I_{m-r} \end{bmatrix}.$$

With $C'' = -(r-1)C'$ it is obvious that

$$T^{-1} Q^{-1}(\omega)T = \begin{bmatrix} r^{-1} f(\omega)^{-1} I_r & -rf(\omega)^{-2}C'' \\ O & f(\omega)^{-1} I_{m-r} \end{bmatrix}$$

which proves that K is invertible since $G_\lambda = \lambda^{-\sigma_0(r-1)} S$ and hence

$$G_\lambda^{(\alpha)}(\omega) = \lambda^{-\sigma_0(r-1)} S^{(\alpha)}(\omega) = \lambda^{-\sigma_0(r-1)} Q(\omega)c_\alpha, \ |\alpha| = 1. \tag{2.23}$$

We turn to the proof of the last assertion. We note that $T^{-1}F(\omega)Q^{-1}(\omega)T = T^{-1}F(\omega)TT^{-1}Q^{-1}(\omega)T$ is equal to

$$\begin{bmatrix} r^{-1} f(\omega)I_r & C' \\ O & O \end{bmatrix} \begin{bmatrix} rf(\omega)^{-1} I_r & r(r-1)f(\omega)^{-2}C' \\ O & f(\omega)^{-1} I_{m-r} \end{bmatrix}$$
$$= \begin{bmatrix} I_r & rf(\omega)^{-1}C' \\ O & O \end{bmatrix} = rf(\omega)^{-1}T^{-1}F(\omega)T.$$

This shows that $F(\omega)Q^{-1}(\omega) = rf(\omega)^{-1}F(\omega)$ and hence the assertion because $N_\lambda = \lambda^{-\sigma_0(r-1)}F$, $K = \lambda^{-\sigma_0(r-1)}Q$ by (2.23) and $\mathrm{Ord}\, f(\omega)^{-1} = a$. □

2.6 Proof of Key Proposition

Since the proof of Proposition 2.1 is fairly long we first explain our strategy for proving the proposition. Let N, a differential operator of order m' with $C^\infty(\Omega, \mathbb{C}^m)$ coefficients and $\varphi(y, x, \lambda) \in \mathscr{A}(W)[[\tilde{\lambda}]]$ verify the required properties in Proposition 2.1. That is with

$$G(x, \xi) = L(x, \xi)N(x, \xi)$$

we have

$$G_\lambda(z, \varphi_x) = 0, \; G_\lambda{}^{(\alpha)}(z, \varphi_x) = c_\alpha(z, \lambda)K(z, \lambda), \;\; \forall|\alpha| = 1$$

where $K(z, \lambda)$ is invertible. Note that $L(x, D)N(x, D) = G(x, D) + H(x, D)$ where

$$H(x, \xi) = -i \sum_{|\alpha|=1} L^{(\alpha)}(x, \xi)N_{(\alpha)}(x, \xi)$$

and that

$$(L_\lambda(z, D) + bB)N_\lambda(z, D) = G_\lambda(z, D) + H_\lambda(z, D) + bBN_\lambda(z, D) \tag{2.24}$$

where $B \in M_m(\mathbb{C})$ and $b \in \mathbb{C}$. Assume that

$$N_\lambda(z, \varphi_x(z, \lambda))K(z, \lambda)^{-1} \neq O(1)$$

so that some entry $c(z, \lambda)$ of $N_\lambda(z, \varphi_x(z, \lambda))K(z, \lambda)^{-1}$ can be written

$$c(z, \lambda) = \lambda^\nu(c'(z) + o(1)), \;\; c'(z) \neq 0$$

with some $\nu > 0$. Choosing $B \in M_m(\mathbb{C})$ and $b \in \mathbb{C}$ suitably then for $P = L + bB$ we look for u_λ such that the a priori estimates in Proposition 1.2 is incompatible as $\lambda \to \infty$, that is $\lambda^{\bar{\sigma} p}|P_\lambda u_\lambda|_{C^p(W')}$ decays faster than $|u_\lambda|_{C^0(W')}$ as $\lambda \to \infty$. We look for u_λ in the form $N_\lambda \mathscr{U}_\lambda$ so that, as the first step, we are led to find $\mathscr{U}_\lambda$ verifying

$$(L_\lambda + bB)N_\lambda \mathscr{U}_\lambda = O(\lambda^{-k})$$

for any $k \in \mathbb{N}$ and that $N_\lambda \mathscr{U}_\lambda$ does not decay so fast.

We set

$$E_0(z, \lambda) = \exp\{i\varphi(z, \lambda)\lambda^\tau\}$$

where $\tau > 0$ will be determined later. Let $G(x, D)$ be a differential operator of order $m' + 1$ with $C^\infty(\Omega, \mathbb{C}^m)$ coefficients. We recall that (see, for example [5])

$$E_0{}^{-1} G_\lambda(x, D) E_0 = \lambda^{\tau(m'+1)}\big[G_\lambda(x, \varphi_x) + \lambda^{-\tau}\{\sum_{|\alpha|=1} G_\lambda^{(\alpha)}(x, \varphi_x) D^\alpha$$

$$+ i \sum_{|\alpha|=2} G_\lambda^{(\alpha)}(x, \varphi_x) D^\alpha \varphi/\alpha!\} + O(\lambda^{-2\tau})\big]$$

where $O(\lambda^{-2\tau})$ denotes a differential operator such that $\lambda^{2\tau} O(\lambda^{-2\tau})$ is a differential operator with coefficients which are bounded in C^∞ as $\lambda \to \infty$. Set

$$\sum_{|\alpha|=1} c_\alpha(z, \lambda) D^\alpha = l(z, D, \lambda) = \sum_{j=0} l_j(z, D)\lambda^{-\epsilon j},$$

$$G'(z, \lambda) = i \sum_{|\alpha|=2} G_\lambda{}^{(\alpha)}(z, \varphi_x) D^\alpha/\alpha!$$

where $l_j(z, \xi)$ are real and that

$$l_0(z, \xi) = \sum_{j=0}^{n} a_j(z)\xi_j$$

with $a_0(z) \neq 0$ by the hypothesis. Then we have

$$\begin{aligned} &E_0^{-1}\{(L_\lambda + bB)N_\lambda\}E_0 \\ &\quad = \lambda^{\tau m'}\{K(z, \lambda) l(z, D, \lambda) + G'(z, \lambda) + H_\lambda(z, \varphi_x) \\ &\qquad + bBN_\lambda(z, \varphi_x) + O(\lambda^{-\tau+m'+1})\}. \end{aligned}$$

Since $K(z, \lambda)$ is invertible one can write the right-hand side as

$$\lambda^{\tau m'} K(z, \lambda)\{l(z, D, \lambda) + K^{-1}G' + K^{-1}H_\lambda + bK^{-1}BN_\lambda + O(\lambda^{-\tau+g'})\} \tag{2.25}$$

where $g' \leq m' + 1 + \operatorname{Ord} K^{-1}$. We now assume that the assertion of Proposition 2.1 were not true. Then denoting $N_\lambda(z, \varphi_x)K^{-1}(z, \lambda) = (c_{ij}(z, \lambda))$ there are i, j such that

$$c_{ij}(z, \lambda) - \lambda^\nu c_{ij}(z) = o(\lambda^\nu)$$

with some $\nu \in \mathbb{Q}_+$ and $c_{ij}(z)$ which is not identically zero and hence we can choose an open set U in which we have $c_{ij}(z) \neq 0$. We now choose $B = (b_{k\ell}) \in M_m(\mathbb{C})$ so that

$$\begin{cases} b_{ji} = \beta \in \mathbb{C} \\ b_{k\ell} = 0, \ (k,\ell) \neq (j,i) \end{cases}$$

then it is clear that

$$\operatorname{Tr} BN_\lambda K^{-1} - \beta c_{ij}(z)\lambda^\nu = o(\lambda^\nu).$$

Let us set

$$A(b,z,\lambda) = -K^{-1}(G' + H_\lambda + bBN_\lambda).$$

Since $\operatorname{Tr} A = -\operatorname{Tr}\{(G' + H_\lambda + bBN_\lambda)K^{-1}\}$ it follows that

$$\frac{\partial}{\partial b}\operatorname{Tr} A = -\operatorname{Tr} BN_\lambda K^{-1}, \quad \frac{\partial}{\partial b}\operatorname{Tr} A + \beta c_{ij}(z)\lambda^\nu = o(\lambda^\nu). \tag{2.26}$$

In the following we will make several procedures on A under which the trace is invariant modulo $O(1)$. Thus (2.26) implies that the parameter b plays a role of a marker for finding β, that is one can find β by $\partial \operatorname{Tr} A/\partial b$ after several required processes. We rewrite (2.25) as follows

$$\lambda^{\tau m'} K\{l(z,D,\lambda) - A(b,z,\lambda) + \lambda^{-\tau+g'} A'(b,z,D,\lambda)\}$$

with

$$A'(b,z,\xi,\lambda) = \sum_{j=0} A'_j(b,z,\xi)\lambda^{-\epsilon j}, \tag{2.27}$$

where $A'_j(b,z,\xi)$ are polynomials in (b,z,ξ) and of degree less than or equal to $m'+1$ with respect to ξ. Let I be an open interval in $\mathbb{R}$. Note that $A(b,z,\lambda)$ is in $\mathscr{A}(I \times U)\{\{\tilde{\lambda}\}\}$. We now assume the following of which proof we will give the next section.

Lemma 2.15. *One can find $I_1 \times U_1 \subset I \times U$ and an invertible $\Gamma(b,z,\lambda) \in \mathscr{A}(I_1 \times U_1)\{\{\tilde{\lambda}\}\}$ such that*

$$\big(l(z,D,\lambda) - A(b,z,\lambda)\big)\Gamma(b,z,\lambda) = \Gamma(b,z,\lambda)\big(l - \oplus_{j=1}^m \lambda_j(b,z,\lambda) + K'(b,z,\lambda)\big)$$

where $K' \in \mathscr{A}(I_1 \times U_1)[[\tilde{\lambda}]]$ and

$$\operatorname{Tr} A - \sum_{j=1}^{m} \lambda_j = O(1). \tag{2.28}$$

Here $\operatorname{Ord}\Gamma$, $\operatorname{Ord}\Gamma^{-1}$ *are bounded by $c(m)\operatorname{Ord}A$ which is independent of τ.*

Thus Γ has no essential effect on the term $\mathrm{Ord}(\lambda^{-\tau+g'})$ if τ is chosen large enough. In particular one can write with $d = c(m)\mathrm{Ord}A$

$$\lambda^{-\tau+g'}A'(b,z,D,\lambda)\Gamma = \lambda^{-\tau+g'+2d}\Gamma A''(b,z,D,\lambda)$$

where $A''(b,z,\xi,\lambda)$ has the same type development as in (2.27) and hence

$$(l - A + \lambda^{-\tau+g'}A')\Gamma = \Gamma(l - \oplus_1^m \lambda_j + K' + \lambda^{-\tau+g''}A''). \tag{2.29}$$

From (2.26) and (2.28) we can choose $j_0, \beta \in \mathbb{C}$ and an open set $I_2 \times U_2 \subset I_1 \times U_1$ so that

$$\begin{cases} \mathrm{Im}\,\lambda_{j_0}(b,z,\lambda) - \lambda^{\nu_1}c_1(b,z) = o(\lambda^{\nu_1}), & c_1(b,z) \not\equiv 0, \\ \dfrac{\partial}{\partial b}\mathrm{Im}\,\lambda_{j_0}(b,z,\lambda) - \lambda^{\nu_2}c_2(b,z) = o(\lambda^{\nu_2}), & c_2(b,z) \not\equiv 0 \end{cases} \tag{2.30}$$

with $\nu_i \in \mathbb{Q}_+$ since if $\mathrm{Im}\,\lambda_j = O(1)$ for every j we have clearly $\partial(\sum \mathrm{Im}\,\lambda_j)/\partial b = O(1)$ which contradicts (2.26) and (2.28). Moreover we may assume, shrinking $I_2 \times U_2$ if necessary, that for every j we have either

$$\lambda_{j_0}(b,z,\lambda) - \lambda_j(b,z,\lambda) = O(1) \tag{2.31}$$

or

$$\lambda_{j_0}(b,z,\lambda) - \lambda_j(b,z,\lambda) = \lambda^{\nu_j'}(c_j'(b,z) + o(1)) \tag{2.32}$$

with some $\nu_j' \in \mathbb{Q}_+$ and $c_j'(b,z) \not\equiv 0$.

Lemma 2.16. *Write $\theta(b,z,\lambda) = \lambda_{j_0}(b,z,\lambda)$ and assume (2.30). Then for every $(\hat{b},\hat{z}) \in I \times U$ there are a neighborhood $J \times V$ of $(\hat{b},\hat{z})$ and $\psi(b,z,\lambda) \in \mathscr{A}(J \times V)\{\{\tilde{\lambda}\}\}$ such that*

$$l(z,\partial,\lambda)\psi(b,z,\lambda) - \theta(b,z,\lambda) = O(1)$$

where $\partial = (\partial/\partial x_0, \ldots, \partial/\partial x_n)$ and we have with a positive constant c either

$$-\mathrm{Im}\,\psi(b,z,\lambda) \le -c\{x_0 - \hat{x}_0 + |x' - \hat{x}'|^2\}\lambda^{\nu_1}$$

in $J \times V$ and $x_0 \ge \hat{x}_0$ for large λ or

$$-\mathrm{Im}\,\psi(b,z,\lambda) \le -c\{\hat{x}_0 - x_0 + |x' - \hat{x}'|^2\}\lambda^{\nu_1}$$

in $J \times V$ and $x_0 \le \hat{x}_0$ for large λ.

Proof. Recall that

$$\theta(b,z,\lambda)=\sum_{j=-\mu}\theta_j(b,z)\lambda^{-\epsilon j}$$

where we may assume that $\mathrm{Im}\,\theta_{-\mu'}(b,z)$ vanishes nowhere in $J\times V$ with $\nu_1=-\mu'\epsilon$ by (2.30). Set

$$\psi(b,z,\lambda)=\sum_{j=-\mu}\psi_j(b,z)\lambda^{-\epsilon j}.$$

Then the equation $l(z,\partial,\lambda)\psi(b,z,\lambda)=\theta(b,z,\lambda)$ is written as

$$l_0(z,\partial)\psi_p(b,z)=\theta_p(b,z)-\sum_{i+j=p,j\le p-1}l_i(z,\partial)\psi_j(b,z).$$

Since θ_j is real for $-\mu\le j<-\mu'$ by (2.30) we solve ψ_j to be real in $J\times V$ for $-\mu\le j<-\mu'$. We next consider

$$l_0(z,\partial)\psi_{-\mu'}(b,z)=\theta_{-\mu'}(b,z)-\sum_{i+j=-\mu',j\le-\mu'-1}l_i(z,\partial)\psi_j(b,z)=f_{-\mu'}(b,z).$$

Note that $\mathrm{Im}\,f_{-\mu'}(b,z)=\mathrm{Im}\,\theta_{-\mu'}(b,z)\neq 0$ in $I_2\times U_2$. We solve $\psi_{-\mu'}$ with initial condition

$$\psi_{-\mu'}(b,y,\hat{x}_0,x')=i|x'-\hat{x}'|^2.$$

Therefore it follows that

$$\mathrm{Im}\,\psi_{-\mu'}(b,z)\ge c\{x_0-\hat{x}_0+|x'-\hat{x}'|^2\}$$

with a constant $c>0$ for $x_0\ge\hat{x}_0$ if $\mathrm{Im}\,\theta_{-\mu'}\ge c'>0$ and

$$\mathrm{Im}\,\psi_{-\mu'}(b,z)\ge c\{\hat{x}_0-x_0+|x'-\hat{x}'|^2\}$$

for $x_0\le\hat{x}_0$ if $\mathrm{Im}\,\theta_{-\mu'}\le -c'<0$. We next solve $\psi_j(b,z)$, $j\ge-\mu'+1$ with initial condition

$$\psi_j(b,y,\hat{x}_0,x')=0$$

so that

$$-\mathrm{Im}\,\psi_j(b,z)\le C\{|x_0-\hat{x}_0|+|x'-\hat{x}'|^2\}$$

for $j \geq -\mu' + 1$ with a constant $C > 0$. Remarking that

$$-c|x_0 - \hat{x}_0| + C|x_0 - \hat{x}_0|\lambda^{-\epsilon} \leq -c|x_0 - \hat{x}_0|/2$$

for large λ we get the desired result. □

Set

$$E_1(b, z, \lambda) = \exp(i\psi(b, z, \lambda)).$$

From Lemma 2.16 it follows that

$$E_1^{-1} l(z, D, \lambda) E_1 = l(z, D, \lambda) + \tilde{\lambda}_{j_0}$$

where $\tilde{\lambda}_{j_0} - \lambda_{j_0} = O(1)$. Then from (2.29) we have

$$(l + \tilde{\lambda}_{j_0} I - A + \lambda^{-\tau+g'} A')\Gamma = \Gamma(l + \lambda_{j_0} I - \oplus_1^m \lambda_j + K'' + \lambda^{-\tau+g''} A'')$$

with $K'' \in \mathscr{A}(I_2 \times U_2)[[\tilde{\lambda}]]$.

Lemma 2.17. *For every $s \in \mathbb{N}$ there are an open neighborhood $I_s \times U_s \subset I_2 \times U_2$ of $(\hat{b}, \hat{z})$ and a vector $\mathscr{U}_s(b, z, \lambda) \in \mathscr{A}(I_s \times U_s)[[\tilde{\lambda}]]$ with $\sigma_0(\mathscr{U}_s)(\hat{b}, \hat{z}) \neq 0$ such that*

$$(l + \lambda_{j_0} I - \oplus\lambda_j + K'' + \lambda^{-\tau+g''} A'')\mathscr{U}_s = O(\lambda^{-s}).$$

Proof. Exchanging rows and corresponding columns, if necessary, we may assume from (2.31) and (2.32) that

$$-\lambda_{j_0} I + \oplus\lambda_j - K'' - \lambda^{-\tau+g''} A'' = R = (R_{ij})_{1 \leq i,j \leq 2}$$

where

$$R_{ij}(b, z, D, \lambda) = R'_{ij}(b, z, \lambda) + R''_{ij}(b, z, D, \lambda) \text{ for } (ij) = (11), (12), (21),$$
$$R_{22} = \oplus r_j(b, z, \lambda) + R'_{22}(b, z, \lambda) + R''_{22}(b, z, D, \lambda)$$

with $R'_{ij}(b, z, \lambda) \in \mathscr{A}(I_2 \times U_2)[[\tilde{\lambda}]]$ and scalar $r_j(b, z, \lambda) \in \mathscr{A}(I_2 \times U_2)\{\{\tilde{\lambda}\}\}$ such that $\operatorname{Ord} r_j > 0$ where $R''_{ij}(b, z, \xi, \lambda) = \lambda^{-\epsilon} \sum_{k=0} R_{ijk}(b, z, \xi)\lambda^{-\epsilon k}$. We may assume that $\sigma_p(r_j)(b, z)$ never vanish in $I_2 \times U_2$. Let $\Lambda = I \oplus \{\oplus\lambda^{-\kappa_j}\}$ where $\operatorname{Ord} r_j = \kappa_j \in \mathbb{Q}_+$. It is enough to solve

$$\Lambda(l - R)\mathscr{U}_s = O(\lambda^{-s}).$$

Setting $\mathscr{U}_s = {}^t(\mathscr{U}_s^I, \mathscr{U}_s^{II})$ this equation reduces to

$$\begin{cases} (l - R'_{11})\mathscr{U}_s^I - \{\lambda^{-\epsilon} R'''_{11}\mathscr{U}_s^I + R'''_{12}\mathscr{U}_s^{II}\} = O(\lambda^{-s}), \\ \{\oplus\lambda^{-\kappa_j} r_j\}\mathscr{U}_s^{II} - \lambda^{-\epsilon}\{R'''_{21}\mathscr{U}_s^I + R'''_{22}\mathscr{U}_s^{II}\} = O(\lambda^{-s}) \end{cases} \tag{2.33}$$

where

$$R'''_{ij}(b, z, \xi, \lambda) = \sum_{k=0} R'''_{ijk}(b, z, \xi)\lambda^{-\epsilon k}.$$

Note that $R'''_{ijk}(b, z, \xi)$ are polynomials in ξ of degree less than or equal to $m' + 1$. Set

$$\mathscr{U}_s^I = \sum_{j=0} \mathscr{U}_{sj}^I(b, z)\lambda^{-\epsilon j}, \quad \mathscr{U}_s^{II} = \sum_{j=1} \mathscr{U}_{sj}^{II}(b, z)\lambda^{-\epsilon j}.$$

Then it is clear that the system (2.33) is equivalent to

$$(l_0(z, D) - \sigma_0(R_{11})(b, z))\mathscr{U}_{sp}^I = F^I(\mathscr{U}_{sj}^I, \ j \le p-1; \mathscr{U}_{sj}^{II}, \ j \le p-1),$$
$$\{\oplus\sigma_p(r_j)(b, z)\}\mathscr{U}_{sp}^{II} = F^{II}(\mathscr{U}_{sj}^I, \ j \le p-1; \mathscr{U}_{sj}^{II}, \ j \le p-1).$$

Since $\{\oplus\sigma_p(r_j)(b, z)\}$ is non singular these equations can be solved successively with the condition $\sigma_0(\mathscr{U}_s^I)(\hat{b}, \hat{z}) \neq 0$ while $\mathscr{U}_{sp}^{II}$ can be obtained by

$$\mathscr{U}_{sp}^{II}(b, z) = \{\oplus\sigma_p(r_j)(b, z)\}^{-1} F^{II}$$

which proves the assertion. □

Recall that

$$E^{-1}(L_\lambda + bB)N_\lambda E\Gamma = \lambda^{\tau m'} K\Gamma(l + \lambda_{j_0} I - \oplus\lambda_j + K'' + \lambda^{-\tau + g''} A'')$$

with $E = E_0 E_1$. By Lemma 2.17 there are an open set $I_s \times U_s \subset I_2 \times U_2$ and $\mathscr{U}_s(b, z, \lambda) \in \mathscr{A}(I_s \times U_s)[[\tilde{\lambda}]]$ where $\sigma_0(\mathscr{U}_s)(b, z)$ dose not vanish on $I_s \times U_s$ such that

$$E^{-1}(L_\lambda + bB)N_\lambda E\Gamma\mathscr{U}_s = O(\lambda^{-s})$$

which proves that $N_\lambda E\Gamma\mathscr{U}_s$ is an asymptotic solution to $L_\lambda + bB$. Here we must check that $N_\lambda E\Gamma\mathscr{U}_s$ is non trivial which is never obvious because N is a differential operator. To examine this let us write

$$E^{-1}(L_\lambda + bB)E\{E^{-1}N_\lambda E\Gamma\}\mathscr{U}_s = O(\lambda^{-s})$$

and we note that

$$E^{-1}N_\lambda E = \lambda^{\tau m'}\{N_\lambda(z,\varphi_x) + \lambda^{-\tau+g_0}N'(b,z,D,\lambda)\},$$
$$N'(b,z,\xi,\lambda) = \sum_{j=0} N'_j(b,z,\xi)\lambda^{-\epsilon j}$$

where $N'_j(b,z,\xi)$ is a polynomial in ξ of degree m' and g_0 is independent of τ. Thus to show that $E^{-1}N_\lambda E\Gamma\mathscr{U}_s$ is non trivial it is enough to prove

Proposition 2.7. *Set $\mathscr{V}_s = N_\lambda(z,\varphi_x)\Gamma\mathscr{U}_s$. Then there are positive constants c_i independent of τ and s such that*

$$\operatorname{Ord}\mathscr{V}_s > c_1 \ \text{ if } \ \tau > c_2 \ \text{ and } \ s > c_3.$$

Proof. We first recall that $\mathscr{U}_s$ satisfies

$$(l + \tilde{\lambda}_{j_0}I - A + \lambda^{-\tau+g'}A')\Gamma\mathscr{U}_s = O(\lambda^{-s}).$$

Let $\lambda^{g'}A'\Gamma = O(\lambda^{g_1})$ and note that g_1 is independent of τ because so is $\operatorname{Ord}\Gamma$. Let us denote

$$T(z,\lambda) = -K^{-1}(z,\lambda)(G'(z,\lambda) + H_\lambda(z,\varphi_x))$$

and $\kappa = \max(\operatorname{Ord}A, \operatorname{Ord}T, 0)$. Let $\tau > c_2 = 3c(m)\kappa + g_1$, $\ s > c_3 = 3c(m)\kappa$. Then it is clear from the definition of c_i, $i = 1, 2$ that

$$(l + \tilde{\lambda}_{j_0}I - A)\Gamma\mathscr{U}_s = O(\lambda^{-c_3}). \tag{2.34}$$

Since $\operatorname{Ord}\Gamma\mathscr{U}_s \geq -c(m)\kappa$, (2.34) implies that $\tilde{\lambda}_{j_0}$ is an "eigenvalue" of A.

We now suppose that $\operatorname{Ord}\mathscr{V}_s = O(\lambda^{-g_2})$ with $g_2 > c_1 = c_3 + \operatorname{Ord}K^{-1}$. Then it is clear from (2.34) that

$$(l(z,D,\lambda) + \tilde{\lambda}_{j_0}(b,z,\lambda)I - T(z,\lambda))\Gamma\mathscr{U}_s = O(\lambda^{-c_3})$$

since $A\Gamma\mathscr{U}_s = T\Gamma\mathscr{U}_s - bK^{-1}B\mathscr{V}_s$. This implies that $\tilde{\lambda}_{j_0}$ is also an "eigenvalue" of T. We show below that these are not compatible. From Lemma 2.15 there are an open set $V \subset U_s$ and an invertible $\Delta(z,\lambda) \in \mathscr{A}(V)\{\{\tilde{\lambda}\}\}$ with $\operatorname{Ord}\Delta$, $\operatorname{Ord}\Delta^{-1} \leq c(m)\operatorname{Ord}T$ such that

$$(l - T)\Delta = \Delta(l - \oplus_1^m a_j(z,\lambda) + T')$$

with $T' \in \mathscr{A}(V)[[\tilde{\lambda}]]$. Hence we have

$$(l + \tilde{\lambda}_{j_0}I - T)\Delta = \Delta(l + \tilde{\lambda}_{j_0}I - \oplus a_j(z,\lambda) + T').$$

Setting $\mathscr{W}_s = \Delta^{-1}\Gamma\mathscr{U}_s$ it follows that

$$O(\lambda^{-c_3}) = (l + \tilde{\lambda}_{j_0} I - T)\Gamma\mathscr{U}_s = \Delta(l + \tilde{\lambda}_{j_0} I - \oplus a_j(z,\lambda) + T')\mathscr{W}_s$$

so that $(l + \tilde{\lambda}_{j_0} I - \oplus a_j(z,\lambda) + T')\mathscr{W}_s = O(\lambda^{-2c(m)\kappa})$. Since $\mathscr{U}_s = \Gamma^{-1}\Delta\mathscr{W}_s$ we have $0 \le \mathrm{Ord}\,\mathscr{U}_s \le \mathrm{Ord}\mathscr{W}_s + 2c(m)\kappa$ so that $\mathrm{Ord}\mathscr{W}_s \ge -2c(m)\kappa$. Then we can write

$$\mathscr{W}_s = \lambda^{\kappa'}\mathscr{W}_s'$$

with $\sigma_0(\mathscr{W}_s') \neq 0$, $\kappa' \le 2c(m)\kappa$. Therefore we get

$$(l + \tilde{\lambda}_{j_0} I - \oplus a_j(z,\lambda))\mathscr{W}_s' = O(1).$$

This shows that there is k such that $\tilde{\lambda}_{j_0}(b,z,\lambda) - a_k(z,\lambda) = O(1)$. But this contradicts the fact

$$\frac{\partial \mathrm{Im}\tilde{\lambda}_{j_0}}{\partial b} = \lambda^{\nu_2} c_2(b,z), \quad \frac{\partial a_k}{\partial b} = 0$$

because a_k is independent of b. Thus we see that $\mathrm{Ord}\{N_\lambda(z,\varphi_x)\Gamma\mathscr{U}\} \ge -c_1$ which is the desired conclusion. □

Now we complete the proof of Proposition 2.1. We set $\mathscr{U}_s' = E^{-1}N_\lambda E\Gamma\mathscr{U}_s$ and choose τ so that $\tau > c_2$ and

$$\mathrm{Ord}\{\lambda^{-\tau+g_0} N'(b,z,D,\lambda)\Gamma\mathscr{U}_s\} < -c_3.$$

We fix such τ. Then from Proposition 2.7 and (2.34) it follows that $\mathrm{Ord}\,\mathscr{U}_s' > \tau m' - 3c(m)\kappa$ which is independent of s and

$$(L_\lambda + bB)E\mathscr{U}_s' = EO(\lambda^{-s}).$$

Since $E = \exp(i\varphi + i\psi)$ and $\mathrm{Im}(\varphi + \psi) = \mathrm{Im}\,\psi$ using Proposition 1.2 and Lemma 2.16 the rest of the proof is just a repetition of that of Theorem 1.1.

2.7 Proof of Key Proposition, Asymptotic Diagonalization

Let W be an open set in $\mathbb{R}^N$ with a system of local coordinates z. In this section we give a proof of Lemma 2.15 which implies that one can diagonalize any $H(z,\lambda) \in \mathscr{A}(W)\{\{\tilde{\lambda}\}\}$ modulo $O(1)$ $(\lambda \to \infty)$, which we call asymptotic diagonalization.

Proposition 2.8. *Let $R(z,\xi,\lambda) = \sum_{j=0} R_j(z,\xi)\lambda^{\epsilon j}$ with $\epsilon \in \mathbb{Q}_+$ where $R_j(z,\xi)$ are polynomials in ξ of degree 1 and let $H(z,\lambda)$ be in $\mathscr{A}(W)\{\{\tilde{\lambda}\}\}$. Then for every open set $V \subset W$ there are an open set $U \subset V$ and $\Gamma(z,\lambda) \in \mathscr{A}(U)\{\{\tilde{\lambda}\}\}$ which is invertible such that* Ord Γ, Ord $\Gamma^{-1} \le c(m)$Ord$H$ *satisfying*

$$(R(z,D,\lambda) - H(z,\lambda))\Gamma(z,\lambda) = \Gamma(z,\lambda)(R(z,D,\lambda) - \oplus_{j=1}^{m}\lambda_j(z,\lambda) + K(z,\lambda))$$

with $\lambda_j(z,\lambda) \in \mathscr{A}(U)\{\{\tilde{\lambda}\}\}$, $K(z,\lambda) \in \mathscr{A}(U)[[\tilde{\lambda}]]$ and $\sum_{j=1}^{m}\lambda_j(z,\lambda) -$ Tr$H(z,\lambda) = O(1)$ *where $c(m)$ is a constant depending only on m, the size of matrices.*

Remark. Proposition 2.8 does not assert the usual diagonalization process. For example, we will make a procedure such as

$$\begin{bmatrix} \lambda & 0 \\ 0 & 1 \end{bmatrix}^{-1} \begin{bmatrix} 1 & 1 \\ 0 & 1 \end{bmatrix} \begin{bmatrix} \lambda & 0 \\ 0 & 1 \end{bmatrix} = \begin{bmatrix} 1 & 0 \\ 0 & 1 \end{bmatrix} + \lambda^{-1} \begin{bmatrix} 1 & 1 \\ 0 & 1 \end{bmatrix}.$$

In this example, with

$$A = \begin{bmatrix} 1 & 1 \\ 0 & 1 \end{bmatrix}, \quad E = \begin{bmatrix} 1 & 0 \\ 0 & 1 \end{bmatrix}, \quad \Gamma = \begin{bmatrix} 1 & 1 \\ 0 & 1 \end{bmatrix}$$

we have rank$(A - E) = 1$, rank $(\sigma_0(\Gamma^{-1}A\Gamma) - E) = 0$ while

$$\mathrm{Tr}\,\sigma_0(\Gamma^{-1}A\Gamma^{-1}) = \mathrm{Tr}\,A.$$

This invariance of Tr A (mod $O(1)$) holds in general in our asymptotic diagonalization process which plays an important role in the application (to the proof of Theorem 2.1).

To prove this proposition we prepare several notations and lemmas. Let us recall

$$J(r) = \begin{bmatrix} 0 & 1 & & \\ & \ddots & \ddots & \\ & & \ddots & 1 \\ & & & 0 \end{bmatrix} \in M_r(\mathbb{C}), \quad J(r_1,\dots,r_s) = \oplus_{j=1}^{s} J(r_j).$$

Definition 2.4. Let $J = J(r_1,\dots,r_s) \in M_m(\mathbb{C})$, $r_1 \ge \dots \ge r_s(\ge 1)$. We say that $K \in M_m(\mathbb{C})$ is a Sylvester matrix associated to J if K is of the following form; let $K = (K_{ij})_{1\le i,j\le s}$ be the blocking corresponding to that of J and let $K_{ij} = (k_{pq}^{ij})$, $1 \le p \le r_i$, $1 \le q \le r_j$, then $k_{pq}^{ij} = 0$ possibly except for $k_{r_i t}^{ij}$, $1 \le t \le \min(r_i, r_j)$. We denote by $S(J)$ the set of all Sylvester matrices associated to J.

Definition 2.5. Let $J(p_1,\dots,p_s)$, $J(q_1,\dots,q_t) \in M_m(\mathbb{C})$ where $p_1 \geq \cdots \geq p_s$, $q_1 \geq \cdots \geq q_t$. We say $J(p_1,\dots,p_s) \geq J(q_1,\dots,q_t)$ if $(p_1,\dots,p_s,0,\dots,0) \geq (q_1,\dots,q_t,0,\dots,0)$ as elements in $\mathbb{N}^m$ which is equipped with lexicographic order.

The following lemma can be found in Chap. 3 in [31,40].

Lemma 2.18. *Let $J = J(p_1,\dots,p_s) \in M_m(\mathbb{C})$ and $K \in S(J)$. Assume that $J+K$ is similar to $J' = J(q_1,\dots,q_t)$. Then we have $J' \geq J$. If $J+K$ is similar to J then $K = O$.*

Proof. Set

$$\begin{cases} g(K_{ij}) = k^{ij}_{p_i\mu}\lambda^{\mu-1} + \cdots + k^{ij}_{p_i 1}, \quad \mu = \min(p_i,p_j) \quad \text{when} \quad i \neq j, \\ g(K_{ii}) = -\lambda^{p_i} + k^{ii}_{p_i p_i}\lambda^{p_i-1} + \cdots + k^{ii}_{p_i 1}. \end{cases} \tag{2.35}$$

Then it is clear that $(J+K) - \lambda I_m$ is equivalent to $I_p \oplus G(\lambda)$, $p = m - s$ where $G(\lambda) = (g(K_{ij}))_{1\leq i,j\leq s}$. Assume that $J+K$ is similar to J'. Since $J' - \lambda I_m$ is equivalent to $I_q \oplus G'(\lambda)$ with $G'(\lambda) = \oplus_1^t \lambda^{q_i}$ then $I_p \oplus G(\lambda)$ is equivalent to $I_q \oplus G'(\lambda)$. If $q < p$ then the p-th determinant divisor of $I_q \oplus G'(\lambda)$ has the form λ^k $(k \geq 1)$ which contradicts the fact that the p-th determinant divisor of $I_p \oplus G(\lambda)$ is equal to 1. Thus we have $q \geq p$. Hence $G(\lambda)$ is equivalent to $I_{q-p} \oplus (\oplus_1^t \lambda^{q_j})$. Setting $q_{t+1} = \cdots = q_s = 0$, $G(\lambda)$ is then equivalent to $\oplus_1^s \lambda^{q_j}$. Let us denote by d'_k the degree of the k-th determinant divisor $D'_k(\lambda)$ of $G'(\lambda)$. Then it is obvious that $d'_j = q_s + \cdots + q_{s-j+1}$ and $D'_j(\lambda) = \lambda^{d'_j}$, $(1 \leq j \leq s)$. From (2.35) it is easy to see that there is a k-th principal minor of $G(\lambda)$ of the form

$$\epsilon\lambda^{p_s+\cdots+p_{s-k+1}} + \text{lower order} \quad (\epsilon = \pm 1).$$

Since this k-th minor is divisible by $D'_k(\lambda)$ it follows that

$$d'_k = q_s + \cdots + q_{s-k+1} \leq d_k = p_s + \cdots + p_{s-k+1}.$$

Since $p_j = d_{s-j+1} - d_{s-j}$, $q_j = d'_{s-j+1} - d'_{s-j}$, $d_s = d'_s$ it is clear that

$$(p_1,\dots,p_s,0,\dots,0) \leq (q_1,\dots,q_t,0,\dots,0)$$

and this proves the first assertion.

Now we assume that $J+K$ is similar to J and hence $G(\lambda)$ is equivalent to $\oplus_1^s \lambda^{p_j}$. Since $g(K_{sj})$, $g(K_{js})$ is divisible by λ^{p_s} it follows that $g(K_{sj}) = g(K_{js}) = 0$, $j \neq s$ and $g(K_{ss}) = -\lambda^{p_s}$. This shows that $G(\lambda) = G_1(\lambda) \oplus \{-\lambda^{p_s}\}$. Then it follows that $(g(K_{ij}))_{1\leq i,j\leq s-1}$ is equivalent to $\oplus_1^{s-1}\lambda^{p_j}$. By induction we get $g(K_{ij}) = 0$, $i \neq j$ and $g(K_{ii}) = -\lambda^{p_i}$. This shows that $K = O$ and hence the second assertion. □

The following lemma is found in [3,28,62].

Lemma 2.19. *Let* $H(z,\lambda) \in \mathscr{A}(W)\{\{\lambda^{-\epsilon}\}\}$ *be such that* $\sigma_p(H)(z) = J(r_1,\dots,r_s) = J \in M_m(\mathbb{C})$. *Then there is an invertible* $\Gamma(z,\lambda) \in \mathscr{A}(W)[[\lambda^{-\epsilon}]]$ *with* $\sigma_0(\Gamma) = I_m$ *such that*

$$(R(z,D,\lambda) + H(z,\lambda))\Gamma(z,\lambda) = \Gamma(z,\lambda)(R(z,D,\lambda) + \tilde{H}(z,\lambda))$$

where $\tilde{H}(z,\lambda) = \sum \tilde{H}_j(z)\lambda^{-j\epsilon} \in \mathscr{A}(W)\{\{\lambda^{-\epsilon}\}\}$ *with* $\sigma_p(\tilde{H})(z) = J$, $\mathrm{Ord}\tilde{H} = \mathrm{Ord}H$ *and* $\tilde{H}_j(z) \in \mathscr{A}(W, S(J))$ *for* $-\epsilon j < \mathrm{Ord}\tilde{H}$.

Although it is clear from the proof that $\mathrm{Tr}H(z,\lambda) - \mathrm{Tr}\tilde{H}(z,\lambda) = O(1)$ we give a direct proof by computation.

Lemma 2.20. *Assume that* $H(z,\lambda) = \sum_{j=s} H_j\lambda^{-\epsilon j}$ *and* $\tilde{H}(z,\lambda) = \sum_{j=s} \tilde{H}_j \lambda^{-\epsilon j}$ *are in* $\mathscr{A}(W)\{\{\lambda^{-\epsilon}\}\}$ *and*

$$(R(z,D,\lambda) - H(z,\lambda))\Gamma(z,\lambda) = \Gamma(z,\lambda)(R(z,D,\lambda) - \tilde{H}(z,\lambda))$$

with an invertible $\Gamma(z,\lambda) = \sum_{j=0}\Gamma_j\lambda^{-\epsilon j} \in \mathscr{A}(W)[[\lambda^{-\epsilon}]]$ *with* $\det\Gamma_0 \neq 0$. *Then it follows that*

$$\mathrm{Tr}H_j = \mathrm{Tr}\tilde{H}_j$$

for all $j < 0$.

Proof. We first note that

$$\sum_{i+j=p} H_{s+i}\Gamma_j - \Gamma_j\tilde{H}_{s+i} = C_{s+p},\ p = 0,1,\dots,\ C_q = O,\ q < 0.$$

Denoting $\Gamma^{-1} = \sum_{j=0}\Gamma'_j\lambda^{-j\epsilon}$ we consider

$$\sum_{i=0}^{l}\sum_{p=i}^{l}\Gamma'_{l-p}(H_{i+s}\Gamma_{p-i} - \Gamma_{p-i}\tilde{H}_{i+s}) = \sum_{p=0}^{l}\Gamma'_{l-p}C_{s+p}.$$

Here we recall that $\sum_{p=i}^{l}\Gamma'_{l-p}\Gamma_{p-i} = \delta_{l-i,0}I_m$ and hence

$$-\tilde{H}_{l+s} + \sum_{i=0}^{l}\sum_{q+r=l-i}\Gamma'_r H_{i+s}\Gamma_q = \sum_{p=0}^{l}\Gamma'_{l-p}C_{p+s}.$$

When $l+s<0$ the right hand side is zero and then for $l+s<0$ we have

$$\mathrm{Tr}\tilde{H}_{l+s} = \sum_{i=0}^{l}\sum_{q+r=l-i}\mathrm{Tr}(\Gamma_q\Gamma'_r H_{i+s})$$

$$= \mathrm{Tr}(\sum_{i=0}^{l}(\sum_{q+r=l-i}\Gamma_q\Gamma'_r)H_{i+s}) = \mathrm{Tr}H_{l+s}.$$

□

Let $H(z,\lambda) = \sum H_j(z)\lambda^{-\epsilon j} \in \mathscr{A}(W)\{\{\lambda^{-\epsilon}\}\}$ and $\sigma_p(H)(z) = \oplus_{i=1}^{r} B_i(z)$ where

$$B_j(z) = \begin{bmatrix} \lambda_j(z) & \epsilon_{j1} & & \\ & \ddots & \ddots & \\ & & \ddots & \epsilon_{jr_j-1} \\ & & & \lambda_j(z) \end{bmatrix} \in M_{r_j}(\mathbb{C}),\ \epsilon_{ji} = 0 \text{ or } 1$$

and $\lambda_i(z) - \lambda_j(z) \neq 0$ in W if $i \neq j$. The following lemma is classical. We refer to Wasow [71] for a proof.

Lemma 2.21. *Let $H(z,\lambda)$ be as above. Then there are an invertible $\Gamma(z,\lambda) \in \mathscr{A}(W)[[\lambda^{-\epsilon}]]$ with $\Gamma_0 = I_m$ and $\tilde{H}(z,\lambda) = \sum \tilde{H}_j\lambda^{-\epsilon j}$ with $\sigma_p(\tilde{H}) = \sigma_p(H)$ such that*

$$(R(z,D,\lambda) + H(z,\lambda))\Gamma(z,\lambda) = \Gamma(z,\lambda)(R(z,D,\lambda) + \tilde{H}(z,\lambda))$$

where $\tilde{H}_j(z)$ are block diagonal

$$\tilde{H}_j(z) = \oplus_{i=1}^{r}\tilde{H}_{ji}(z).$$

Definition 2.6. Let $J = J(r_1,\ldots,r_s) \in M_m(\mathbb{C})$. Then a matrix

$$\Lambda(\lambda,\epsilon) = \oplus_{j=1}^{s}\Lambda_{r_j}(\lambda,\epsilon),\ \Lambda_j(\lambda,\epsilon) = \oplus_{i=1}^{j}\lambda^{-i\epsilon} \in M_j(\mathbb{C}),\ \epsilon \in \mathbb{Q}_+$$

is called a shearing matrix associated to J (see Wasow [71] and the references given there).

We note that $\Lambda(\lambda,\epsilon) \in \mathscr{A}(W)\{\{\lambda^{-\epsilon}\}\}$, $\mathrm{Ord}\Lambda(\lambda,\epsilon) = -\epsilon$ and $\Lambda(\lambda,\epsilon)$ is invertible such that $\Lambda^{-1} = \oplus_1^s\Lambda_{r_j}(\lambda,-\epsilon)$, $\mathrm{Ord}\Lambda^{-1} = \max_j(r_j\epsilon)$. It is clear that

$$\Lambda(\lambda,\epsilon)^{-1}J\Lambda(\lambda,\epsilon) = \lambda^{-\epsilon}J.$$

For $A = (a_{ij}) \in M_m(\mathbb{C})$ we have

$$\Lambda(\lambda,\epsilon)^{-1}A\Lambda(\lambda,\epsilon) = (a_{ij}\lambda^{t_{ij}\epsilon})$$

where $t_{ij} \in \mathbb{Z}$ are independent of A.

Lemma 2.22. *Let $H(z,\lambda) \in \mathscr{A}(W)\{\{\tilde{\lambda}\}\}$ be such that* $\mathrm{Ord} H > 0$ *and* $\sigma_p(H)(z) = J = J(r_1, \ldots, r_s)$. *Then there are a shearing matrix $\Lambda(z,\theta)$ with* $\mathrm{Ord}\Lambda$, $\mathrm{Ord}\Lambda^{-1} \leq m\,\mathrm{Ord} H$ *and $\tilde{H}(z,\lambda) \in \mathscr{A}(W)\{\{\tilde{\lambda}\}\}$ with*

$$\Lambda(\lambda,\theta)^{-1} H(z,\lambda)\Lambda(\lambda,\theta) = \tilde{H}(z,\lambda)$$

such that we have either $\mathrm{Ord} H > \mathrm{Ord}\tilde{>}0$ *and* $\sigma_p(\tilde{H})(z) = J + K(z)$, $K(z) \not\equiv 0$ *or* $\mathrm{Ord}\tilde{H}(z,\lambda) \leq 0$.

Proof. Let $H(z,\lambda) = \lambda^{-t\epsilon}(J + H'(z,\lambda))$, $H'(z,\lambda) = (h'_{ij}(z,\lambda)) \in \mathscr{A}(W)[[\lambda^{-\epsilon}]]$. Note that with some $q_{ij} \in \mathbb{N}_+$ we have

$$h'_{ij}(z,\lambda) = \lambda^{-\epsilon q_{ij}}(a_{ij}(z) + O(\lambda^{-\epsilon})),\ a_{ij}(z) \not\equiv 0.$$

Thus we get

$$\Lambda(\lambda,\delta)^{-1} H'\Lambda(\lambda,\delta) = \{\lambda^{-\epsilon q_{ij}+\delta t_{ij}}(a_{ij}(z) + O(\lambda^{-\epsilon}))\}.$$

Let $\theta(\epsilon)$ be

$$\theta(\epsilon) = \min_{i,j,1+t_{ij}>0} (\epsilon q_{ij}/(1+t_{ij}), -t\epsilon).$$

By definition we have $0 < \theta(\epsilon) \leq -t\epsilon = \mathrm{Ord} H$ and $-\epsilon q_{ij} + \theta(\epsilon)t_{ij} \leq -\theta(\epsilon)$. Taking $\theta = \theta(\epsilon)$ it follows that $\Lambda(\lambda,\theta)^{-1} H'\Lambda(\lambda,\theta) = O(\lambda^{-\theta})$ and hence

$$\Lambda(\lambda,\theta)^{-1} H\Lambda(\lambda,\theta) = O(\lambda^{-\theta-t\epsilon})$$

where $\mathrm{Ord}\Lambda$, $\mathrm{Ord}\Lambda^{-1} \leq m\,\mathrm{Ord} H$. Here note that $\Lambda^{-1}H\Lambda \in \mathscr{A}(W)\{\{\tilde{\lambda}\}\}$ because $\theta \in \mathbb{Q}_+$. If $-\theta - t\epsilon > 0$, from the definition of θ, there are k, l with $-\epsilon q_{kl} + t_{kl}\theta = -\theta$ so that we have

$$\Lambda(\lambda,\theta)^{-1} H'\Lambda(\lambda,\theta) = \lambda^{-\theta}(K(z) + O(\lambda^{-\epsilon})),\quad K(z) \not\equiv O$$

which proves the assertion. □

Proposition 2.9. *Let $H(z,\lambda) \in \mathscr{A}(W)\{\{\tilde{\lambda}\}\}$ with* $\mathrm{Tr} H(z,\lambda) = O(1)$. *Then for every open set $V \subset W$ there are an open set $U \subset V$ and an invertible $\Gamma(z,\lambda) \in \mathscr{A}(U)\{\{\tilde{\lambda}\}\}$ with* $\mathrm{Ord}\,\Gamma$, $\mathrm{Ord}\,\Gamma^{-1} \leq c(m)\mathrm{Ord} H$ *such that we have either*

$$(R+H)\Gamma = \Gamma(R+K) \text{ with } K \in \mathscr{A}(U)[[\tilde{\lambda}]]$$

or

$$(R+H)\Gamma = \Gamma(R+\tilde{H}) \text{ with } \tilde{H} \in \mathscr{A}(U)\{\{\tilde{\lambda}\}\}$$

where $\mathrm{Ord}H \geq \mathrm{Ord}\tilde{H} > 0$, $\mathrm{Tr}\tilde{H} = O(1)$ *and* $\sigma_p(\tilde{H})(z)$ *has a non zero eigenvalue for every* $z \in U$.

To prove this proposition we need the following lemma.

Lemma 2.23. *Let* $H(z, \lambda) \in \mathscr{A}(U)\{\{\tilde{\lambda}\}\}$ *where* $\sigma_p(H)(z) = J = J(p_1, \dots, p_s)$ *and* $\mathrm{Tr}H(z, \lambda) = O(1)$. *Then there are an open set* $V \subset U$ *and an invertible* $\Gamma(z, \lambda) \in \mathscr{A}(V)\{\{\tilde{\lambda}\}\}$ *with* $\mathrm{Ord}\,\Gamma$, $\mathrm{Ord}\,\Gamma^{-1} \leq m\,\mathrm{Ord}H$ *such that* $(R(z, D, \lambda) + H(z, \lambda))\Gamma(z, \lambda)$ *is equal to one of the followings*

(a) $R(z, D, \lambda) + K(z, \lambda)$, $K(z, \lambda) \in \mathscr{A}(V)[[\tilde{\lambda}]]$,
(b) $R(z, D, \lambda) + H^1(z, \lambda)$, $H^1(z, \lambda) \in \mathscr{A}(V)\{\{\tilde{\lambda}\}\}$ *with* $\mathrm{Ord}H \geq \mathrm{Ord}H^1 > 0$, $\mathrm{Tr}H^1 = O(1)$ *where* $\sigma_p(H^1)(z)$ *has a non zero eigenvalue for every* $z \in V$,
(c) $R(z, D, \lambda) + H^1(z, \lambda)$, $H^1(z, \lambda) \in \mathscr{A}(V)\{\{\tilde{\lambda}\}\}$ *with* $\mathrm{Ord}H \geq \mathrm{Ord}H^1 > 0$ *and* $\sigma_p(H^1)(z) = J_1 = J(q_1, \dots, q_t) > J$, $\mathrm{Tr}H^1(z, \lambda) = O(1)$.

Proof. By Lemma 2.19 there is $\Gamma'(z, \lambda) \in \mathscr{A}(U)[[\tilde{\lambda}]]$, $\Gamma'_0 = I$ satisfying $(R + H)\Gamma' = \Gamma'(R + H')$ where $H' \in \mathscr{A}(U)\{\{\tilde{\lambda}\}\}$ verifies the requirements in Lemma 2.19 and $\mathrm{Tr}H' = O(1)$. From Lemma 2.22 there is a shearing matrix $\Lambda(\lambda, \theta)$ associated to J with $\mathrm{Ord}\Lambda$, $\mathrm{Ord}\Lambda^{-1} \leq m\,\mathrm{Ord}H$ such that $H'\Lambda = \Lambda H''$. Since $R\Lambda = \Lambda R$ we get

$$(R + H)\Gamma'\Lambda = \Gamma'\Lambda(R + H'').$$

If $\mathrm{Ord}H'' = 0$ we have the case (a). Otherwise $\mathrm{Ord}H = \mathrm{Ord}H' > \mathrm{Ord}H'' > 0$ and $\sigma_p(H'')(z) = J + C(z)$, $C(z) \not\equiv 0$. Assume that $\mathrm{Ord}H'' > 0$. If $\sigma_p(H'')(z)$ has a non zero eigenvalue at some $z \in U$ then there is an open set $V \subset U$ on which $\sigma_p(H'')(z)$ has a non zero eigenvalue. This is the case (b). In these two cases it is enough to take $\Gamma = \Gamma'\Lambda$. It is clear that $\mathrm{Ord}\,\Gamma^{-1} = \mathrm{Ord}\Lambda^{-1}\Gamma'^{-1} \leq m\,\mathrm{Ord}H$. We turn to the remaining case, that is $\mathrm{Ord}H'' > 0$ and $\sigma_p(H'')(z)$ is nilpotent for every $z \in U$. Then one can choose an open set $V \subset U$ so that there is $N(z) \in \mathscr{A}(V, M_m(\mathbb{C}))$ which verifies that (see [27, 72])

$$N^{-1}(z)\sigma_p(H'')(z)N(z) = J_1 = J(q_1, \dots, q_t).$$

Since $\sigma_p(H'')(z) = J + C(z)$ and we can assume $C(z) \neq 0$ at every $z \in V$, shrinking V if necessary, it follows from Lemma 2.18 that $J_1 > J$. Now we arrive at

$$(R + H)\Gamma'\Lambda N = \Gamma'\Lambda N(R + H''')$$

where $H''' \in \mathscr{A}(V)\{\{\tilde{\lambda}\}\}$, $\sigma_p(H''') = J_1$, $\mathrm{Tr}H''' = O(1)$. Taking $\Gamma = \Gamma'\Lambda N$ it is clear that $\mathrm{Ord}\,\Gamma$, $\mathrm{Ord}\,\Gamma^{-1} \leq m\,\mathrm{Ord}H$. This completes the proof. □

Proof of Proposition 2.9. If $\sigma_p(H)(z)$ has a non zero eigenvalue at some $z \in U$ we can choose an open set $V \subset U$ in which $\sigma_p(H)(z)$ has a non zero eigenvalue. In this case it is enough to take $\Gamma = I$. If $\sigma_p(H)(z)$ is nilpotent for every $z \in U$ repeating the same argument in the proof of Lemma 2.23 we can assume that $\sigma_p(H)(z) = J = J(p_1, \dots, p_s)$. Now we apply Lemma 2.23 repeatedly. If we arrive at the case (c) we again apply Lemma 2.23. Since the case (c) occurs only finitely many times which is bounded by a constant $c_1(m)$ depending only on m, we arrive at the cases (a) or (b) after at most $c_1(m)$ times and this proves the assertion except for the estimate of Ord Γ. The desired Γ is given by

$$\Gamma = \Gamma^1\Gamma^2 \cdots \Gamma^s, \quad \operatorname{Ord}\Gamma^i \le m \operatorname{Ord} H, \ s \le c_1(m).$$

Then it is clear that Ord Γ, Ord $\Gamma^{-1} \le c_1(m)m$ Ord$H$ which completes the proof. □

Corollary 2.5. *Let $H(z, \lambda) \in \mathscr{A}(W)\{\{\tilde{\lambda}\}\}$, Ord$H > 0$. Then for every open set $V \subset W$ there are an open set $U \subset V$ and an invertible $\Gamma(z, \lambda) \in \mathscr{A}(U)\{\{\tilde{\lambda}\}\}$ with* Ord Γ, Ord $\Gamma^{-1} \le c(m)$Ord$H$ *such that we have either*

$$(R + H)\Gamma = \Gamma(R + K) \ \text{ with } \ K(z, \lambda) \in \mathscr{A}(U)[[\tilde{\lambda}]]$$

or

$$(R + H)\Gamma = \Gamma(R + \tilde{H}) \text{ with } \tilde{H}(z, \lambda) \in \mathscr{A}(U)\{\{\tilde{\lambda}\}\}, \ \operatorname{Ord}H \ge \operatorname{Ord}\tilde{H} > 0$$

where $\mathrm{Tr}H - \mathrm{Tr}\tilde{H} = O(1)$ *and* $\sigma_p(\tilde{H})(z)$ *has a non zero eigenvalue at every* $z \in U$.

Proof. Write $H(z, \lambda) = \phi(z, \lambda)I + H'(z, \lambda)$, $\phi = \mathrm{Tr}H/m$ so that $\mathrm{Tr}H' = 0$. Since $\phi\Gamma = \Gamma\phi$ the assertion is an immediate consequence of Proposition 2.9. □

Lemma 2.24. *Let $H(z, \lambda) \in \mathscr{A}(W)\{\{\tilde{\lambda}\}\}$. Then for every open set $V \subset W$ there are an open set $U \subset V$ and an invertible $\Gamma(z, \lambda) \in \mathscr{A}(U)\{\{\tilde{\lambda}\}\}$ with* Ord Γ, Ord $\Gamma^{-1} \le c(m)$Ord$H$ *such that*

$$(R + H)\Gamma = \Gamma(R + \tilde{H}), \ \tilde{H} \in \mathscr{A}(U)\{\{\tilde{\lambda}\}\}, \ \mathrm{Tr}H - \mathrm{Tr}\tilde{H} = O(1)$$

and if Ord $\tilde{H} > 0$ *then either $\tilde{H}$ is block diagonal with more than two blocks or $\tilde{H}$ is diagonal modulo $O(1)$.*

Proof. Set $H = \phi I + H'$, $\phi(z, \lambda) = \mathrm{Tr}H/m$ so that $\mathrm{Tr}H' = 0$. It follows from Proposition 2.9 that there are an open set $U \subset V$ and an invertible $\Gamma' \in \mathscr{A}(U)\{\{\tilde{\lambda}\}\}$ with Ord Γ', Ord $\Gamma'^{-1} \le c(m)$Ord$H$ such that

$$(R + H')\Gamma' = \Gamma'(R + H''), \ \mathrm{Tr}H'' = O(1).$$

If Ord$H'' \le 0$ then $\phi I + H''$ is diagonal modulo $O(1)$. Otherwise $\sigma_p(H'')(z)$ has a non zero eigenvalue at every $z \in U$. Hence we can choose an open set $U_1 \subset U$ and $N(z) \in \mathscr{A}(U_1, M_m(\mathbb{C}))$ satisfying

$$N(z)^{-1}\sigma_p(H'')(z)N(z) = \oplus_1^s B_j$$

where

$$B_j(z) = \begin{bmatrix} \lambda_j(z) & \epsilon_{ji} & & \\ & \ddots & \ddots & \\ & & \ddots & \epsilon_{jr_j-1} \\ & & & \lambda_j(z) \end{bmatrix} \in M_{r_j}(\mathbb{C}),\ \epsilon_{jk} = 0 \text{ or } 1$$

and $\lambda_i(z) \neq \lambda_j(z)$ in U_1 if $i \neq j$. Since $\mathrm{Tr}H'' = O(1)$ we have necessarily $s \geq 2$. By Lemma 2.21 we can take $\Gamma'' \in \mathscr{A}(U_1)[[\tilde{\lambda}]]$ with $\Gamma_0'' = I$ which satisfies

$$(R + H'')\Gamma'' = \Gamma''(R + H'''),\ \mathrm{Ord}H''' = O(1)$$

where H''' is block diagonal. Setting $\Gamma = \Gamma' N \Gamma''$ we have the desired conclusion with $U = U_1$ because $\phi\Gamma = \Gamma\phi$. □

Proof of Proposition 2.8. We first note that if $H \in \mathscr{A}(W)\{\{\tilde{\lambda}\}\}$ is block diagonal then Lemma 2.24 holds for each block. Applying Lemma 2.24 to H, we get H'. If H' is diagonal modulo $O(1)$ nothing to be proved. Otherwise H' is block diagonal with more than two blocks. In the latter case we again apply Lemma 2.24 to each block. Then after at most m times repetition of this procedure we arrive at the assertion because in each step the number of blocks modulo $O(1)$ increases at least by one. □

2.8 Involutive Characteristics

It is important to note that (2.14) or (2.6) holds for any choice of $\sigma, \delta \in \mathbb{Q}_+^{n+1}$ provided $(\sigma,\delta) \in I(h,\rho)$. In the proof of Theorem 2.1 we eventually take $\sigma = (\bar{\sigma},\ldots,\bar{\sigma},1)$, $\delta = (\bar{\delta},\ldots,\bar{\delta})$, $\bar{\delta} = 1 - \bar{\sigma}$, $3\bar{\delta} > 1$ but it is expected that if we can choose another $(\sigma,\delta) \in I(h,\rho)$ then the result will be improved. This case actually happens. Let ρ be a characteristic of order r for $h(x,\xi)$. From Lemma 1.11 we see that $h_\rho(X)$ is a hyperbolic polynomial with respect to θ. Then we can define the linearity space $\Lambda(h_\rho)$ of h_ρ and we can improve Theorem 2.1 when $\Lambda(h_\rho)$ is involutive. We start with defining the linearity space $\Lambda(h_\rho)$.

Definition 2.7 (see [2,16,35]). The hyperbolic cone $\Gamma(h_\rho)$ of $h_\rho(X)$ is defined as the connected component of the set

$$\{X = (x,\xi) \in \mathbb{R}^{n+1} \times \mathbb{R}^{n+1} \mid h_\rho(X) \neq 0\}$$

containing $\theta = (0,\ldots,0,1,0,\ldots,0)$. The propagation cone $C(h_\rho)$ of h_ρ is the set

$$\{X \in \mathbb{R}^{n+1} \times \mathbb{R}^{n+1} \mid \sigma(X, Y) \le 0, Y \in \Gamma(h_\rho)\}$$

where σ is the symplectic 2 form given by $\sigma = \sum_{j=0}^{n} d\xi_j \wedge dx_j$ so that

$$\sigma((x, \xi), (y, \eta)) = \langle \xi, y \rangle - \langle x, \eta \rangle.$$

The linearity space $\Lambda(h_\rho)$ of h_ρ is defined as

$$\{X \in \mathbb{R}^{n+1} \times \mathbb{R}^{n+1} \mid h_\rho(tX + Y) = h_\rho(Y), t \in \mathbb{R}, Y \in \mathbb{R}^{n+1} \times \mathbb{R}^{n+1}\}.$$

It is clear that

$$\Gamma(h_\rho) + \Lambda(h_\rho) \subset \Gamma(h_\rho).$$

Remark that $\Lambda(h_\rho)$ is the maximal linear subspace Λ such that $\Gamma(h_\rho) + \Lambda \subset \Gamma(h_\rho)$. Indeed we have

Lemma 2.25. *Let $\langle X \rangle$ be the linear subspace spanned by X. If*

$$\Gamma(h_\rho) + \langle X \rangle \subset \overline{\Gamma(h_\rho)}$$

then $X \in \Lambda(h_\rho)$ where $\overline{\Gamma}$ denotes the closure of Γ.

Proof. Let $Y \in \Gamma(h_\rho, \theta)$ and consider $h_\rho(sY + tX) = h_\rho((s - \epsilon)Y + \epsilon Y + tX) = 0$ for small $\epsilon > 0$. Since $\epsilon Y + tX \in \overline{\Gamma}$ and h_ρ is hyperbolic with respect to θ it follows (see [2, 19]) that

$$h_\rho(sY + tX) = 0 \Longrightarrow s \le \epsilon$$

and hence $s \le 0$. On the other hand we can write

$$h_\rho(sY + tX) = h_\rho(Y) \prod (s - \lambda_j(tX)),$$
$$h_\rho(sY - tX) = (-1)^{rh}_\rho(-sY + tX) = (-1)^{rh}_\rho(Y) \prod (-s - \lambda_j(tX))$$

from which we conclude that $\lambda_j(tX) = 0$. Thus we get

$$h_\rho(sY + tX) = h_\rho(Y) \prod s.$$

Choosing $s = 1$ one has $h_\rho(Y + tX) = h_\rho(Y)$ for all $Y \in \Gamma(h_\rho, \theta)$. Since $\Gamma(h_\rho)$ is an open cone then we have the assertion. □

Definition 2.8. We say that a linear subspace $E \subset \mathbb{R}^{n+1} \times \mathbb{R}^{n+1}$ is involutive if

$$E^{\sigma} = \{X \mid \sigma(X,Y) = 0, \forall Y \in E\} \subset E.$$

Here we note

Lemma 2.26. *$\Lambda(h_\rho)$ is involutive if and only if*

$$C(h_\rho) \subset \Lambda(h_\rho).$$

Proof. Let us write $C = C(h_\rho, \theta)$, $\Lambda = \Lambda(h_\rho)$ and $\Gamma = \Gamma(h_\rho)$. Assume that Λ is involutive, that is $\Lambda^\sigma \subset \Lambda$. Since $\Gamma + \Lambda \subset \Gamma$ by Lemma 2.25 we have

$$\sigma(X, Y + tZ) \le 0, \quad \forall t \in \mathbb{R},\ \forall Y \in \Gamma,\ \forall Z \in \Lambda$$

for $X \in C$ which shows that $\sigma(X,Z) = 0$ and hence $X \in \Lambda^\sigma \subset \Lambda$. That is $C \subset \Lambda$.

Conversely assume that $C \subset \Lambda$. Suppose $\Lambda^\sigma \not\subset \Lambda$. Take $X \in \Lambda^\sigma$ such that $X \notin \Lambda$. Thanks to Lemma 2.25 it follows that $\Gamma + \langle X \rangle \notin \overline{\Gamma}$. Thus there exist $Y \in \Gamma$ and $s \in \mathbb{R}$ such that $Y + sX \notin \overline{\Gamma}$. From the Hahn-Banach theorem there is $W \in \mathbb{R}^{2(n+1)}$ such that

$$\sigma(W, Z) \le 0, \ \ \forall Z \in \Gamma, \quad \sigma(W, Y + sX) > 0.$$

From the first inequality we see $W \in C$ and hence $\sigma(W, X) = 0$ because $C \subset \Lambda$ by the assumption. This shows

$$\sigma(W, Y + sX) = \sigma(W, Y) \le 0$$

which is a contradiction. □

If ρ is a simple characteristic, then $h_\rho(X) = \langle \nabla_{x,\xi} h(\rho), X\rangle$ and clearly $\Lambda(h_\rho)$ is the hyperplane defined by $\langle \nabla_{x,\xi} h(\rho), (x,\xi)\rangle = 0$ where

$$\nabla_{x,\xi} h = (\partial h/\partial x_0, \ldots, \partial h/\partial x_n, \partial h/\partial \xi_0, \ldots, \partial h/\partial \xi_n)$$

and the propagation cone $C(h_\rho)$ is

$$\mathbb{R}_+ \cdot H_h(\rho) = \Big\{ c\big(\frac{\partial h}{\partial \xi}(\rho), -\frac{\partial h}{\partial x}(\rho)\big) \mid c \ge 0 \Big\}$$

where H_h denotes the Hamilton vector field of h. We have

Theorem 2.2. *Assume that $A_j(x)$ are real analytic in Ω and $0 \in \Omega$. Let $\rho = (0, \bar{\xi})$, $\bar{\xi} \in \mathbb{R}^{n+1} \setminus \{0\}$ be a characteristic of order r for $h(x, \xi)$. Suppose that $\Lambda(h_\rho)$ is involutive. Then if L is strongly hyperbolic near the origin then every $(m-1)$-th minor of L vanishes of order $r-1$ at ρ, that is for any $(m-1)$-th minor $q(x, \xi)$ of $L(x, \xi)$ we have*

$$\partial_x^\beta \partial_\xi^\alpha q(\rho) = 0, \quad \forall |\alpha + \beta| < r - 1.$$

Corollary 2.6. *Assume that $A_j(x)$ are real analytic in Ω and $0 \in \Omega$. Let $\rho = (0, \bar{\xi})$, $\bar{\xi} \in \mathbb{R}^{n+1} \setminus \{0\}$ be a characteristic of order r for $h(x, \xi)$ with involutive $\Lambda(h_\rho)$. If L is strongly hyperbolic near the origin then we have*

$$\dim \operatorname{Ker} L(\rho) = r.$$

Proof. The assertion follows from the same argument as in the proof of Corollary 2.1. □

From Corollary 2.6 we have

Corollary 2.7. *Assume that $A_j(x)$ are real analytic in Ω and $0 \in \Omega$. Let $\rho = (0, \bar{\xi})$, $\bar{\xi} \in \mathbb{R}^{n+1} \setminus \{0\}$ be a characteristic of order r for $h(x, \xi)$ satisfying $h_\rho(X) = cl(X)^r$ with a linear function $l(X)$ and a constant $c \neq 0$. If L is strongly hyperbolic near the origin we have*

$$\dim \operatorname{Ker} L(\rho) = r.$$

Proof. The assertion follows from Corollary 2.6 since $C(h_\rho) = cH_l \cdot \mathbb{R}_+ \subset \Lambda = \{X \mid l(X) = 0\}$. □

We now consider the case that $h(x, \xi)$ has the form

$$h(x, \xi) = \prod_{j=1}^{s} (\xi_0 - \lambda_j(x, \xi'))^{m_j} = \prod_{j=1}^{s} g_j(x, \xi)^{m_j} \tag{2.36}$$

near $x = 0$ and for any $\xi' \neq 0$ where m_j are constants independent of (x, ξ') and $\lambda_j(x, \xi')$ are smooth and different from each other. Let us fix $(\bar{x}, \bar{\xi}')$, $\bar{\xi}' \neq 0$. Set

$$\rho_j = (\bar{x}, \lambda_j(\bar{x}, \bar{\xi}'), \bar{\xi}') = (\bar{x}, \bar{\lambda}_j, \bar{\xi}')$$

so that ρ_j is a characteristic of $h(x, \xi)$ of order m_j. It is clear that

$$h_{\rho_j}(x, \xi) = c\langle \nabla_{x,\xi} g_j(\rho_j), X \rangle^{m_j}, \quad c = \prod_{k \neq j} g_k(\rho_j)^{m_k}$$

where $X = (x, \xi)$. Then from Corollary 2.7 it follows that

$$\dim \operatorname{Ker} L(\rho_j) = m_j, \quad j = 1, \ldots, s.$$

This shows that $A(\bar{x}, \xi')$ is diagonalizable. Thus we have proved

Corollary 2.8 ([27]). *Assume that $A_j(x)$ are real analytic in $\Omega \ni 0$ and $h(x,\xi)$ verifies (2.36). If L is strongly hyperbolic near the origin then $A(x,\xi')$ is diagonalizable for any x close to the origin and for any ξ'.*

This corollary holds without the analyticity assumption on $A_j(x)$, see [27].

2.9 Localization at Involutive Characteristics

In this section we prove Theorem 2.2. The key point is that if $\Lambda(h_\rho)$ is involutive then we have $(\sigma, \delta) \in I(h, \rho)$ when σ_j $(j < n)$ and δ_j are enough close to $1/2$. We consider the following change of coordinates

$$\tilde{y} = A\tilde{x}, \quad y_n = x_n + q(\tilde{x})/2, \quad \tilde{x} = (x_0, \ldots, x_{n-1})$$

where A is an $n \times n$ non singular matrix and $q(\tilde{x}) = \langle Q\tilde{x}, \tilde{x}\rangle$ is a quadratic form in $\tilde{x}$.

Lemma 2.27. *Let $\tilde{y} = A\tilde{x}$, $y_n = x_n + q(\tilde{x})/2$ be as above. We write $(y,\eta) = (\tilde{y}, y_n, \tilde{\eta}, \eta_n)$ the corresponding new coordinates in $\mathbb{R}^{n+1} \times \mathbb{R}^{n+1}$. Then in these new coordinates h_ρ is given by*

$$h_\rho(y, \tilde{\eta}) = h_\rho(A^{-1}\tilde{y}, y_n, {}^t\!A\tilde{\eta} + QA^{-1}\tilde{y})$$

where ${}^t\!A$ denotes the transposed of A.

Proof. We first note that

$$\tilde{x} = A^{-1}\tilde{y} = \varphi(y), \quad x_n = y_n - q(A^{-1}\tilde{y})/2 = \varphi_n(y),$$
$$\tilde{\xi} = {}^t\!A\tilde{\eta} + QA^{-1}\tilde{y}\eta_n = \psi(y,\eta), \quad \xi_n = \eta_n.$$

Recall that

$$h(x,\xi) = \xi_n^{m-r} h_\rho(x\xi_n, \tilde{\xi}) + O(|\xi_n|^{m-r-1}(|\xi_n x| + |\tilde{\xi}|)^{r+1}) \text{ as } |x| + |\tilde{\xi}| \to 0.$$

and

$$\xi_n^{m-r} h_\rho(x\xi_n, \tilde{\xi}) = \lim_{\lambda\to\infty} \lambda^{-\gamma(a,r)} h(T_\lambda(x,\xi)), \; T_\lambda(x,\xi) = (\lambda^{-b}x, \lambda^a\tilde{\xi}, \lambda\xi_n)$$

where $b = 1 - a$, $a > b > 0$. Since $\varphi(T_\lambda(y,\eta)) = \lambda^{-b}A^{-1}\tilde{y}$, $\varphi_n(T_\lambda(y,\eta)) = \lambda^{-b}(y_n + o(1))$, $\psi(T_\lambda(y,\eta)) = \lambda^a({}^t\!A\tilde{\eta} + QA^{-1}\tilde{y}\eta_n + o(1))$ it is clear that

$$\lim_{\lambda\to\infty} \lambda^{-\gamma(a,r)}(\lambda\eta_n)^{m-r} h_\rho(\varphi(T_\lambda(y,\eta))\lambda\eta_n, \varphi_n(T_\lambda(y,\eta))\lambda\eta_n, \psi(T_\lambda(y,\eta)))$$
$$= \eta_n^{m-r} h_\rho(A^{-1}\tilde{y}\eta_n, y_n\eta_n, {}^tA\tilde{\eta} + QA^{-1}\tilde{y}\eta_n).$$

From this we get the assertion. □

Here we note that the symplectic map:

$$(\tilde{x}, \tilde{\xi}) \to (A^{-1}\tilde{x}, {}^tA\tilde{\xi} + QA^{-1}\tilde{x})$$

with a non singular A and a symmetric Q generates a group G. In fact with $T_i(\tilde{x}, \tilde{\xi}) = (A_i^{-1}\tilde{x}, {}^tA_i\tilde{\xi} + Q_iA_i^{-1}\tilde{x})$, $i = 1, 2$ we see

$$\begin{aligned} T_2T_1(\tilde{x}, \tilde{\xi}) &= T_2(A_1^{-1}\tilde{x}, {}^tA_1\tilde{\xi} + Q_1A_1^{-1}\tilde{x}) \\ &= (A_2^{-1}A_1^{-1}\tilde{x}, {}^tA_2({}^tA_1\tilde{\xi} + Q_1A_1^{-1}\tilde{x}) + Q_2A_2^{-1}A_1^{-1}\tilde{x}) \\ &= (A^{-1}\tilde{x}, {}^tA\tilde{\xi} + \{{}^tA_2Q_1A_2 + Q_2\}A^{-1}\tilde{x}) \end{aligned}$$

where $A = A_1A_2$ and $Q = {}^tA_2Q_1A_2 + Q_2$ is symmetric. This group G is generated by linear change of coordinates $\tilde{x}$;

$$T_A : (\tilde{x}, \tilde{\xi}) \mapsto (A^{-1}\tilde{x}, {}^tA\tilde{\xi}), \quad A \text{ is non singular}$$

and

$$S_Q : (\tilde{x}, \tilde{\xi}) \mapsto (\tilde{x}, \tilde{\xi} + Q\tilde{x}), \quad Q \text{ is symmetric}$$

since we have

$$S_QT_A(\tilde{x}, \tilde{\xi}) = S_Q(A^{-1}\tilde{x}, {}^tA\tilde{\xi}) = (A^{-1}\tilde{x}, {}^tA\tilde{\xi} + QA^{-1}\tilde{x})$$

(see for example [18]). We use this lemma to reduce h_ρ to a more convenient form in the case that $\Lambda(h_\rho)$ is involutive.

Proposition 2.10. *Assume that $\Lambda(h_\rho)$ is an involutive subspace. Then we can find a new system of local coordinates y; $\tilde{y} = A\tilde{x}$, $y_n = x_n + q(\tilde{x})/2$ preserving the x_0 coordinate, an integer $k \in \mathbb{N}$ and a homogeneous polynomial q of degree r such that*

$$h_\rho(y, \tilde{\eta}) = q(\eta_0, \dots \eta_k, y_{k+1}, \dots, y_n).$$

Proof. To simplify notations we set

$$h_\rho(x, \tilde{\xi}) = p(x, \tilde{\xi}), \quad \Lambda(h_\rho) = \Lambda.$$

Let Λ be given by

$$\Lambda = \{(x, \tilde{\xi}) \mid l_j(x, \tilde{\xi}) = 0,\ 0 \le j \le N\}$$

where $l_j(x, \tilde{\xi})$ are linearly independent linear forms in $(x, \tilde{\xi})$. Here we note that $p(x, \tilde{\xi})$ is a polynomial of degree r in $l(x, \tilde{\xi}) = (l_0(x, \tilde{\xi}), \ldots, l_N(x, \tilde{\xi}))$. To see this, take linear forms $k_1(x, \tilde{\xi}), \ldots, k_M(x, \tilde{\xi})$ so that $\ell_1, \ldots, \ell_N, k_1, \ldots, k_M$ is a new coordinate system in $\mathbb{R}^{2n+1}$. With $z = (x, \tilde{\xi})$ we have

$$p(z) = \sum_{|\alpha+\beta|=r} c_{\alpha\beta}\ell(z)^{\alpha}k(z)^{\beta} = q(\ell, k).$$

Note that

$$p(z + w) = q(\ell(z + w), k(z + w)) = q(\ell(z), k(z) + k(w))$$

which is equal to $p(z) = q(\ell(z), k(z))$. Since $\Lambda \ni w \mapsto k(w) \in \mathbb{R}^M$ is surjective it follows that q is independent of k and hence the result. □

We prepare two lemmas.

Lemma 2.28. $\Lambda = \{(x, \xi) \mid \ell_j(x, \xi) = 0, j = 0, 1, \ldots, N\}$ *is involutive if and only if the Poisson bracket* $\{\ell_i, \ell_j\}$ *vanishes for any pair* ℓ_i, ℓ_j

$$\{\ell_i, \ell_j\} = \sum_{p=0}^{n}\Big(\frac{\partial \ell_i}{\partial \xi_p}\frac{\partial \ell_j}{\partial x_p} - \frac{\partial \ell_i}{\partial x_p}\frac{\partial \ell_j}{\partial \xi_p}\Big) = 0, \quad \forall i, j.$$

Proof. Define H_{ℓ_j} by $\ell_j(x, \xi) = \sigma((x, \xi), H_{\ell_j})$. Since $(x, \xi) \in \Lambda$ if and only if $\sigma((x, \xi), H_{\ell_j}) = 0$ for all j it is clear that

$$\Lambda^{\sigma} = \text{span}\{H_{\ell_j}, j = 0, 1, \ldots, N\}.$$

Then $\Lambda^{\sigma} \subset \Lambda$ is equivalent to $H_{\ell_j} \in \Lambda$ for all j, that is

$$\ell_i(H_{\ell_j}) = \sigma(H_{\ell_j}, H_{\ell_i}) = \{\ell_j, \ell_i\} = 0$$

which proves the assertion. □

Lemma 2.29. *Let* $\{(x, \xi) \mid \ell_j(x, \xi) = 0, j = 1, \ldots, p\}$ *be a linear subspace where* ℓ_j *are linearly independent linear forms in* (x, ξ) *with* $x = (x_1, \ldots, x_p)$ *and* $\xi = (\xi_1, \ldots, \xi_p)$. *Assume that* $\{\ell_i, \ell_j\} = 0$ *for all* i, j. *Then there exists* $T \in G$ *such that*

$$\begin{aligned}&\{(x, \xi) \mid \ell_j(T(x, \xi)) = 0, j = 1, \ldots, p\}\\ &= \{(x, \xi) \mid \xi_1 = \cdots = \xi_k = 0, x_{k+1} = \cdots = x_p = 0\}.\end{aligned}$$

Proof. Let $\ell_1 = a_1 x_1 + \cdots + a_p x_p + b_1 \xi_1 + \cdots + b_p \xi_p$. Assume that $b_j \neq 0$ with some j. By repeated applications of S_Q with

$$Q = \mathrm{diag}(0, \ldots, 0, -a_j/b_j, 0, \ldots, 0)$$

one can assume that

$$\ell_1 = c_1 \xi_1 + \cdots + c_l \xi_l + c_{l+1} x_{l+1} + \cdots + c_p x_p$$

where $c_j \neq 0$. Let Q' be $p \times p$ matrix with the first row

$$(0, \ldots, 0, -c_1^{-1} c_{l+1}, \ldots, -c_1^{-1} c_p)$$

and zero other entries and put $Q = Q' + {}^t Q'$. Then it is clear that

$$\tilde{\ell}_1(x, \xi) = \ell_1(x, \xi + Qx) = c_1 \xi_1 + \cdots + c_l \xi_l.$$

Choose a non singular A such that $c_1({}^t A\xi)_1 + \cdots + c_l({}^t A\xi)_l = \xi_1$ so that we have

$$\tilde{\ell}_1(A^{-1}x, {}^t A\xi) = \xi_1.$$

If $b_1 = \cdots = b_p = 0$ taking a non singular A such that $a_1(A^{-1}x)_1 + \cdots + a_p(A^{-1}x)_p = x_1$ we have

$$\tilde{\ell}_1(x, \xi) = \ell_1(A^{-1}x, {}^t A\xi) = x_1.$$

Thus one can assume that either $\ell_1 = \xi_1$ or $\ell_1 = x_1$.

Let $\ell_1 = \xi_1$. Considering $\ell_j - \alpha_j \ell_1$ we may assume that ℓ_j, $j \geq 2$ are independent of ξ_1. From $\{\ell_1, \ell_j\} = 0$ it follows that ℓ_j is independent of x_1 and hence

$$\ell_j = \ell_j(x_2, \ldots, x_p, \xi_2, \ldots, \xi_p), \quad j \geq 2.$$

When $\ell_1 = x_1$, repeating a similar argument we conclude that

$$\ell_j = \ell_j(x_2, \ldots, x_p, \xi_2, \ldots, \xi_p), \quad j \geq 2.$$

By induction on p there is a $T \in G$ such that either $\ell_i(T(x, \xi)) = \xi_i$ or $\ell_i(T(x, \xi)) = x_i$, $j = 1, \ldots, p$. Renumbering the indices if necessary we get the desired assertion. □

Remark. In the proof we have used the following fact several times

$$\{\ell(T(x, \xi)), \tilde{\ell}(T(x, \xi))\} = \{\ell, \tilde{\ell}\}(T(x, \xi)).$$

We turn back to the proof of the proposition. Note that $\partial l_j/\partial\xi_0 \neq 0$ with some j because p is a polynomial in ξ_0 of degree r. Thus we may assume that $l_0 = \xi_0$ and l_j, $1 \le j \le N$ are independent of ξ_0 considering $\ell_j - c_j\ell_0$, $j = 1, \dots, N$ with suitable c_j. Since Λ is involutive l_j, $1 \le j \le N$ are independent of x_0 and then

$$l_j = l_j(x', \tilde{\xi}'), \ \ \tilde{\xi}' = (\xi_1, \dots, \xi_{n-1}), \ \ 1 \le j \le N.$$

Set $\Lambda' = \{(x', \tilde{\xi}') \mid l_j(x', \tilde{\xi}') = 0, 1 \le j \le N\}$ and assume first that l_j are independent of x_n. Since Λ' is involutive thanks to Lemma 2.29 there is a symplectic map $T : (\tilde{x}', \tilde{\xi}') \to (A\tilde{x}', {}^tA\tilde{\xi}' + QA^{-1}\tilde{x}')$ with a non singular A and a symmetric Q such that

$$T^{-1}(\Lambda') = \{\xi_1 = \cdots = \xi_k = 0, x_{k+1} = \cdots = x_N = 0\}$$

with some $k \in \mathbb{N}$. Clearly T preserves the x_0 coordinate. As remarked before $T \circ p$ is a polynomial in $(\xi_0, \dots, \xi_k, x_{k+1}, \dots, x_N)$ we have the desired conclusion.

We next assume that $\partial l_j/\partial x_n \neq 0$ with some j. Without restrictions one may assume that

$$\frac{\partial l_N}{\partial x_n} \neq 0, \ \frac{\partial l_j}{\partial x_n} = 0, \ 1 \le j \le N-1.$$

Setting $\Lambda'' = \{l_j(\tilde{x}', \tilde{\xi}') = 0, 1 \le j \le N-1\}$ and applying Lemma 2.29 we can assume that $l_j = \xi_j$, $j = 1, \dots, k$ and $l_{k+j} = x_{k+j}$, $j = 1, \dots, N-1-k$. Hence considering $\ell_N - \sum_{j=1}^{N-1} c_j\ell_j$ we may assume that

$$l_N = l_N(x_1, \dots, x_k, x_N, \dots, x_n, \xi_{k+1}, \dots, \xi_{n-1}).$$

On the other hand since the Poisson brackets $\{l_N, l_j\}$ of l_N and l_j vanish it follows that

$$l_N = l_N(x_N, \dots, x_n, \xi_N, \dots, \xi_{n-1}).$$

If $N = n$ then $l_N = l_N(x_n)$ and the proof is complete. If $N < n$ then applying Lemma 2.29 again there is a $T \in G$ preserving the $(x_1, \dots, x_{N-1}, \xi_0, \dots, \xi_{N-1})$ coordinates such that

$$l_N(T(x, \xi)) = x_N + ax_n \ \text{ or } \ \xi_N + ax_n.$$

Exchanging the x_N and x_{k+1} coordinates in the latter case we have the desired assertion since p is a polynomial in l_j, $0 \le j \le N$. □

When $\Lambda(h_\rho)$ is involutive one can choose $(\sigma, \delta) \in I(h, \rho)$ so that $h_\rho^{\sigma,\delta}$ has a simpler form. From Proposition 2.10 we can assume that

$$h_\rho(x, \tilde{\xi}) = q(\xi_0, \dots, \xi_k, x_{k+1}, \dots, x_n) \tag{2.37}$$

with a homogeneous polynomial q and an integer $k \in \mathbb{N}$. Let us write

$$\beta = (\beta_0, \beta', \beta'', \beta_n) = (\beta_0, \beta_1, \dots, \beta_k, \beta_{k+1}, \dots, \beta_{n-1}, \beta_n)$$
$$\alpha = (\tilde{\alpha}, \alpha_n) = (\alpha_0, \alpha', \alpha'', \alpha_n) = (\alpha_0, \alpha_1, \dots, \alpha_k, \alpha_{k+1}, \dots, \alpha_{n-1}, \alpha_n).$$

From Lemma 2.2 and (2.37) it follows that

$$h^{(\tilde{\alpha})}_{(\beta)}(0, e_n) \neq 0,\ |\tilde{\alpha} + \beta| = r \Longrightarrow \beta_0 = 0, \beta' = 0, \alpha'' = 0.$$

Take $\epsilon > 0$, $\epsilon_0 > 0$ so that

$$0 < \epsilon_0 < \epsilon < 1/2(2r+1),\ \ \epsilon(r-2) < \epsilon_0 r \tag{2.38}$$

and set

$$\begin{cases} \sigma_0 = \delta_0 = 1/2 - \epsilon_0, \sigma_j = \delta_j = 1/2 - \epsilon, 1 \le j \le k, \\ \sigma_j = \delta_j = 1/2 + \epsilon, k+1 \le j \le n-1, \\ \sigma_n = 1, \delta_n = 1/2 + \epsilon. \end{cases} \tag{2.39}$$

Lemma 2.30. *Let* σ, δ *be as in (2.39) and* $y = (0, \dots, 0, y_n)$. *Assume that (2.37). Then we have* $(\sigma, \delta) \in I(h, \rho)$ *and*

$$\bar{h}^{\sigma,\delta}_{\rho}(y, x, \xi) = c\xi_n^{m-r}\xi_0^r$$

with a constant $c \neq 0$.

Proof. Recall that $\bar{h}^{\sigma,\delta}_{\rho}(y, x, \xi)$ is defined as the limit of

$$\lambda^{-\gamma(\sigma_0, r)} h(\lambda^{-\delta}y + \lambda^{-\sigma}x, \lambda^{\sigma}\xi) = \sum_{j \ge r} I_j$$

as $\lambda \to \infty$ where

$$I_j = \lambda^{-\gamma} \sum_{|\tilde{\alpha}+\beta|=j} h^{(\tilde{\alpha})}_{(\beta)}(0, \lambda\xi_n)(\lambda^{-\delta}y + \lambda^{-\sigma}x)^{\beta}(\lambda^{\tilde{\sigma}}\tilde{\xi})^{\tilde{\alpha}}/(\tilde{\alpha}!\beta!).$$

A general term in I_j is

$$(y_n + \lambda^{-1/2+\epsilon}x_n)^{\beta_n} x'^{\beta'} x''^{\beta''} x_0^{\beta_0} \xi_0^{\alpha_0} \xi'^{\alpha'} \xi''^{\alpha''} \xi_n^{m-|\tilde{\alpha}|} \lambda^{F(\tilde{\alpha},\beta)}$$

where

$$F(\tilde{\alpha}, \beta) = \epsilon_0 r + r/2 - |\tilde{\alpha} + \beta|/2 - \epsilon(\beta_n - |\beta'| + |\beta''|)$$
$$+\epsilon_0(\beta_0 - \alpha_0) - \epsilon(|\alpha'| - |\alpha''|).$$

For $|\tilde{\alpha}+\beta| = l$ we have

$$F(\tilde{\alpha},\beta) \le -(l-r)/2 + \epsilon|\tilde{\alpha}+\beta| + \epsilon r \le -(l-r)/2 + \epsilon(l+r)$$

because $\epsilon_0 < \epsilon$. This is negative when $l > r$ since $\epsilon < 1/2(2r+1)$. Thus we get

$$\bar{h}_\rho^{\sigma,\delta}(y,x,\xi) = \lim_{\lambda\to\infty} I_r.$$

When $|\tilde{\alpha}+\beta| = r$ and $h_{(\beta)}^{(\tilde{\alpha})}(0,e_n) \neq 0$ so that $\beta_0 = 0$, $\beta' = 0$, $\alpha'' = 0$ we have

$$F(\tilde{\alpha},\beta) = (\epsilon_0 - \epsilon)(r - \alpha_0).$$

Then $F(\tilde{\alpha},\beta) \le 0$ and $F(\tilde{\alpha},\beta) = 0$ if and only if $\alpha_0 = r$. This completes the proof. □

Let $q(x,\xi)$ be a homogeneous polynomial in ξ of degree $m-1$ with coefficients in $C^\infty(U)$ where U is an open set in $\mathbb{R}^{n+1}$. Let us set

$$\lambda^{-\gamma(\sigma_0,r)} q(\lambda^{-\delta}y + \lambda^{-\sigma}x, \lambda^{\sigma}\xi) = \lambda^{-\gamma} q_\lambda(y,x,\xi) = \sum_j q_j(y,x,\xi)\lambda^{-\epsilon j}. \quad (2.40)$$

We now study $\sigma_p(\lambda^{-\gamma}q_\lambda)$.

Lemma 2.31. *Let σ, δ be as in (2.39). Assume that one of the following conditions holds*

(a) $\mathrm{Ord}(\lambda^{-\gamma}q_\lambda) \le 0$,
(b) *the degree of $\sigma_p(\lambda^{-\gamma}q_\lambda)$ with respect to ξ_0 is at least $r-1$.*

Then we have

$$q_{(\beta)}^{(\tilde{\alpha})}(0,e_n) = 0, \quad |\tilde{\alpha}+\beta| < r-1.$$

Proof. Suppose that the assertion were not true. Then there would be $\tilde{\alpha}$, β with $|\tilde{\alpha}+\beta| = s \le r-2$ such that $q_{(\beta)}^{(\tilde{\alpha})}(0,e_n) \neq 0$. We can assume that $q_{(\beta)}^{(\tilde{\alpha})}(0,e_n) = 0$, $|\tilde{\alpha}+\beta| < s$ as we observed before. Set

$$S = \{(\tilde{\alpha},\beta);\, |\tilde{\alpha}+\beta| = s,\ q_{(\beta)}^{(\tilde{\alpha})}(0,e_n) \neq 0\},\ \kappa = \max_{(\tilde{\alpha},\beta)\in S} F(\tilde{\alpha},\beta). \quad (2.41)$$

Since $F(\tilde{\alpha},\beta) - 1 \ge \epsilon_0 r + (r-l)/2 - 1 - \epsilon l \ge \epsilon_0 r - \epsilon(r-2)$, if $|\tilde{\alpha}+\beta| = l \le r-2$, it is clear that $\kappa - 1 > 0$ by (2.38). As in the proof of Lemma 2.30 we write the right-hand side of (2.40) as a sum of I_j's where I_j is a sum of such terms;

$$q_{(\beta)}^{(\tilde{\alpha})}(0,e_n)(y_n + \lambda^{-1/2+\epsilon}x_n)^{\beta_n} x'^{\beta'} x''^{\beta''} x_0^{\beta_0} \xi_0^{\alpha_0} \xi'^{\alpha'} \xi''^{\alpha''} \xi_n^{m-1-|\tilde{\alpha}|} \lambda^{F(\tilde{\alpha},\beta)-1}/(\tilde{\alpha}!\beta!)$$

over $|\tilde{\alpha}+\beta| = j$. Note that $I_s = \lambda^{\kappa-1}(J(y,x,\xi)+o(1))$ where $J(y,x,\xi)$ is a polynomial in (y,x,ξ) and of degree less than or equal to s with respect to ξ_0 and is not identically zero because S is not empty. Here we note that

$$\min_{|\tilde{\alpha}+\beta|=s} F(\tilde{\alpha},\beta) > \max_{|\tilde{\alpha}+\beta|=j} F(\tilde{\alpha},\beta),\ j \geq s+1,\ s \leq r.$$

This shows that $\sigma_p(\lambda^{-\gamma} q_\lambda)(y,x,\xi) = J(y,x,\xi)$ is a polynomial of degree s with respect to ξ_0 and $\mathrm{Ord}(\lambda^{-\gamma} q_\lambda) = \kappa - 1 > 0$ so that both (a) and (b) fail. This proves the assertion. □

Proof of Theorem 2.2. Take σ, δ as in Lemma 2.31. Applying Lemma 2.31 instead of Lemma 2.3 in the argument of the proof of Theorem 2.1 we get the desired result. □

2.10 Concluding Remarks

The formulation of necessary conditions taken here is partly motivated by the studies on sufficient conditions for strong hyperbolicity of first order systems in [56] and partly comes from the methods used to study necessary conditions for hyperbolicity of systems in [47,49]. Theorems 2.1 and 2.2 are found in [50]. For further necessary conditions for strong hyperbolicity involving spectral structure of the characteristics we refer to [32,48,49].

A first order system with constant coefficients is strongly hyperbolic if and only if it is uniformly diagonalizable [30]. Some related results are in [6]. For first order systems which depend only on the time variable, some necessary conditions for strong hyperbolicity were given from different point of views, see for example [38, 40, 45, 73]. For strongly hyperbolic systems with characteristics of constant multiplicity, we refer to [27].

For hyperbolic systems with characteristics of constant multiplicity, under the constant rank condition, the question (A) is discussed in [60, 61]. See also [24]. Without this constant rank condition the same question is studied in [10] in the case of multiplicity 2. Assuming that the coefficients are real analytic, without the constant rank condition, the question (A) is solved in [39,40,67] for any multiplicity.

We have some results without assuming the analyticity of $A_j(x)$.

Proposition 2.11. *Let $A_j(x) \in C^\infty(\Omega)$ and $0 \in \Omega$. Let $\rho = (0,\bar{\xi})$, $\bar{\xi} \in \mathbb{R}^{n+1} \setminus \{0\}$ be a characteristic of order r for h. Assume that $h_\rho(x,\xi_0,\xi')$ has $r-2$ simple zeros at some (x,ξ'). Then if L is strongly hyperbolic near the origin then every $(m-1)$-th minor of $L(x,\xi)$ vanishes of order $r-2$.*

Proof. Take σ, δ as in Lemma 2.3 with $3\bar{\delta} > 1$ and $\bar{\sigma} > \bar{\delta}$. Assume that $h_\rho(y,\xi_0,\tilde{\xi}')$ has $r-2$ simple zeros at some $(y,\tilde{\xi}')$. Then from Lemma 2.2 we see that at least $r-2$ roots $\omega^j(z,\xi',\lambda)$ are simple roots of $h_\lambda(z,\xi)$ when $(z,\xi') \in V$ where V is an open set in $\mathbb{R}^{2(n+1)+n}$. We may assume that $\omega^j(z,\xi',\lambda)$, $1 \le j \le r-2$ are simple. Let $\varphi(z,\lambda)$ be characteristic functions to $\omega^j(z,\xi',\lambda)$ with $(z, \varphi_{x'}(z,\lambda)) \in V$. It is clear that

$$h_\lambda^{(\alpha)}(z, \varphi_x(z,\lambda)) = \lambda^\gamma(c_\alpha^j(z) + o(1)), \ |\alpha| = 1, \ 1 \le j \le r-2$$

with $c^j_{(1,0,\ldots,0)}(z) \neq 0$. Let q be any $(m-1)$-th minor of L then from Corollary 2.3 it follows that

$$q_\lambda(z, \varphi_x(z,\lambda)) = O(\lambda^\gamma), \ 1 \le j \le r-2.$$

Assume $\mathrm{Ord}(\lambda^{-\gamma} q_\lambda) > 0$ then from Lemma 2.8 it follows that $\sigma_p(\lambda^{-\gamma} q_\lambda)(z,\xi)$ is divisible by $\prod_{j=1}^{r-2}(\xi_0 - \omega_0^j(z,\xi'))$ and hence the degree of $\sigma_p(\lambda^{-\gamma} q_\lambda)$ with respect to ξ_0 is greater than or equal to $r-2$. Then by Lemma 2.3 we get the desired result. □

Corollary 2.9. *Let $A_j(x) \in C^\infty(\Omega)$ and $0 \in \Omega$. Let $\rho = (0, \bar{\xi})$, $\bar{\xi} \in \mathbb{R}^{n+1} \setminus \{0\}$ be a characteristic of order r for h. Assume that the irreducible factorization of h_ρ has no multiple factor. Then we have the same conclusion as in Proposition 2.11.*

We make some comments about Theorem 2.1 and Corollary 2.1. From Corollary 2.1 it follows that if L is strongly hyperbolic and ρ is a characteristic of order m then

$$0 \le \operatorname{rank} L(\rho) \le [\frac{m}{2}].$$

This condition on the rank of $L(\rho)$ is sharp in the following sense. Let us denote by Σ the set of characteristics of order m and we assume that Σ is a smooth manifold. Since $h_\rho(x,\xi)$ is independent of directions $T_\rho\Sigma$ so that $h_\rho(x,\xi)$ is a polynomial on $T_\rho\mathbb{R}^{n+1}/T_\rho\Sigma$.

Theorem 2.3 ([56]). *Assume that* rank $L = [m/2]$ *and* $C(h_\rho) \cap \Lambda(h_\rho) = \{0\}$ *on* Σ. *We also assume that h_ρ is strictly hyperbolic on $T_\rho\mathbb{R}^{n+1}/T_\rho\Sigma$. If every $(m-1)$-th minor of L vanishes of order $m-2$ then $L(x,D)$ is (microlocally near ρ) strongly hyperbolic.*

It should be remarked that when $m = 2$ then h_ρ is always strictly hyperbolic on $T_\rho\mathbb{R}^{n+1}/T_\rho\Sigma$.

Recall that a scalar operator P with principal symbol p is strongly hyperbolic if and only if every multiple characteristic is at most double and at every double characteristic ρ the following condition holds

$$C(p_\rho) \cap \Lambda(p_\rho) = \{0\}$$

(see [29]). From these facts it is very natural to conjecture

Conjecture. *Let ρ be a multiple characteristic of order r. If $L(x,D)$ is strongly hyperbolic near the origin then one of the following conditions holds*

(i) $C(h_\rho) \cap \Lambda(h_\rho) = \{0\}$,
(ii) $\dim \operatorname{Ker} L(\rho) = r$.

In order to prove the conjecture, thanks to Lemma 2.26 and Corollary 2.7, it is enough to show

$$C(h_\rho) \not\subset \Lambda(h_\rho), \ \ C(h_\rho) \cap \Lambda(h_\rho) \neq \{0\} \Longrightarrow \dim \operatorname{Ker} L(\rho) = r.$$

By [49] the conjecture is true for $r = 2$. It is also true for the case $n = 1$ because in this case we can conclude $C(h_\rho) \subset \Lambda(h_\rho)$ from $C(h_\rho) \cap \Lambda(h_\rho) \neq \{0\}$.

For the case (ii), introducing nondegenerate characteristics, we discuss sufficient conditions for $L(x,D)$ to be strongly hyperbolic in Chapter 4.

Problem. In Definition 1.2 requiring the C^∞ well-posedness both in the future and in the past seems to be surplus. Prove: if the Cauchy problem for $P+Q$ is C^∞ well posed in the future near the origin for any differential operator Q of order $q-1$ with $C^\infty(\Omega, M_m(\mathbb{C}))$ coefficients then Theorem 2.1 holds.

Problem. In the proofs of Theorems 2.1 and 2.2 the analyticity of $A_j(x)$ in Ω is required only when the set $\{X \mid h_\rho(X) = 0\}$ is not reduced hence it seems that this analyticity assumption could be dropped.

Problem. In the proofs of Theorems 2.1 and 2.2 we use essentially only two choices of $(\sigma,\delta) \in \mathbb{Q}_+^{n+1} \times \mathbb{Q}_+^{n+1}$, while many other choices are possible. Which kind results we could obtain by other choices?

We state one such result. Let ρ be a multiple characteristic of order m with involutive $\Lambda(\rho)$. If $L(x,D)$ is strongly hyperbolic near the origin we have $\dim \operatorname{Ker} L(\rho) = m$ by Theorem 2.2 which implies $L(\rho) = O$. Then one can define the "localization" of $L_\rho(x,\xi)$ at ρ by (for the precise definition and several properties of localization, see Sect. 4.1)

$$L(\rho + \mu(x,\xi)) = \mu(L_\rho(x,\xi) + O(\mu)), \quad \mu \to 0$$

which should be considered as a first approximation of $L(x,\xi)$ on $\operatorname{Ker} L(\rho)$. Here $\det L_\rho(x,\xi)$ is a hyperbolic polynomial with respect to $\theta = (0,\ldots,0,1,0,\ldots,0)$ (see Sect. 4.1).

Theorem 2.4 ([51]). *Assume that L is strongly hyperbolic near the origin. Let κ be a characteristic of order r of L_ρ. Then any $(m-1)$-th minor of L_ρ vanishes of order at least $r-2$ at κ.*

Chapter 3
Two by Two Systems with Two Independent Variables

Abstract In this chapter we study the C^∞ well-posedness of the Cauchy problem for 2×2 systems with two independent variables with real analytic coefficients. For such a system L the characteristic set is given by zeros of some nonnegative real analytic function. We define pseudo-characteristic curves for L as the real part of the zeros of nonnegative functions associated to the system and we give a necessary and sufficient condition for the Cauchy problem for L to be C^∞ well posed in terms of pseudo-characteristic curves and Newton polygons. In particular we can characterize strongly hyperbolic 2×2 systems with two independent variables. This gives another proof of the strong hyperbolicity of the 2×2 system discussed in Sect. 1.3. By checking this necessary and sufficient condition we provide many instructive examples. For instance, we see that there are examples which are strictly hyperbolic apart from the initial line with polynomial coefficients such that the Cauchy problem is not C^∞ well posed for any lower order term.

3.1 Reduction to Almost Diagonal Systems

Let us study a 2×2 system

$$Lu = \partial_t u - A(t, x)\partial_x u + B(t, x)u$$

where $t, x \in \mathbb{R}$ and $A(t, x)$, $B(t, x)$ are 2×2 matrices which are real analytic near the origin of $\mathbb{R}^2$. Moreover we assume that

$$A(t, x) \text{ is real valued.}$$

We study the following Cauchy problem

T. Nishitani, *Hyperbolic Systems with Analytic Coefficients*, Lecture Notes in Mathematics 2097, DOI 10.1007/978-3-319-02273-4_3,

$$\begin{cases} Lu = f, & f = 0 \text{ in } t < \tau \\ u = 0 \text{ in } t < \tau. \end{cases} \tag{CP}$$

Here we recall Theorem 1.1 in Sect. 1.4.

Proposition 3.1. *If (CP) is C^∞ well posed near the origin then all eigenvalues of $A(t,x)$ are real when (t,x) varies near the origin.*

We next remark that one can always assume that the trace of $A(t,x)$ is zero.

Lemma 3.1. *In a new system of local coordinates*

$$s = t, \quad y = \phi(t,x), \quad \phi(0,x) = x$$

one can assume that $\mathrm{Tr}\, A(t,x) \equiv 0$ *where* $\phi(t,x)$ *verifies*

$$\frac{\partial \phi}{\partial t} = \frac{1}{2}\mathrm{Tr}\, A(t,x)\frac{\partial \phi}{\partial x}, \quad \phi(0,x) = x.$$

Proof. Easy. □

In what follows we assume that $\mathrm{Tr}\, A(t,x) \equiv 0$ and hence

$$A(t,x) = \begin{bmatrix} a_{11} & a_{12} \\ a_{21} & -a_{11} \end{bmatrix}.$$

Let us denote

$$h(t,x) = -\det A(t,x) = a_{11}^2 + a_{12}a_{21}.$$

Note that if all eigenvalues of $A(t,x)$ are real then

$$h(t,x) \geq 0$$

and vice versa.

Let us take

$$T = \begin{bmatrix} 1 & i \\ i & 1 \end{bmatrix}.$$

Note that if the Cauchy problem for L is C^∞ well posed then so is for $T^{-1}LT$ and vice versa. Thus it is enough to study $T^{-1}LT$

$$L^\sharp = T^{-1}LT = \partial_t - A^\sharp(t,x)\partial_x + B^\sharp(t,x)$$

where $A^\sharp(t,x) = T^{-1}A(t,x)T$ and $B^\sharp(t,x) = T^{-1}B(t,x)T$. More precisely

$$A^\sharp(t,x) = \begin{bmatrix} \dfrac{i(a_{12}-a_{21})}{2} & \dfrac{a_{12}+a_{21}}{2} + ia_{11} \\ \dfrac{a_{12}+a_{21}}{2} - ia_{11} & -\dfrac{i(a_{12}-a_{21})}{2} \end{bmatrix} = \begin{bmatrix} c^\sharp & a^\sharp \\ \overline{a^\sharp} & -c^\sharp \end{bmatrix} = \begin{bmatrix} c^\sharp & a^\sharp \\ \overline{a^\sharp} & \overline{c^\sharp} \end{bmatrix}$$

where

$$a^\sharp = \frac{a_{12}+a_{21}}{2} + ia_{11}, \quad c^\sharp = \frac{i(a_{12}-a_{21})}{2} \tag{3.1}$$

and $\bar{a}$ denotes the complex conjugate of a.

Lemma 3.2. *We have*

$$|a^\sharp| \geq |c^\sharp|, \quad 4|a^\sharp|^2 \geq \mathrm{Tr}\,(A\,{}^t\!A) = \sum_{i,j=1}^{2} a_{ij}(t,x)^2, \quad |a^\sharp|^2 \geq h.$$

In particular we have $a^\sharp(t,x) = 0 \iff A(t,x) = O$.

Proof. Note that

$$h = |a^\sharp|^2 - |c^\sharp|^2$$

by (3.1). Since $h \geq 0$ it follows that $|a^\sharp|^2 \geq |c^\sharp|^2$ and $|a^\sharp|^2 \geq h$. Observing that

$$A^\sharp(A^\sharp)^* = T^{-1}A\,{}^tAT$$

we have $\mathrm{Tr}\,(A\,{}^tA) = \mathrm{Tr}\,(A^\sharp(A^\sharp)^*) = 2(|c^\sharp|^2 + |a^\sharp|^2) \leq 4|a^\sharp|^2$. □

Let us put

$$M = \partial_t + A^\sharp\partial_x + C + {}^{co}B^\sharp - A_x^\sharp$$

where ${}^{co}B^\sharp$ stands for the cofactor matrix of $B^\sharp$, $A_x^\sharp = \partial_x A^\sharp$ and $C = (c_{ij})$ will be determined later. We use M in order to reduce $L^\sharp$ to a second order 2×2 almost diagonal system with singular coefficients. Note that

$$\begin{aligned} L^\sharp M &= \partial_t^2 - h\partial_x^2 + (A_t^\sharp - A^\sharp C + \mathrm{Tr}\,(AB))\partial_x \\ &\quad + (B^\sharp + {}^{co}B^\sharp + C - A_x^\sharp)\partial_t + L^\sharp(C + {}^{co}B^\sharp - A_x^\sharp) \end{aligned}$$

because, for instance, we have

$$B^\sharp A^\sharp - A^{\sharp co}B^\sharp = (B^\sharp A^\sharp) + {}^{co}(B^\sharp A^\sharp) = \mathrm{Tr}\,(A^\sharp B^\sharp) = \mathrm{Tr}\,(AB).$$

We now want to choose C so that we have

$$A_t^\sharp - A^\sharp C + \mathrm{Tr}\,(AB) = \text{diagonal matrix}.$$

Let us examine $A_t^\sharp - A^\sharp C + \mathrm{Tr}\,(AB)$ which is

$$\begin{bmatrix} \partial_t c^\sharp - c^\sharp c_{11} - a^\sharp c_{21} + \mathrm{Tr}\,(AB) & \partial_t a^\sharp - c^\sharp c_{12} - a^\sharp c_{22} \\ \partial_t \overline{a^\sharp} - \overline{c^\sharp} c_{21} - \overline{a^\sharp} c_{11} & \partial_t \overline{c^\sharp} - \overline{c^\sharp} c_{22} - \overline{a^\sharp} c_{12} + \mathrm{Tr}\,(AB) \end{bmatrix}.$$

We want to choose $C = (c_{ij})$ so that

$$\begin{cases} \partial_t a^\sharp - c^\sharp c_{12} - a^\sharp c_{22} = 0, \\ \partial_t \overline{a^\sharp} - \overline{c^\sharp} c_{21} - \overline{a^\sharp} c_{11} = 0 \end{cases}$$

that is

$$c_{11} = \frac{\partial_t \overline{a^\sharp}}{\overline{a^\sharp}} - \frac{\overline{c^\sharp}}{\overline{a^\sharp}} c_{21}, \quad c_{22} = \frac{\partial_t a^\sharp}{a^\sharp} - \frac{c^\sharp}{a^\sharp} c_{12}. \tag{3.2}$$

Lemma 3.3. *Assume that $C = (c_{ij})$ verifies (3.2). Then with*

$$\begin{cases} Y = \overline{a^\sharp} \partial_t c^\sharp - c^\sharp \partial_t \overline{a^\sharp} + \overline{a^\sharp} \mathrm{Tr}\,(AB), \\ Z = a^\sharp \partial_t \overline{c^\sharp} - \overline{c^\sharp} \partial_t a^\sharp + a^\sharp \mathrm{Tr}\,(AB) \end{cases}$$

we have

$$L^\sharp M = (\partial_t^2 - h\partial_x^2) I + Q\partial_x + R\partial_t + S$$

where

$$Q = \begin{bmatrix} Y/\overline{a^\sharp} - hc_{21}/\overline{a^\sharp} & 0 \\ 0 & Z/a^\sharp - hc_{12}/a^\sharp \end{bmatrix}$$

with $R = C - A_x^\sharp + B^\sharp + {}^{co}B^\sharp$ and $S = L^\sharp({}^{co}B^\sharp - A_x^\sharp)$.

Proof. We study $(2,2)$-th entry of $A_t^\sharp - A^\sharp C + \mathrm{Tr}\,(AB)$;

$$\partial_t \overline{c^\sharp} - \overline{c^\sharp} \left(\frac{\partial_t a^\sharp}{a^\sharp} - \frac{c^\sharp}{a^\sharp} c_{12} \right) - \overline{a^\sharp} c_{12} + \mathrm{Tr}\,(AB)$$
$$= \frac{1}{a^\sharp} \{ a^\sharp \partial_t \overline{c^\sharp} - \overline{c^\sharp} \partial_t a^\sharp - (|a^\sharp|^2 - |c^\sharp|^2) c_{12} + a^\sharp \mathrm{Tr}\,(AB) \} = \frac{1}{a^\sharp}(Z - hc_{12}).$$

We can examine the other entries similarly. □

In what follows we choose $c_{12} = c_{21} = 0$ so that

$$C = \begin{bmatrix} \dfrac{\partial_t \overline{a^\sharp}}{\overline{a^\sharp}} & 0 \\ 0 & \dfrac{\partial_t a^\sharp}{a^\sharp} \end{bmatrix}.$$

Then we see that

$$L^\sharp C = \begin{bmatrix} \partial_t(\dfrac{\partial_t \overline{a^\sharp}}{\overline{a^\sharp}}) & 0 \\ 0 & \partial_t(\dfrac{\partial_t a^\sharp}{a^\sharp}) \end{bmatrix} + \begin{bmatrix} c^\sharp \partial_x(\dfrac{\partial_t \overline{a^\sharp}}{\overline{a^\sharp}}) & a^\sharp \partial_x(\dfrac{\partial_t a^\sharp}{a^\sharp}) \\ \overline{a^\sharp} \partial_x(\dfrac{\partial_t \overline{a^\sharp}}{\overline{a^\sharp}}) & \overline{c^\sharp} \partial_x(\dfrac{\partial_t a^\sharp}{a^\sharp}) \end{bmatrix}.$$

Lemma 3.4. *Let us define*

$$D^\sharp = c^\sharp \partial_t a^\sharp - a^\sharp \partial_t c^\sharp = a^\sharp \partial_t \overline{c^\sharp} - \overline{c^\sharp} \partial_t a^\sharp.$$

Then we have $Z = D^\sharp + a^\sharp \mathrm{Tr}\,(AB)$, $\overline{Y} = D^\sharp + a^\sharp \mathrm{Tr}\,(A\bar{B})$.

Proof. It is clear since A is real valued by assumption. □

Lemma 3.5. *Let us put*

$$M = \partial_t + A^\sharp \partial_x + A^\sharp_x + {}^{co}B^\sharp + \tilde{C}$$

where

$$\tilde{C} = -\begin{bmatrix} \dfrac{\partial_t a^\sharp}{a^\sharp} & 0 \\ 0 & \dfrac{\partial_t \overline{a^\sharp}}{\overline{a^\sharp}} \end{bmatrix}.$$

Then we have

$$ML^\sharp = (\partial_t^2 - h\partial_x^2)I - h_x\partial_x + \tilde{Q}\partial_x + \tilde{R}\partial_t + \tilde{S}$$

where

$$\tilde{Q} = \begin{bmatrix} Z/a^\sharp & 0 \\ 0 & Y/\overline{a^\sharp} \end{bmatrix}, \quad \tilde{R} = \tilde{C} + A^\sharp_x + B^\sharp + {}^{co}B^\sharp, \ \tilde{S} = MB^\sharp.$$

Proof. Noting $A^\sharp_x A^\sharp + A^\sharp A^\sharp_x = h_x$ the proof is similar to that of Lemma 3.3. □

Remark that $(\partial_t^2 - h\partial_x^2)I + Q\partial_x$ and $(\partial_t^2 - h\partial_x^2)I + \tilde{Q}\partial_x - h_x\partial_x$ are

$$\begin{bmatrix} \partial_t^2 - h\partial_x^2 + (\overline{\dfrac{D^\sharp + a^\sharp \mathrm{Tr}(A\bar{B})}{a^\sharp}})\partial_x & 0 \\ 0 & \partial_t^2 - h\partial_x^2 + (\dfrac{D^\sharp + a^\sharp \mathrm{Tr}(AB)}{a^\sharp})\partial_x \end{bmatrix}$$

and

$$\begin{bmatrix} \partial_t^2 - h\partial_x^2 + (\dfrac{D^\sharp + a^\sharp \mathrm{Tr}(AB)}{a^\sharp} - h_x)\partial_x & 0 \\ 0 & \partial_t^2 - h\partial_x^2 + (\overline{\dfrac{D^\sharp + a^\sharp \mathrm{Tr}(A\bar{B})}{a^\sharp}} - h_x)\partial_x \end{bmatrix}$$

respectively. Thus our system is, essentially, reduced to a second order 2×2 "almost" diagonal system with singular coefficients in front of ∂_x.

3.2 Nonnegative Real Analytic Functions

From the observations made in the preceding section it is plausible to expect that the zeros of not only h but also of $a^\sharp$ plays an important role. In this subsection we make precise studies about zeros of a nonnegative real analytic function defined near the origin.

Lemma 3.6. *Let $F(t,x)$ be a nonnegative real analytic function defined near the origin. Then there is a real valued $f(t,x)$ defined in V (a neighborhood of the origin) such that $f(t,x)$ is real analytic in $V \setminus (0,0)$ continuous in V, unique up to a non zero factor such that*

$$f(t,x)^2 = F(t,x), \quad f(t,x) = x^n \prod_{j=1}^{l}(t - t_j(x)) \prod_{j=l+1}^{m} |t - t_j(x)|\Phi(t,x) \tag{3.3}$$

where $\Phi(0,0) \neq 0$ and $t_j(x)$ is obtained as the restriction to $\mathbb{R}$ of

$$t_j(z) = \sum_{k \geq 0} C_{jk} z^{k/p_j}, \quad z = x + iy, \quad (p_j \in \mathbb{N})$$

and hence one can write $t_j(x)$ in $0 < \pm x < \delta$

$$t_j(x) = \sum_{k \geq 0} C_{jk}^{\pm}(\pm x)^{k/p_j}$$

as convergent Puiseux series. Here $\mathrm{Im}\, t_j(x) \neq 0$ *for* $0 < |x| < \delta$ *with some* $\delta > 0$ *for* $j \geq l+1$ *(for* $1 \leq j \leq l$ *it may happen* $\mathrm{Im}\, t_j(x) = 0$ *in* $0 < |x| < \delta$*).*

Proof. Note that one can write

$$F(t,x) = x^{2n} g_1^{l_1} \cdots g_\nu^{l_\nu} h_1^{m_1} \cdots h_\mu^{m_\mu} \bar{h}_1^{m_1} \cdots \bar{h}_\mu^{m_\mu} \Phi$$

where g_i are real, that is $\bar{g}_i = g_i$ and $\bar{h}_i \neq h_i$ and $\Phi(0,0) \neq 0$. Here we denote $\bar{h}(t,x) = \overline{h(\bar{t},\bar{x})}$ and one can assume that $g_i, h_j, \bar{h}_k$ are Weierstrass polynomials in t. Indeed from the Weierstrass preparation theorem, taking $F(t,x) \geq 0$ into account, $F(t,x)$ is written as

$$F(t,x) = x^{2n}(t^{2r} + f_1(x)t^{2r-1} + \cdots + f_{2r}(x))\Phi(t,x)$$

where $f_i(0) = 0$, $\Phi(0,0) \neq 0$. Let us factorize $F(t,x)$ as the product of irreducible factors;

$$F = x^{2n} g_1^{l_1} \cdots g_\nu^{l_\nu} k_1^{m_1} \cdots k_p^{m_p} \Phi$$

with $\bar{k}_i \neq k_i$. Since $\bar{F} = F$ we have

$$\bar{F} = x^{2n} g_1^{l_1} \cdots g_\nu^{l_\nu} \bar{k}_1^{m_1} \cdots \bar{k}_p^{m_p} \Phi = x^{2n} g_1^{l_1} \cdots g_\nu^{l_\nu} k_1^{m_1} \cdots k_p^{m_p} \Phi.$$

On the other hand from the uniqueness of the factorization $\bar{k}_j^{m_j}$ coincides with some $\bar{k}_i^{m_i}$. This proves the assertion. Taking $\delta > 0$ small enough, we may suppose that the resultant of any pair among $g_i, h_j, \bar{h}_k$ is different from zero in $0 < |x| < \delta$. We also may assume that the discriminant of every $g_i, h_j, \bar{h}_k$ is different from zero in $0 < |x| < \delta$. Factorize

$$h_i = \prod_{k=1}^{n(i)} (t - t_k(x))$$

then we have $\mathrm{Im}\, t_j(x) \neq 0$ for $x \in \mathbb{R}$, $0 < |x| < \delta$ since otherwise we would have $\bar{h}_i(t_k(\hat{x}), \hat{x}) = h_i(t_k(\hat{x}), \hat{x}) = 0$ with some $\hat{x} \in \mathbb{R}$, $0 < |\hat{x}| < \delta$ where $\mathrm{Im}\, t_k(\hat{x}) = 0$ which contradicts the assumption that the resultant of h_i and $\bar{h}_i$ is different from zero in $x \in \mathbb{R}$, $0 < |x| < \delta$. Thus one can write

$$h_i \bar{h}_i = \prod_{k=1}^{n(i)} |t - t_k(x)|^2 = \left(\prod_{k=1}^{n(i)} |t - t_k(x)| \right)^2.$$

We turn to g_i. Let us write

$$g_i = \prod_{k=1}^{n(i)} (t - t_k(x)).$$

If there is a $x \in \mathbb{R}$, $0 < |x| < \delta$ such that $\mathrm{Im}\, t_k(x) = 0$ with some k then l_i is even (recall that the discriminant of g_i is different from zero) because $F(t,x) \geq 0$. Hence one can write

$$g_i^{l_i} = \left(\prod_{k=1}^{n(i)} (t - t_k(x))^{l_i/2} \right)^2 .$$

Finally if $\mathrm{Im}\, t_k(x) \neq 0$ for all $x \in \mathbb{R}$, $0 < |x| < \delta$ and for all k then, since g_i is real, $\overline{t_k(x)}$ is also a root of $g_i = 0$ so that $\overline{t_k(x)}$ coincides with some $t_i(x)$ and

$$g_i = \prod (t - t_k(x))(t - \overline{t_k(x)}) = \left(\prod |t - t_k(x)| \right)^2 .$$

This proves the assertion. □

3.3 Well-Posedness and Pseudo-Characteristic Curves

In this section we state a necessary and sufficient condition for C^∞ well-posedness of (CP). We start with

Definition 3.1. Let $F(t,x)$ be a nonnegative real analytic function defined near the origin so that we have (3.3) by Lemma 3.6. We denote by $\mathscr{C}^+(F)$ the set of functions $\mathrm{Re}\, t_j(x)$, $j = 1,\dots,m$ which are restrictions of $\mathrm{Re}\, t_j(z)$ in $x > 0$ and by $\mathscr{C}^-(F)$ the set of $\mathrm{Re}\, t_j(x)$, $j = 1,\dots,m$ which are restrictions of $\mathrm{Re}\, t_j(z)$ in $x < 0$. In particular we call the curves $t = \phi(x)$, $\phi \in \mathscr{C}^\pm(h|a^\sharp|^2)$ pseudo-characteristic curves of the reference system $L = \partial_t - A(t,x)\partial_x$.

Definition 3.2. Let $F(t,x)$ be a nonnegative real analytic function defined near the origin and let $\phi(x) \in \mathscr{C}^\pm(F)$. For any real analytic function $B(t,x)$ defined near the origin and for any $\phi \in \mathscr{C}^\pm(F)$ we define the Newton polygon $\Gamma(B_\phi)$ which is a subset in $\mathbb{R}^2 = \{(t,x) \mid t, x \in \mathbb{R}\}$ as follows; we set

$$B_\phi(t,x) = B(t + \phi(x), x)$$

then $B_\phi(t,x)$ is defined for $\pm x > 0$ and one can express $B_\phi(t,x)$ by the Puiseux series expansion;

$$B_\phi(t,x) = \sum_{i,k \geq 0} B_{ik}^\pm t^i (\pm x)^{k/p}$$

with some $p \in \mathbb{N}$. We define $\Gamma(B_\phi)$ by

$$\Gamma(B_\phi) = \text{convex hull of } \{ \bigcup_{B_{ik}^\pm \neq 0} ((i, \frac{k}{p}) + \mathbb{R}_+^2) \}$$

where $\mathbb{R}^2_+ = \{(t,x) \mid t \geq 0, x \geq 0\}$. We define $\Gamma(B_\phi) = \emptyset$ if $B \equiv 0$.

Then we have

Theorem 3.1. *Notations being as above. In order that the Cauchy problem* (CP) *for L is C^∞ well posed near the origin it is necessary and sufficient that*

$$\Gamma(tZ_\phi) = \Gamma(t[D^\sharp + a^\sharp \mathrm{Tr}\,(AB)]_\phi) \subset \frac{1}{2}\Gamma([h|a^\sharp|^2]_\phi),$$

$$\Gamma(tY_\phi) = \Gamma(t[D^\sharp + a^\sharp \mathrm{Tr}\,(A\bar{B})]_\phi) \subset \frac{1}{2}\Gamma([h|a^\sharp|^2]_\phi)$$

for any pseudo-characteristic curve $t = \phi(x)$ of $\partial_t - A(t,x)\partial_x$.

Applying Theorem 3.1 we can provide many instructive examples.

Example 3.1. Let us consider a scalar second order hyperbolic operator with two independent variables with real analytic coefficients

$$Pv = \partial_t^2 v - \alpha(t,x)\partial_x^2 v + \beta(t,x)\partial_x v$$

where $\alpha(t,x) \geq 0$. We study the Cauchy problem

$$\begin{cases} Pu = f, & f = 0 \ \text{in} \ t < \tau \\ u = 0 & \text{in} \ t < \tau. \end{cases} \tag{3.4}$$

If we set $u_1 = \partial_x v$, $u_2 = \partial_t v$, $u = {}^t(u_1, u_2)$, then the equation is reduced to the following system;

$$Lu = \partial_t u - \begin{bmatrix} 0 & 1 \\ \alpha & 0 \end{bmatrix} \partial_x u + \begin{bmatrix} 0 & 0 \\ \beta & 0 \end{bmatrix} u = \begin{bmatrix} 0 \\ f \end{bmatrix}.$$

If the Cauchy problem (CP) for P is C^∞ well posed then so is for L and vice versa. It is clear that $a^\sharp = (1+\alpha)/2$, $D^\sharp = i\partial_t\alpha/2$ and $h = \alpha$. Since $a^\sharp(0,0) \neq 0$ the pseudo-characteristic curves of L coincides with $t = \phi(x)$, $\phi \in \mathscr{C}^\pm(\alpha)$ which are introduced in [47] for the scalar operator P. Since

$$\Gamma(tD^\sharp_\phi) = \Gamma(t[\partial_t a]_\phi) = \Gamma(t\partial_t \alpha_\phi) \subset \Gamma(\alpha_\phi) \subset \frac{1}{2}\Gamma(\alpha_\phi)$$

the conditions in Theorem 3.1 are reduced to

$$\Gamma(t[\mathrm{Tr}\,(AB)]_\phi) = \Gamma(t\beta_\phi) \subset \frac{1}{2}\Gamma(\alpha_\phi)$$

for any $\phi \in \mathscr{C}^\pm(\alpha)$. This is exactly the same condition obtained in [47] for P.

We summarize

Corollary 3.1. *In order that the Cauchy problem (3.4) is C^∞ well posed it is necessary and sufficient*

$$\Gamma(t\beta_\phi) \subset \frac{1}{2}\Gamma(\alpha_\phi) \tag{3.5}$$

for any $\phi \in \mathscr{C}^\pm(\alpha)$.

Example 3.2. We exhibit A with $h(t,x) > 0$ outside the initial line $t = 0$ for which no $B(t,x)$ could be taken so that (CP) is C^∞ well posed. Let us consider

$$A(t,x) = \begin{bmatrix} x^2 - t^4/2 & x^2 + xt^2 \\ -x^2 + xt^2 & -(x^2 - t^4/2) \end{bmatrix}.$$

It is easy to see that

$$h = t^8/4,\ c^\sharp = ix^2,\ a^\sharp = xt^2 + i(x^2 - t^4/2),\ D^\sharp = 2ix^3t + 2x^2t^3.$$

Suppose that $B(t,x) = (b_{ij}(t,x))$ is given. Then $a^\sharp \mathrm{Tr}\,(AB)$ has the form

$$C_{40}x^4 + C_{32}x^3t^2 + C_{24}x^2t^4 + C_{16}xt^6 + C_{08}t^8$$

where $C_{ij}(t,x)$ are linear combinations of b_{ij}. Note that we have

$$h|a^\sharp|^2 = t^8(x^4 + t^8/4)/4 = x^4t^8/4 + t^{16}/16.$$

Taking $\phi = 0$ we easily see that

$$\Gamma(t[D^\sharp + a^\sharp \mathrm{Tr}\,(AB)]) \not\subset \frac{1}{2}\Gamma(h|a^\sharp|^2)$$

because $D^\sharp + a^\sharp \mathrm{Tr}\,(AB)$ has the form

$$2ix^3t + 2x^2t^3 + C_{40}x^4 + C_{32}x^3t^2 + C_{24}x^2t^4 + C_{16}xt^6 + C_{08}t^8$$

where $2ix^3t$ could not be canceled out. This proves that for any $B(t,x)$, the Cauchy problem (CP) is not C^∞ well posed.

Such a strange example was given in [37] for the first time where h vanishes identically, while in this example the eigenvalues of $A(t,x)$ are $\pm t^4/2$ which implies that L is strictly hyperbolic apart from the initial line $t = 0$.

Example 3.3. Let us consider

$$A(t,x) = \begin{bmatrix} a_{11}(t,x) & a_{12}(t,x) \\ a_{21}(t,x) & -a_{11}(t,x) \end{bmatrix}$$

where $h(t,x) = a_{11}^2(t,x) + a_{12}(t,x)a_{21}(t,x) \equiv 0$. That is, the zero eigenvalue is folded. Since $h \equiv 0$ it is clear that the conditions in Theorem 3.1 are reduced to

$$D^\sharp + a^\sharp \mathrm{Tr}\,(AB) \equiv 0, \quad D^\sharp + a^\sharp \mathrm{Tr}\,(A\bar{B}) \equiv 0. \tag{3.6}$$

If we factor out the common factor $K(t,x)$ among $a_{ij}(t,x)$ one can write

$$A(t,x) = \begin{bmatrix} K\sigma\rho & K\sigma^2 \\ -K\rho^2 & -K\sigma\rho \end{bmatrix}$$

where ρ and σ are relatively prime. It is clear that $c^\sharp = iK(\sigma^2+\rho^2)/2$ and $a^\sharp = K(\sigma^2-\rho^2)/2 + iK\sigma\rho$ and hence $D^\sharp = K(\rho\partial_t\sigma - \sigma\partial_t\rho)a^\sharp$. Let us write

$$B(t,x) = \begin{bmatrix} b_{11} & b_{12} \\ b_{21} & b_{22} \end{bmatrix}.$$

Since $\mathrm{Tr}\,(AB) = K[b_{21}\sigma^2 - b_{12}\rho^2 + (b_{11}-b_{22})\sigma\rho]$ it follows that

$$D^\sharp + a^\sharp \mathrm{Tr}\,(AB) = a^\sharp K\big[\rho\partial_t\sigma - \sigma\partial_t\rho + b_{21}\sigma^2 - b_{12}\rho^2 + (b_{11}-b_{22})\sigma\rho\big].$$

Thus the conditions (3.6) are equivalent to

$$\rho\partial_t\sigma - \sigma\partial_t\rho + b_{21}\sigma^2 - b_{12}\rho^2 + (b_{11}-b_{22})\sigma\rho \equiv 0$$

which is the Levi condition obtained in [67].

Example 3.4. Let us consider

$$A(t,x) = \psi(t,x)\begin{bmatrix} 0 & 1 \\ t^4 & 0 \end{bmatrix}.$$

For this A we have

$$a^\sharp = \psi(1+t^4)/2, \quad c^\sharp = i\psi(1-t^4)/2, \quad D^\sharp = 2it^3\psi^2, \quad h = t^4\psi^2.$$

Since $\Gamma(tD^\sharp_\phi) \subset \Gamma([h|a^\sharp|^2]_\phi)/2$ is clear then the condition

$$\Gamma(t[\mathrm{Tr}\,(AB)]_\phi) \subset \frac{1}{2}\Gamma(h_\phi) \tag{3.7}$$

is necessary and sufficient for the C^∞ well-posedness of (CP). Let us write

$$B = \begin{bmatrix} b_{11} & b_{12} \\ b_{21} & b_{22} \end{bmatrix}$$

then the condition (3.7) is reduced to $\Gamma(t[\psi b_{21}]_\phi) \subset \Gamma(h_\phi)/2$ which is equivalent to $b_{21}(0,x) = 0$. Note that the condition is independent of ψ while if we consider the scalar second order operator having the same characteristic roots as $A(t,x)$

$$Pv = \partial_t^2 v - \psi(t,x)^2 t^4 \partial_x^2 v + \beta(t,x)\partial_x v$$

then the condition (3.5) is $\Gamma(t\beta_\phi) \subset \Gamma([\psi t^2]_\phi)$ which clearly depends on ψ.

3.4 Strongly Hyperbolic 2 × 2 Systems

We now give a necessary and sufficient condition in order that $\partial_t - A(t,x)\partial_x$ is strongly hyperbolic.

Theorem 3.2. *For $\partial_t - A(t,x)\partial_x$ to be strongly hyperbolic near the origin it is necessary and sufficient that*

$$\Gamma(tD_\phi^\sharp) \subset \frac{1}{2}\Gamma([h|a^\sharp|^2]_\phi), \quad \Gamma(t[a_{ij}]_\phi) \subset \frac{1}{2}\Gamma(h_\phi)$$

for any pseudo-characteristic curve $t = \phi(x)$ of $\partial_t - A(t,x)\partial_x$.

Example 3.5. Let

$$A(t,x) = \begin{bmatrix} a_{11}(t,x) & a_{12}(t,x) \\ a_{21}(t,x) & -a_{11}(t,x) \end{bmatrix}$$

be symmetric, that is $a_{12}(t,x) = a_{21}(t,x)$. In this case we have

$$D^\sharp = 0, \quad |a_{11}| \le |a^\sharp|, \quad |a_{12}| = |a_{21}| \le |a^\sharp|, \quad h = |a^\sharp|^2. \tag{3.8}$$

Since $D^\sharp = 0$ the conditions in Theorem 3.2 are reduced to

$$\Gamma(t[a_{ij}]_\phi) \subset \frac{1}{2}\Gamma(h_\phi)$$

which is clearly satisfied because $|a_{ij}|^2 \le h$ by (3.8).

Here we take the strongly hyperbolic 2 × 2 system which is not symmetrizable studied in Sect. 1.3.

Example 3.6. Let us consider

$$A(t,x) = \psi(t,x)\begin{bmatrix} 0 & 1 \\ t^2 & 0 \end{bmatrix}.$$

In this case we have

$$a^\sharp = \psi(1+t^2)/2,\ \ c^\sharp = i\psi(1-t^2)/2,\ \ D^\sharp = i\psi^2 t,\ \ h = t^2\psi^2.$$

Note that

$$\Gamma(tf_\phi) = \Gamma(t) + \Gamma(f_\phi) \subset \Gamma([tf]_\phi)$$

because $\Gamma(t) \subset \Gamma(t_\phi)$. Then remarking $|ta_{ij}|^2 \le Ch$, $|tD^\sharp|^2 \le Ch|a^\sharp|^2$ we see that

$$\Gamma(tD^\sharp_\phi) \subset \Gamma([tD^\sharp]_\phi) \subset \frac{1}{2}\Gamma([h|a^\sharp|^2]_\phi),$$

$$\Gamma(t[a_{ij}]_\phi) \subset \Gamma([ta_{ij}]_\phi) \subset \frac{1}{2}\Gamma(h_\phi)$$

and hence the assumptions in Theorem 3.2 are verified. Thus $\partial_t - A(t,x)\partial_x$ is strongly hyperbolic for any ψ. We remark that $A(0,x)$ is not symmetrizable if $\psi(0,x) \neq 0$ and this example provides a class of strongly hyperbolic systems which are not symmetrizable.

Definition 3.3. We say that a $m \times m$ system of first order differential operators

$$L = \partial_t + \sum_{j=1}^{n} A_j(x)\partial_{x_j}$$

is uniformly diagonalizable if there is $C > 0$ such that for any (x,ξ'), $|\xi'| = 1$ there exists a $m \times m$ non singular matrix $H(x,\xi')$ with $\|H^{-1}(x,\xi')\|$, $\|H(x,\xi')\| \le C$ such that

$$H^{-1}(x,\xi')\Big(\sum_{j=1}^{n} A_j(x)\xi_j\Big)H(x,\xi')$$

is diagonal where $\|H(x,\xi')\| = \mathrm{Tr}\,(H^{\,t}H)$ and the smoothness of $H(x,\xi')$ with respect to (x,ξ') is not required.

Example 3.7. Assume that $\partial_t - A(t,x)\partial_x$ is uniformly diagonalizable, that is for every (t,x) near the origin there is a $U(t,x)$ such that

$$U^{-1}AU = \begin{bmatrix} \alpha(t,x) & 0 \\ 0 & -\alpha(t,x) \end{bmatrix}$$

is a diagonal matrix where $\|U(t,x)\|$, $\|U(t,x)^{-1}\| \le C$ with some $C > 0$ independent of (t,x). Here we remark that the smoothness of $\alpha(t,x)$ is not assumed. Note that $A = U\mathrm{diag}(\alpha,-\alpha)U^{-1}$ shows $\|A\|^2 \le \|U\|^2\|U^{-1}\|^2(2\alpha^2) \le 2C^4\alpha^2$. On the other hand, since $\alpha^2 = h = -\det A$, we have

$$\sum_{i,j=1}^{2} a_{ij}(t,x)^2 \le 2C^4 h.$$

Thus we have $|c^\sharp|^2 \le 2|a^\sharp|^2 \le C'h$ and hence

$$\Gamma(tD^\sharp_\phi) \subset \frac{1}{2}\Gamma([h|a^\sharp|^2]_\phi).$$

Indeed since $\Gamma(t[\partial_t c^\sharp]_\phi) = \Gamma(t\partial_t c^\sharp_\phi) \subset \Gamma(c^\sharp_\phi) \subset \Gamma(h_\phi)/2$ we have

$$\begin{aligned}\Gamma(t[a^\sharp\partial_t c^\sharp]_\phi) &= \Gamma(a^\sharp_\phi) + \Gamma(t[\partial_t c^\sharp]_\phi) \subset \Gamma(a^\sharp_\phi) + \Gamma(c^\sharp_\phi)\\ &\subset \frac{1}{2}\Gamma([|a^\sharp|^2]_\phi) + \frac{1}{2}\Gamma(h_\phi) = \frac{1}{2}\Gamma([h|a^\sharp|^2]_\phi)\end{aligned}$$

and similarly $\Gamma(t[c^\sharp\partial_t a^\sharp]_\phi) \subset \Gamma([h|a^\sharp|^2]_\phi)/2$. On the other hand $|a_{ij}|^2 \le 4|a^\sharp|^2 \le C''h$ proves

$$\Gamma(t[a_{ij}]_\phi) \subset \Gamma([a_{ij}]_\phi) \subset \frac{1}{2}\Gamma(h_\phi)$$

and hence the conditions in Theorem 3.2 are verified.

Corollary 3.2. *Let $A(t,x)$ be a 2×2 real analytic real valued matrix. If $\partial_t - A(t,x)\partial_x$ is uniformly diagonalizable then $\partial_t - A(t,x)\partial_x$ is strongly hyperbolic.*

3.5 Nonnegative Functions and Newton Polygons

Let $F(t,x)$ be a nonnegative real analytic function and let $B(t,x)$ be any real analytic function defined near the origin. Comparing the two Newton polygons $\Gamma(B_\phi)$ and $\Gamma(F_\phi)$ with $\phi \in \mathscr{C}^\pm(F)$ provides useful ingredients to study the behavior $|B|$ against F (Proposition 3.2 below). We first recall that one can write

$$f(t,x)^2 = F(t,x), \quad f(t,x) = x^n \prod_{j=1}^{l}(t - t_j(x)) \prod_{j=l+1}^{m} |t - t_j(x)|\Phi(t,x). \tag{3.9}$$

Since $f(t,x)^2 = F(t,x)$ it is clear that

$$|\partial_t f| = |\partial_t \sqrt{F}|, \quad |\partial_x f| = |\partial_x \sqrt{F}|, \quad \frac{\partial_t \sqrt{F}}{\sqrt{F}} = \frac{\partial_t f}{f}.$$

Definition 3.4. Let $F(t,x)$ be a nonnegative real analytic function defined near the origin so that we have (3.9); We set

$$t^*(x) = \Big(\sum_{j=1}^{m} |t_j(x)|^2\Big)^{1/2}.$$

We may assume, after shrinking δ if necessary, that

$$\begin{aligned}
&\operatorname{Re} t_{\mu_1}(x) \le \operatorname{Re} t_{\mu_2}(x) \le \cdots \le \operatorname{Re} t_{\mu_m}(x), \quad 0 < x < \delta,\\
&\operatorname{Re} t_{\nu_1}(x) \le \operatorname{Re} t_{\nu_2}(x) \le \cdots \le \operatorname{Re} t_{\nu_m}(x), \quad -\delta < x < 0.
\end{aligned}$$

Then we put

$$\sigma_j(x) = \begin{cases} \operatorname{Re} t_{\mu_j}(x), & x > 0,\\ \operatorname{Re} t_{\nu_j}(x), & x < 0 \end{cases}$$

and define

$$\begin{aligned}
&s_j(x) = \frac{1}{2}\{\sigma_j(x) + \sigma_{j+1}(x)\}, \quad j = 1, 2, \ldots, m-1,\\
&s_0(x) = -3t^*(x), \quad s_m(x) = 3t^*(x).
\end{aligned}$$

We also define

$$\begin{aligned}
\tilde{\omega}_j &= \{(t,x) \mid |x| < \delta, s_{j-1}(x) \le t \le s_j(x)\}, \quad j = 1, \ldots, m,\\
\tilde{\omega}(T) &= \{(t,x) \mid |x| < \delta, s_m(x) \le t \le T\}.
\end{aligned}$$

Note that the curve $t = \sigma_j(x)$ divides $\tilde{\omega}_j$ into two regions.

Lemma 3.7. *Let $F(t,x)$ be as above and $f(t,x)$ be as in (3.9). Then there are $c_i > 0$ such that*

$$\frac{c_1}{t - t^*(x)} \le \frac{c_1}{t - 2t^*(x)} \le \frac{f_t}{f} \le \frac{c_2}{t - t^*(x)}$$

in $\tilde{\omega}(T)$ (taking $T > 0$ small enough).

Proof. Recall that

$$\frac{f_t}{f} = \sum_{j=1}^{l} \frac{1}{t - t_j(x)} + \sum_{j=l+1}^{m} \frac{t - \operatorname{Re} t_j(x)}{|t - t_j(x)|^2} + \frac{\Phi_t}{\Phi}.$$

Since

$$\sum_{j=1}^{l}\frac{1}{t-t_j(x)}=\sum_{j=1}^{l}\frac{t-\operatorname{Re}t_j(x)+i\operatorname{Im}t_j(x)}{|t-t_j(x)|^2}=\sum_{j=1}^{l}\frac{t-\operatorname{Re}t_j(x)}{|t-t_j(x)|^2}$$

because the left-hand side is real we get

$$\frac{f_t}{f}=\sum_{j=1}^{m}\frac{t-\operatorname{Re}t_j(x)}{|t-t_j(x)|^2}+\frac{\Phi_t}{\Phi}. \tag{3.10}$$

Hence we get

$$\frac{f_t}{f}\geq\sum_{j=1}^{m}\frac{t-\operatorname{Re}t_j(x)}{|t-t_j(x)|^2}-C.$$

On the other hand noting that

$$t-2t^*(x)\geq\frac{t}{3}\geq\frac{1}{4}(t+|t_j(x)|)\geq\frac{1}{4}|t-t_j(x)|$$

in $\tilde{\omega}(T)$ we have

$$\frac{1}{|t-t_j(x)|}\geq\frac{1}{4(t-2t^*(x))}.$$

Since $t-\operatorname{Re}t_j(x)\geq t-2t^*(x)$ it follows that

$$\frac{f_t}{f}\geq\frac{1}{6}\sum_{j=1}^{m}\frac{1}{t-2t^*(x)}-C\geq\frac{c_1}{t-2t^*(x)}$$

taking T small enough because $0\leq t-2t^*(x)\leq T$ in $\tilde{\omega}(T)$ which implies

$$-\frac{TC}{t-2t^*(x)}\leq -C.$$

We turn to the right-hand side. Note

$$|t-t_j(x)|\geq t-|t_j(x)|\geq t-t^*(x)$$

and hence by (3.10) one has

$$\frac{f_t}{f}\leq\sum_{j=1}^{m}\frac{1}{|t-t_j(x)|}+C\leq\frac{1}{t-t^*(x)}+C.$$

Using $C\leq CT/(t-t^*(x))$ we have the desired assertion. □

Lemma 3.8. *Let $F(t,x)$ be as above and $f(t,x)$ be given by Lemma 3.6. Then there is a $C > 0$ such that*

$$\partial_t\left(\frac{f_t}{f}\right) \leq C$$

in $\tilde{\omega}(T)$.

Proof. From (3.10) one has

$$\partial_t\left(\frac{f_t}{f}\right) = -\sum_{j=1}^{l}\frac{1}{(t-t_j(x))^2} - \sum_{j=l+1}^{m}\frac{(t-\operatorname{Re}t_j(x))^2-(\operatorname{Im}t_j(x))^2}{|t-t_j(x)|^4} + \partial_t\left(\frac{\Phi_t}{\Phi}\right).$$

Here we note that

$$\operatorname{Re}\frac{1}{(t-t_j(x))^2} = \frac{(t-\operatorname{Re}t_j(x))^2-(\operatorname{Im}t_j(x))^2}{|t-t_j(x)|^4}.$$

This shows that

$$\partial_t\left(\frac{f_t}{f}\right) = -\sum_{j=1}^{m}\frac{(t-\operatorname{Re}t_j(x))^2-(\operatorname{Im}t_j(x))^2}{|t-t_j(x)|^4} + \partial_t\left(\frac{\Phi_t}{\Phi}\right).$$

In $\tilde{\omega}(T)$ we see that

$$t-\operatorname{Re}t_j(x) \geq 3|t_j(x)| - \operatorname{Re}t_j(x) \geq 2|t_j(x)| \geq |\operatorname{Im}t_j(x)|$$

and hence $(t-\operatorname{Re}t_j(x))^2-(\operatorname{Im}t_j(x))^2 \geq 0$. This gives

$$\partial_t\left(\frac{f_t}{f}\right) \leq \partial_t\left(\frac{\Phi_t}{\Phi}\right) \leq C$$

and hence the result. □

Proposition 3.2. *Assume that*

$$\Gamma(tB_\phi) \subset \frac{1}{2}\Gamma(F_\phi), \quad \forall\phi \in \mathscr{C}^{\pm}(F).$$

Then there is $C > 0$ such that (taking T small enough)

$$|(t-\sigma_j(x))B(t,x)| \leq C|f(t,x)| \ \textit{for}\ (t,x) \in \tilde{\omega}_j,\ j = 1,\dots,m,$$

$$|(t-s_m(x))B(t,x)| \leq C|f(t,x)| \ \textit{for}\ (t,x) \in \tilde{\omega}(T),\ \textit{if}\ n \geq 1,$$

$$|B(t,x)| \leq C|\partial_t f(t,x)| \ \textit{for}\ (t,x) \in \tilde{\omega}(T),\ \textit{if}\ n = 0.$$

Proof. We give the proof in Sect. 3.7. □

Definition 3.5. Let $f(x)$, $g(x)$ be two functions defined near the origin or in a half neighborhood of the origin. Then we denote $f \sim g$ if and only if there exists some $C > 0$ such that

$$C^{-1}|f(x)| \le |g(x)| \le C|f(x)|.$$

Lemma 3.9. *Let $n = 0$. Then there is a $C > 0$ such that*

$$\sup_{0 \le t \le t^*(x)} |f(t,x)| \le C|x|.$$

Proof. It is enough to show that $|F(t,x)| \le C|x|^2$ for $0 \le t \le t^*(x)$. By definition there is j such that

$$F(t_j(x), x) = 0, \quad t_j(x) \sim t^*(x).$$

If $g_i(t_j(x), x) = 0$ with $l_i \ge 2$ for some i then one gets

$$|g_i(t,x)|^{l_i} \le C|x|^2, \quad 0 \le t \le t^*(x).$$

To see this note that $g_i(t_j(x), x) = t_j(x)^{n(i)} + O(|x|) = 0$ and hence we have $t_j(x)^{n(i)} = O(|x|)$. This gives $g_i(t,x) = O(|x|)$ for $0 \le t \le t^*(x)$. If $h_i(t_j(x), x) = 0$ then it is easy to see that

$$|h_i(t,x)| \le C|x| \quad \text{for } 0 \le t \le t^*(x)$$

and hence $|h_i(t,x)\bar{h}_i(t,x)| \le C|x|^2$ for $0 \le t \le t^*(x)$. Finally let $g_i(t_j(x), x) = 0$ with $l_i = 1$. Since $g_i(t,x) \ge 0$ then $g_i(t,x) = t^{2\bar{m}} + d_1(x)t^{2\bar{m}-1} + \cdots + d_{2\bar{m}}(x)$ where $d_{2\bar{m}}(x) = O(|x|^2)$. On the other hand since every root $t(x)$ of $g_i(t,x) = 0$ is a branch of

$$\sum_{i \ge 1} C_i (x^{1/2\bar{m}})^i$$

then it follows that $C_1 = 0$ and hence every root is $O(|x|^{1/\bar{m}})$. This shows that $d_j(x)(|x|^{1/\bar{m}})^{2\bar{m}-j} = O(|x|^2)$ and hence $g_i(t,x) = O(|x|^2)$ for $0 \le t \le t*(x)$. □

Lemma 3.10. *Let $F(t,x)$ and $f(t,x)$ be as above. Then we have*

$$\sup_{|t| \le T, 0<|x|<\delta} |f_x(t,x)| \le C$$

with some $C > 0$.

Proof. Recall that $f(t,x)$ is real analytic in $V \setminus (0,0)$ satisfying $f(t,x)^2 = F(t,x)$. If $g(t,x)^2 = F(t,x)$ then we have $f(t,x) = g(t,x)$ or $f(t,x) = -g(t,x)$ in $V \setminus (0,0)$. That is $f(t,x)$ is unique up to the sign. Taking this fact into account, we can argue exchanging t and x to conclude that

$$f(t,x) = t^k \prod_{j=1}^{\tilde{l}} (x - s_j(t)) \prod_{j=\tilde{l}+1}^{\tilde{m}} |x - s_j(t)| \Psi(t,x)$$

where $\mathrm{Im}\, s_j(t) \neq 0$ if $j \geq \tilde{l} + 1$. Then it is clear that $f_x(t,x)$ is bounded because

$$\frac{\partial}{\partial x}|x - s_j(t)| = \frac{x - \mathrm{Re}\, s_j(t)}{|x - s_j(t)|}$$

is bounded. □

Lemma 3.11. *Let $F(t,x)$ and $f(t,x)$ be as above. Let $n = 0$. Then for any $K > 0$, there is T_K such that we have*

$$\text{either } \; f_t(t,x) \geq Kf(t,x) > 0 \quad \text{or} \quad -f_t(t,x) \geq -Kf(t,x) > 0$$

in $\tilde{\omega}(T)$ for $0 < T \leq T_K$.

Proof. Recall that

$$f(t,x) = \prod_{j=1}^{l} (t - t_j(x)) \prod_{j=l+1}^{m} |t - t_j(x)| e(t,x).$$

Then it is easy to see

$$\bar{f}_t f + f_t \bar{f} = 2 \sum_{p=1}^{m} (t - \mathrm{Re}\, t_p(x)) \prod_{j \neq p} |t - t_j(x)|^2 e^2 + \prod_{j=1}^{m} |t - t_j(x)|^2 (e^2)_t.$$

On the other hand, by definition we see for $(t,x) \in \tilde{\omega}(T)$

$$t - \mathrm{Re}\, t_k(x) \geq t - t^*(x) \geq \frac{2}{3} t \geq \frac{1}{2}|t - t_j(x)|, \quad k = 1, \ldots, m.$$

Then one has

$$(f^2)_t - Kf^2 \geq \sum_k (1 - CK|t - t_k(x)|)|t - t_k(x)| \prod_{j \neq k} |t - t_j(x)|^2 e^2.$$

Since $\sup_{\tilde{\omega}(T)} |t - t_k(x)| \to 0$ as $T \to 0$ we get the desired result. □

Remark. Although we have only discussed the behavior of $|B|$ and f in $\tilde{\omega}(T)$, similar arguments give corresponding results in $\tilde{\omega}(-T) = \{(t,x) \mid -T \le t \le s_0(x)\}$.

3.6 Behavior Around Pseudo-Characteristic Curves

In this section we study the behaviors of h and a near pseudo-characteristic curves $t = \phi(x)$, $\phi \in \mathscr{C}^{\pm}$ which will be needed when we derive energy estimates in the following sections. To simplify notations let us put

$$a(t,x) = |a^{\sharp}(t,x)|^2.$$

Recall that

$$h(t,x) = x^{2n_1}\big(t^{2m_1} + h_1(x)t^{2m_1-1} + \cdots + h_{2m_1}(x)\big)e(t,x)^2$$

where $e(0,0) \neq 0$, $h_i(0) = 0$. We apply Lemma 3.6 to h to get a real $b(t,x)$ such that

$$b^2(t,x) = h(t,x), \;\; b(t,x) = x^{n_1}\prod_{i=1}^{l_1}(t - t_i(x)) \prod_{i=l_1+1}^{m_1} |t - t_i(x)| e(t,x).$$

Since $|a^{\sharp}(t,x)|^2$ is also a nonnegative real analytic function we apply Lemma 3.6 again to obtain a real $\tilde{b}(t,x)$ such that

$$\tilde{b}^2(t,x) = |a^{\sharp}(t,x)|^2, \;\; \tilde{b}(t,x) = x^{n_2}\prod_{j=1}^{l_2}(t - t_j(x)) \prod_{j=l_2+1}^{m_2} |t - t_j(x)| \tilde{e}(t,x).$$

Remark. By the Weierstrass preparation theorem one can write

$$a^{\sharp}(t,x) = x^{n_2}\big(t^{m_2} + a_1(x)t^{m_2-1} + \cdots + a_{m_2}(x)\big)\Psi(t,x)$$

with $a_i(0) = 0$, $\Psi(0,0) \neq 0$. Repeating the same arguments as in Sect. 3.2 one can express

$$a^{\sharp}(t,x) = x^{n_2} g_1^{\mu_1} \cdots g_p^{\mu_p} h_1^{\nu_1} \cdots h_q^{\nu_q} \Psi$$

where $\bar{g}_i = g_i$ and $\bar{h}_i \neq h_i$. This gives

$$|a^{\sharp}|^2 = x^{2n_2}(g_1^{\mu_1} \cdots g_p^{\mu_p})^2 h_1^{\nu_1} \bar{h}_1^{\nu_1} \cdots h_q^{\nu_q} \bar{h}_q^{\nu_q} |\Psi|^2.$$

Thus $\tilde{b}$ is given by

$$\tilde{b}(t,x) = x^{n_2} g_1^{\mu_1} \cdots g_p^{\mu_p} |h_1|^{\nu_1} \cdots |h_q|^{\nu_q} |\Psi|$$

where

$$g_1^{\mu_1} \cdots g_p^{\mu_p} = \prod_{j=1}^{\ell_2} (t - t_j(x)), \quad |h_1^{\nu_1} \cdots h_q^{\nu_q}| = \prod_{j=\ell_2+1}^{m_2} |t - t_j(x)|.$$

Definition 3.6. With $F(t,x) = h|a^\sharp|^2 = ha$ we define $t^*(x)$, $\sigma_j(x)$, $s_j(x)$, ω_j, $\omega(T)$ according to Definition 3.4. We define $t_h^*(x)$, $\tilde{\omega}_h(T)$ and $t_a^*(x)$, $\tilde{\omega}_a(T)$ by Definition 3.4 with $F = h$ and $F = a$ respectively. Recall that $t = \sigma_j(x)$ are nothing but the pseudo-characteristic curves of $\partial_t - A(t,x)\partial_x$

Remark. Note that $n_1 = 0$ implies $m_1 \geq 1$.

Proof. Let $n_1 = 0$. Note that $|a^\sharp|^2 \geq h$ implies $n_2 = 0$. On the other hand $n_2 = 0$ means $m_2 \geq 1$ because $a^\sharp(0,0) = 0$. Hence $|a^\sharp|^2 \geq h$ again shows that $m_1 \geq 1$. □

Remark. Since one can write

$$t_h^*(x) = |x|^\alpha (C_h + o(|x|)), \quad t_a^*(x) = |x|^\beta (C_a + o(|x|)), \quad C_h,\ C_a > 0$$

with some α, $\beta > 0$ then taking $\delta > 0$ so small one may assume that

$$\text{either} \quad 2t_a^*(x) \geq t_h^*(x) \quad \text{or} \quad 2t_h^*(x) \geq t_a^*(x)$$

holds in $|x| \leq \delta$.

Lemma 3.12. *Let $n_1 = 0$ and $2t_a^*(x) \geq t_h^*(x)$ (resp. $2t_h^*(x) \geq t_a^*(x)$). Then there is a $C > 0$ such that*

$$\frac{b_t}{b} \leq C \frac{\tilde{b}_t}{\tilde{b}} \quad (\textit{resp.}\ \frac{\tilde{b}_t}{\tilde{b}} \leq C \frac{b_t}{b}) \quad \textit{in}\ \tilde{\omega}(T).$$

Proof. Suppose $2t_a^*(x) \geq t_h^*(x)$. Clearly we have $t - 2t_a^*(x) \leq t - t_h^*(x)$ and hence by Lemma 3.7 we have

$$\frac{b_t}{b} \leq \frac{C'}{t - t_h^*(x)} \leq \frac{C'}{t - 2t_a^*(x)} \leq C'' \frac{\tilde{b}_t}{\tilde{b}} \quad in \quad \tilde{\omega}(T)$$

because $\tilde{\omega}(T) \subset \tilde{\omega}_a(T) \cap \tilde{\omega}_h(T)$. The proof for the other case is similar. □

Lemma 3.13. *Let $n_1 = 0$. Then there is a $C > 0$ such that*

$$\left|\frac{\partial_t a^\sharp}{a^\sharp}\right| \leq C \frac{\tilde{b}_t}{\tilde{b}}, \quad \left|\partial_t (\frac{\partial_t a^\sharp}{a^\sharp})\right| \leq C \left(\frac{\tilde{b}_t}{\tilde{b}}\right)^2$$

in $\tilde{\omega}_a(T)$.

Proof. Recall that

$$a^\sharp = x^{n_2} \prod_{j=1}^{m_2} (t - t_j(x))\Psi$$

and note that

$$\frac{\partial_t a^\sharp}{a^\sharp} = \sum \frac{1}{t - t_j(x)} + \frac{\Psi_t}{\Psi}.$$

Since $|t - t_j(x)| \geq t - t_a^*(x)$ in $\tilde{\omega}_a(T)$ we have

$$\left| \frac{\partial_t a^\sharp}{a^\sharp} \right| \leq \frac{c_1}{t - t_a^*(x)} + c_2 \leq \frac{c_3}{t - t_a^*(x)} \leq c_4 \left(\frac{\tilde{b}_t}{\tilde{b}} \right)$$

in $\tilde{\omega}_a(T)$, taking T small enough. Similarly we have

$$\partial_t \left(\frac{\partial_t a^\sharp}{a^\sharp} \right) = -\sum \frac{1}{(t - t_j(x))^2} + \partial_t \left(\frac{\Psi_t}{\Psi} \right)$$

from which it is easy to see that

$$\left| \partial_t (\frac{\partial_t a^\sharp}{a^\sharp}) \right| \leq \frac{c_1}{(t - t_a^*(x))^2} \leq c_2 \left(\frac{\tilde{b}_t}{\tilde{b}} \right)^2$$

in $\tilde{\omega}_a(T)$. □

Lemma 3.14. *There is a $C > 0$ such that*

$$\sup_{0 \leq t \leq t^*(x)} |b(t, x)| \leq C|x|.$$

Proof. If $n_1 \geq 1$ then the assertion is obvious. Let $n_1 = 0$ and hence $m_1, m_2 \geq 1$. When $t^*(x) \sim t_h^*(x)$ then Lemma 3.9 (or rather its proof) proves the lemma. Then we now assume that there is no j such that $h(t_j(x), x) = 0$, $t_j(x) \sim t^*(x)$. In this case we have

$$a^\sharp(t_j(x), x) = 0, \quad t_j(x) \sim t^*(x)$$

with some j. This shows that the line with the slowest steep of $\Gamma(h)$ is steeper than that of $\Gamma(a^\sharp)$ which proves that

$$h_k(x) t^*(x)^{2m_1 - k} = o(t^*(x)^{2m_1}), \quad 1 \leq k \leq 2m_1. \tag{3.11}$$

On the other hand since $a^\sharp \overline{a^\sharp} \geq h$, $(\bar{a})^\sharp = \overline{a^\sharp}$ we see $m_1 \geq m_2$. From $a^\sharp(t_j(x), x) = 0$ it follows that

$$t_j(x)^{m_2} = O(|x|).$$

This shows that $t^*(x)^{2m_1} = O(|x|^2)$ and then from (3.11) we have

$$\sup_{0 \leq t \leq t^*(x)} |h(t,x)| \leq C|x|^2$$

which proves the desired assertion. □

Definition 3.7. To simplify notations let us put

$$\rho_j(t,x) = t - \sigma_j(x), \quad j = 1, \ldots, m, \quad \rho_{m+1}(x) = t - s_m(x).$$

Lemma 3.15. *We have the followings.*

(i) *Let* $n_1 \geq 1$. *Then for* $j = 1, \ldots, m+1$ *we have*

$$\sup_{0 \leq t \leq T, |x| < \epsilon} |b(t,x)\rho_{jx}(t,x)| \to 0, \quad \epsilon \to 0.$$

(ii) *Let* $n_1 = 0$. *Then for* $j = 1, \ldots, m+1$ *we have*

$$\sup_{0 \leq t \leq t^*(x), |x| < \epsilon} |b(t,x)\rho_{jx}(t,x)| \to 0, \quad \epsilon \to 0.$$

(iii) *Let* $n_1 = 0$ *and* $2t_a^*(x) \geq t_h^*(x)$. *Then for* $j = 1, \ldots, m+1$ *we have*

$$\sup_{0 \leq t \leq t^*(x), |x| < \epsilon} |\tilde{b}(t,x)\rho_{jx}(t,x)| \to 0, \quad \epsilon \to 0.$$

Proof. Remarking that $\rho_{jx}(x) = O(|x|^{\sigma-1})$ with some $\sigma > 0$ the assertions (i) and (ii) follow from Lemma 3.14. Noting that $2t_a^*(x) \geq t_h^*(x)$ implies $t^*(x) \sim t_a^*(x)$ the assertion (iii) follows from Lemma 3.9. □

Lemma 3.16. *There is* $C > 0$ *such that*

$$|\rho_j(t,x)b_t(t,x)| \leq C|b(t,x)|$$

in $\tilde{\omega}_j$, $j = 1, \ldots, m$ *and*

$$|\rho_{m+1}(t,x)b_t(t,x)| \leq C|b(t,x)|$$

in $\tilde{\omega}(T)$.

Proof. Recall that

$$\frac{b_t}{b} = \sum_{i=1}^{l_1} \frac{1}{t - t_i(x)} + \sum_{i=l_1+1}^{m_1} \frac{t - \operatorname{Re} t_i(x)}{|t - t_i(x)|^2} + \frac{e_t}{e}.$$

Let $(t, x) \in \tilde{\omega}_j$. Then it is clear that

$$|t - \sigma_j(x)| \le |t - \operatorname{Re} t_\mu(x)| \le |t - t_\mu(x)|$$

for all μ. This shows that

$$\left|\frac{b_t}{b}\right| \le \sum \frac{1}{|t - t_i(x)|} + c \le \frac{c'}{|t - \sigma_j(x)|} + c$$

in $\tilde{\omega}_j$. Taking T small so that

$$c \le \frac{c''}{|t - \sigma_j(x)|}$$

in $\tilde{\omega}_j$ we have the assertion. If $(t, x) \in \tilde{\omega}(T)$, then we see

$$|t - s_m(x)| \le |t - \operatorname{Re} t_\mu(x)| \le |t - t_\mu(x)|$$

for all μ and hence the assertion follows from the same arguments as before. □

3.7 Proof of Proposition 3.2

In this section we give a proof of Proposition 3.2. We may assume that $\mu_j = j$ renumbering the indices if necessary. We fix $1 \le j_0 \le m$. Assume that

$$\operatorname{Re} t_{j_0-k-1}(x) < \operatorname{Re} t_{j_0-k}(x) = \cdots = \operatorname{Re} t_{j_0}(x) = \cdots = \operatorname{Re} t_{j_0+l}(x) < \operatorname{Re} t_{j_0+l+1}(x)$$

in $0 < x < \delta$. Put $\lambda(x) = \operatorname{Re} t_{j_0}(x)$ and

$$\phi^+ = \frac{1}{2}(\operatorname{Re} t_{j_0+l+1}(x) - \lambda(x)), \quad \phi^-(x) = \frac{1}{2}(\lambda(x) - \operatorname{Re} t_{j_0-k-1}(x)).$$

When $\operatorname{Re} t_{j_0}(x) = \operatorname{Re} t_m(x)$ (resp. $\operatorname{Re} t_{j_0} = \operatorname{Re} t_1(x)$) we set

$$\phi^+ = \frac{1}{2}(3t^*(x) - \lambda(x)), \quad (\text{resp. } \phi^- = \frac{1}{2}(\lambda(x) + 3t^*(x))).$$

Lemma 3.17. *For any* $1 \le \nu \le m$

$$|\lambda(x) - t_\nu(x)| + \delta|\phi^\pm(x)| \sim |\lambda(x) + \delta\phi^\pm(x) - t_\nu(x)|$$

holds uniformly in $0 \le \delta \le 1$.

Proof. Let $\operatorname{Re} t_{j_0}(x) < \operatorname{Re} t_m(x)$. Note

$$|\lambda(x) - t_\nu(x) + \delta\phi^+(x)|^2 = (\lambda(x) - \operatorname{Re} t_\nu(x) + \delta\phi^+(x))^2 + (\operatorname{Im} t_\nu(x))^2.$$

If $\operatorname{Re} t_\nu(x) \ge \operatorname{Re} t_{j_0+l+1}(x)$ then

$$\begin{aligned}&|\lambda(x) - \operatorname{Re} t_\nu(x) + \delta\phi^+(x)|\\&\quad \ge |\operatorname{Re} t_\nu(x) - \lambda(x)| - \delta|\phi^+(x)| \ge \frac{1}{2}|\operatorname{Re} t_\nu(x) - \lambda(x)|.\end{aligned}$$

Since $\operatorname{Re} t_\nu(x) - \lambda(x) \ge 2\phi^+(x) \ge 0$ it follows that

$$|\lambda(x) - \operatorname{Re} t_\nu(x) + \delta\phi^+(x)| \sim |\lambda(x) - \operatorname{Re} t_\nu(x)| + \delta|\phi^+(x)|$$

which proves the assertion for ϕ^+. We next assume $\operatorname{Re} t_\nu(x) \le \operatorname{Re} t_{j_0}(x)$. In this case we have

$$|\lambda(x) - \operatorname{Re} t_\nu(x) + \delta\phi^+(x)| = |\lambda(x) - \operatorname{Re} t_\nu(x)| + \delta\phi^+(x)$$

then one gets the assertion. The proof of the other cases are similar. □

Write

$$B(t,x) = x^{\bar n}\tilde B(t,x)\tilde E(t,x)$$

where $\tilde E(0,0) \ne 0$. Recall that our assumption is

$$\Gamma(tx^{\bar n}\tilde B(t+\phi(x),x)) \subset \Gamma(x^n \prod_{\nu=1}^{m} \Lambda_\nu(t+\phi(x),x))$$

where $\Lambda_\nu(t,x) = t - t_\nu(x)$. Let us define $\epsilon(\nu)$, $1 \le \nu \le m$ and ϵ by

$$|\lambda(x) - t_\nu(x)| \sim x^{\epsilon(\nu)}, \quad \phi^+(x) \sim x^\epsilon.$$

Assume that

$$\epsilon(\nu_1) \ge \cdots \ge \epsilon(\nu_l) > \epsilon \ge \epsilon(\nu_{l+1}) \ge \cdots \ge \epsilon(\nu_m).$$

From Lemma 3.17 it follows that

$$\prod_{j=l+1}^{m} |\Lambda_j(\lambda(x)+\delta\phi^+(x),x)| \geq c \prod_{j=l+1}^{m} x^{\epsilon(\nu_j)}$$

with some $c>0$ uniformly in $0\leq\delta\leq 1$. Lemma 3.17 again shows

$$\prod_{j=1}^{l} |\Lambda_j(\lambda(x)+\delta\phi^+(x),x)| \sim \prod_{j=1}^{l}(x^{\epsilon(\nu_j)}+\delta x^{\epsilon}) \geq c\delta^p x^{\epsilon p+\epsilon(\nu_{p+1})+\cdots+\epsilon(\nu_l)}$$

with $c>0$ for $p=0,1,\ldots,l$. Hence, writing

$$tx^{\bar{n}}\tilde{B}(t+\lambda(x),x)=\sum b_j(x)t^j,\quad b_{j+1}(x)=\frac{1}{j!}x^{\bar{n}}\partial_t^j\tilde{B}(\lambda(x),x)$$

the assumption implies that

$$\text{Order}\{x^{\bar{n}}\partial_t^j\tilde{B}(\lambda(x),x)\}\geq n+\sum_{i=j+2}^{m}\epsilon(\nu_i). \tag{3.12}$$

Lemma 3.18. *For $0\leq\delta\leq 1$ we have*

$$|\delta\phi^{\pm}(x)x^{\bar{n}}\tilde{B}(\lambda(x)+\delta\phi^{\pm}(x),x)|\leq C|x^n\prod_{\nu=1}^{m}\Lambda_\nu(\lambda(x)+\delta\phi^{\pm}(x),x)|$$

with C independent of δ.

Proof. Let us write

$$\tilde{B}(\lambda(x)+\delta\phi^+(x),x)=\sum_{j=0}^{\bar{m}}B_j(x)\delta^j,\quad B_j(x)=\frac{1}{j!}\phi^+(x)^j\partial_t^j\tilde{B}(\lambda(x),x).$$

From (3.12) it follows that

$$|x^{\bar{n}}\partial_t^j\tilde{B}(\lambda(x),x)|\leq Cx^{n+\sum_{i=j+2}^{m}\epsilon(\nu_i)}$$

and hence

$$\delta|\phi^+||\delta^j\phi^+(x)^jx^{\bar{n}}\partial_t^j\tilde{B}(\lambda(x),x)|\leq C\delta^{j+1}x^{\epsilon(\nu_{j+2})+\cdots+\epsilon(\nu_m)+(j+1)\epsilon+n}. \tag{3.13}$$

Let $j+2\leq l$ then the right-hand side of (3.13) is bounded by

$$C\delta^{j+1}x^{n+(j+1)\epsilon+\epsilon(\nu_{j+2})+\cdots+\epsilon(\nu_l)}\prod_{i=l+1}^{m}x^{\epsilon(\nu_i)}\leq C|x^n\prod_{\nu=1}^{m}\Lambda_\nu(\lambda(x)+\delta\phi^+(x),x)|.$$

If $j+2>l$ then noting

$$(j+1)\epsilon+\epsilon(\nu_{j+2})+\cdots+\epsilon(\nu_m)\geq l\epsilon+\epsilon(\nu_{l+1})+\cdots+\epsilon(\nu_m)$$

the right-hand side of (3.13) is estimated by

$$C\delta^l x^{n+l\epsilon}\left(\prod_{i=l+1}^{m} x^{\epsilon(\nu_i)}\right)\delta^{j+1-l}\leq C|x^n\prod_{\nu-1}^{m}\Lambda_\nu(\lambda(x)+\delta\phi^+(x),x)|$$

which ends the proof of the assertion for ϕ^+. The proof for ϕ^- is similar. □

Proof of Proposition 3.2. Recall that

$$\tilde{\omega}_{j_0}=\{(t,x)\mid |x|<\delta,\lambda(x)-\phi^-(x)\leq t\leq\lambda(x)+\phi^+(x)\}.$$

Let $(t,x)\in\tilde{\omega}_{j_0}\cap\{t\geq\lambda(x)\}$. Then there is a $0\leq\delta\leq 1$ such that $t=\lambda(x)+\delta\phi^+(x)$. From Lemma 3.18 it follows

$$|(t-\lambda(x))B(t,x)|\leq C|f(t,x)|.$$

In the case $(t,x)\in\tilde{\omega}_{j_0}\cap\{t\leq\lambda(x)\}$ the proof is similar. □

Lemma 3.19. *In $\tilde{\omega}(T)$ with small T we have*

$$|B(t,x)|\leq C\sum_{q=1}^{m}|x^n\prod_{\nu\neq q}\Lambda_\nu(t,x)|,$$

$$|(t-\operatorname{Re}t_m(x))B(t,x)|\leq C|x^n\prod_{\nu=1}^{m}\Lambda_\nu(t,x)|.$$

Proof. Repeating the same proof of Lemma 3.18 we see with $\lambda(x)=\operatorname{Re}t_m(x)$ that

$$|\delta\phi^+(x)x^{\bar{n}}\tilde{B}(\lambda(x)+\delta\phi^+(x),x)|\leq C|x^n\prod_{\nu=1}^{m}\Lambda_\nu(\lambda(x)+\delta\phi^+(x),x)|$$

holds for all $0\leq\delta$. For any $(t,x)\in\tilde{\omega}(T)$, taking $\delta>0$ so that $t=\lambda(x)+\delta\phi^+(x)$ the second inequality follows. Since

$$t-\operatorname{Re}t_m(x)\geq t-|t_m(x)|\geq\frac{2}{3}t\geq\frac{1}{3}|t-t_\nu(x)|,$$

$$|t-t_\nu(x)|\geq t-|t_\nu(x)|\geq\frac{2}{3}t\geq\frac{1}{3}(t-\operatorname{Re}t_m(x))$$

where the first inequality follows from the second one. □

Lemma 3.20. *In $\tilde{\omega}(T)$ with small T we have*

$$\sum_{q=1}^{m} \Big|x^n \prod_{\nu\neq q}^{m} \Lambda_\nu(t,x)\Big| \sim |\partial_t f(t,x)|.$$

Proof. Since it is clear that

$$|\partial_t f(t,x)| \le C \sum_{q=1}^{m} \Big|x^n \prod_{\nu\neq q} \Lambda_\nu(t,x)\Big|$$

it is enough to show the converse. Note that

$$\begin{aligned} f\partial_t f &= x^{2n} \sum_{\nu=1}^{m} (t - \operatorname{Re} t_\nu(x)) \prod_{\mu\neq\nu} |t - t_\mu(x)|^2 |e|^2 \\ &\quad + x^{2n} \prod_{\mu=1}^{m} |t - t_\mu(x)|^2 e\partial_t e. \end{aligned}$$

On the other hand we have $t - \operatorname{Re} t_\nu(x) \ge c|t - t_\nu(x)|$ for $1 \le \nu \le m$ in $\tilde{\omega}(T)$ because

$$t - \operatorname{Re} t_\nu(x) \ge \frac{2}{3}t \ge \frac{t}{3} + \frac{1}{3}|t_\nu(x)| \ge \frac{1}{3}|t - t_\nu(x)|.$$

Thus we see

$$f\partial_t f \ge cx^{2n} \sum_{\nu=1}^{m} |t - t_\nu(x)| \prod_{\mu\neq\nu} |t - t_\mu(x)|^2$$

with $c > 0$. Hence dividing the above inequality by $|f(t,x)|$ we get the desired assertion. □

3.8 Energy Estimates Near Pseudo-Characteristic Curves

Since the existence of analytic solutions with analytic initial data is assured by the Cauchy–Kowalevsky theorem, applying the usual limiting arguments, to prove the sufficiency of C^∞ well-posedness, it is enough to derive energy estimates for analytic solutions to $L^\sharp u = f$. Since u verifies

$$ML^\sharp u = (\partial_t^2 u - h\partial_x^2 + (\tilde{Q} - h_x)\partial_x + \tilde{R}\partial_t + \tilde{S})u = Mf$$

where M is given in Lemma 3.5, we use this equation to get energy estimates. We show that we can obtain weighted energy estimates although $\tilde{R}$ and $\tilde{S}$ are not smooth in general.

Let $D \subset W$ be an open set and $\rho(t,x) \in C^\infty(D)$ where $\rho_t > 0$ in D. Put $p = \partial_t^2 - h(t,x)\partial_x^2$ and note that

$$p - h_x\partial_x = \partial_t^2 - \partial_x h \partial_x.$$

We study the energy form

$$\begin{aligned}(pu - h_x\partial_x u)\overline{\partial_t u} &+ \overline{(pu - h_x\partial_x u)}\partial_t u\\ &= \partial_t G_1(u) + \partial_x G_2(u) - R(u)\end{aligned}$$

where

$$\begin{cases} G_1(u) = |\partial_t u|^2 + h(t,x)|\partial_x u|^2, \\ G_2(u) = -h(\partial_t u\overline{\partial_x u} + \overline{\partial_t u}\partial_x u), \\ R(u) = h_t|\partial_x u|^2. \end{cases}$$

Multiply $e^{-\theta t}\rho^{\pm N}$ to the energy form and integrate over D to get

$$\begin{aligned} 2\int_D & e^{-\theta t}\rho^{\pm N}|pu - h_x\partial_x u||\partial_t u|dxdt \\ \geq & \int_D \left[\partial_t(e^{-\theta t}\rho^{\pm N}G_1(u)) + \partial_x(e^{-\theta t}\rho^{\pm N}G_2(u))\right]dxdt \\ & \mp N\int_D e^{-\theta t}\rho^{\pm N-1}(\rho_t G_1(u) + \rho_x G_2(u))dxdt \\ & +\theta\int_D e^{-\theta t}\rho^{\pm N}G_1(u)dxdt - \int_D e^{-\theta t}\rho^{\pm N}R(u)dxdt \end{aligned}$$

where $\theta > 0$ is a positive parameter and N is an even integer. Note that

$$\begin{aligned} N\int_D |e^{-\theta t}\rho^{\pm N-1}\rho_x G_2(u)|dxdt \leq & \frac{N}{4}\int_D |e^{-\theta t}\rho^{\pm N-1}\rho_t||\partial_t u|^2dxdt \\ & +4N\int_D |e^{-\theta t}\rho^{\pm N-1}h^2\rho_x^2\rho_t^{-1}||\partial_x u|^2dxdt \end{aligned}$$

by the Cauchy–Schwarz inequality. Similarly we have

$$\begin{aligned} 2\int_D & e^{-\theta t}\rho^{\pm N}|pu - h_x\partial_x u||\partial_t u|dxdt \\ \leq & \frac{4}{N}\int_D |e^{-\theta t}\rho^{\pm N+1}\rho_t^{-1}||pu - h_x\partial_x u|^2dxdt \\ & +\frac{N}{4}\int_D |e^{-\theta t}\rho^{\pm N-1}\rho_t||\partial_t u|^2dxdt. \end{aligned}$$

We choose $\pm$ so that $\mp\rho^{\pm N-1}\rho_t > 0$ in D, that is if $\rho > 0$ in D we take ρ^{-N} and if $\rho < 0$ in D then we take ρ^N. Using these inequalities we get

$$\begin{aligned}
\frac{4}{N}\int_D &|e^{-\theta t}\rho^{\pm N+1}\rho_t^{-1}||pu - h_x\partial_x u|^2 dxdt\\
&\geq \int_D \big[\partial_t(e^{-\theta t}\rho^{\pm N}G_1(u)) + \partial_x(e^{-\theta t}\rho^{\pm N}G_2(u))\big]dxdt\\
&+\frac{N}{4}\int_D |e^{-\theta t}\rho^{\pm N-1}\rho_t|(2|\partial_t u|^2 + h(t,x)|\partial_x u|^2)dxdt\\
&+\int_D \sigma(t,x)|e^{-\theta t}\rho^{\pm N-1}||\partial_x u|^2 dxdt + \theta\int_D e^{-\theta t}\rho^{\pm N}G_1(u)dxdt
\end{aligned}$$

where

$$\sigma(t,x) = \frac{3N}{4}h\rho_t - 4Nh^2\rho_x^2\rho_t^{-1} - C|\rho h_t|. \tag{3.14}$$

We turn to $\partial_t u\cdot\bar{u} + \overline{\partial_t u}\cdot u = \partial_t|u|^2$. Multiply $e^{-\theta t}\rho^{\pm N-2}\rho_t^2$ we get

$$\begin{aligned}
C_1N^{-1}\int_D |e^{-\theta t}\rho^{\pm N-1}\rho_t||\partial_t u|^2 dxdt \geq &\int_D \partial_t(e^{-\theta t}\rho^{\pm N-2}\rho_t^2|u|^2)dxdt\\
&+\frac{N}{4}\int_D e^{-\theta t}\rho^{\pm N-3}\rho_t^3|u|^2 dxdt\\
&+\theta\int_D e^{-\theta t}\rho^{\pm N-2}\rho_t^2|u|^2 dxdt.
\end{aligned}$$

Let us put

$$E(u) = |\partial_t u|^2 + h(t,x)|\partial_x u|^2 + cN^2\rho^{-2}\rho_t^2|u|^2 \quad (c = (4C_1)^{-1}) \tag{3.15}$$

and

$$\Gamma(u) = -(e^{-\theta t}\rho^{\pm N}E(u))dx + (e^{-\theta t}\rho^{\pm N}G_2(u))dt. \tag{3.16}$$

Since

$$\int_{\partial D}\Gamma(u) = \int_D \big[\partial_t(e^{-\theta t}\rho^{\pm N}E(u)) + \partial_x(e^{-\theta t}\rho^{\pm N}G_2(u))\big]dxdt$$

where ∂D is oriented by D, we have

Proposition 3.3. *Assume $\rho \in C^\infty(D)$, $\rho \neq 0$, $\rho_t > 0$ in D and N is even. Choose $\pm$ so that $\mp\rho^{\pm N-1}\rho_t > 0$ in D. Then we have*

$$\frac{4}{N}\int_D |e^{-\theta t}\rho^{\pm N-1}\rho_t^{-1}||pu - h_x\partial_x u|^2 dxdt$$
$$\geq \int_{\partial D}\Gamma(u) + \int_D \sigma(t,x)|e^{-\theta t}\rho^{\pm N-1}||\partial_x u|^2 dxdt$$
$$+\frac{N}{4}\int_D |e^{-\theta t}\rho^{\pm N-1}\rho_t|(E(u) - C'N\rho^{-1}\rho_{tt}|u|^2)dxdt$$
$$+\theta\int_D e^{-\theta t}\rho^{\pm N}E(u)dxdt$$

where

$$\sigma(t,x) = \frac{3N}{4}h\rho_t - 4Nh^2\rho_x^2\rho_t^{-1} - C|\rho h_t|.$$

Let $\gamma : [a,b] \ni x \mapsto (\phi(x), x)$ be a space-like curve, that is

$$1 - h(\phi(x), x)\phi'(x)^2 > 0$$

for $x \in [a,b]$. Then form $|\phi'(x)G_2(u)| \leq \phi'(x)^2 h|\partial_t u|^2 + h|\partial_x u|^2$ it follows that

$$\int_\gamma \Gamma(u) \leq 0.$$

We now introduce weight functions and regions associated to $A(t,x)$. From Lemma 3.9 we see

$$\sup_{s_0(x)\leq t\leq s_m(x)} |h(t,x)| \leq C|x|^2$$

with some $C > 0$. Then it is easy to check that one can choose small $\delta > 0$ and $T > 0$ so that the two curves $t = s_0(x)$, $t = s_m(x)$ and all pseudo-characteristic curves $t = \sigma_j(x)$ intersect $|x| = \delta(T - t)$ near the origin. We fix these $T > 0$ and $\delta > 0$.

Definition 3.8. Without restrictions we may assume that $b > 0$, $\tilde{b} > 0$ in $\tilde{\omega}_h(T)$ and $\tilde{\omega}_a(T)$ respectively since otherwise it suffices to take $-b$ and $-\tilde{b}$ instead of b and $\tilde{b}$. We define $\rho_{A,D}(t,x)$ by

$$\begin{cases} \rho_{A,D}(t,x) = \rho_j = t - \sigma_j(x) & \text{if } D = \omega_j, j = 1,2,\ldots,m, \\ \rho_{A,D}(t,x) = \rho_{m+1} = t - s_m(x) & \text{if } D = \omega(T) \text{ and } n_1 \geq 1, \\ \rho_{A,D}(t,x) = b(t,x) & \text{if } D = \omega(T), n_1 = 0 \text{ and } 2t_h^*(x) \geq t_a^*(x), \\ \rho_{A,D}(t,x) = \tilde{b}(t,x) & \text{if } D = \omega(T), n_1 = 0 \text{ and } 2t_a^*(x) \geq t_h^*(x) \end{cases}$$

where we have set

$$\omega_j = \{(t,x) \mid |x| < \delta(T-t), s_{j-1}(x) \le t \le s_j(x)\}, \quad j = 1,\dots,m,$$

$$\omega(T) = \{(t,x) \mid s_m(x) \le t \le s_m(\bar{x})\}$$

where $\bar{x}$ is defined by $s_m(\bar{x}) = T - \delta^{-1}\bar{x}$ and $T > 0$ and $\delta > 0$ are given above.

Lemma 3.21. *Let $D = \omega_j$ or $D = \omega(T)$ and $\rho = \rho_{A,D}$. Then there are $c > 0$, $C > 0$ such that, taking T small, we have*

$$\sigma(t,x) \ge cNb(t,x)^2\rho_t, \quad C\rho^{-2}\rho_t^2 \ge \rho^{-1}\rho_{tt} \quad in \quad D$$

for $N \ge N_0$.

Proof. We first study the case $n_1 \ge 1$. In this case from the definition we see that $\rho_{A,D} = t - \sigma_j$ or $t - s_m$. Note

$$\rho h_t = 2\rho b b_t.$$

From Lemma 3.16 we have $|\rho b_t| \le Cb$ in D and hence $|\rho h_t| \le Cb^2$ in D. On the other hand from Lemma 3.15 we see that

$$\sup_{0\le t\le t^*(x),|x|<\epsilon} |b(t,x)\rho_x| \to 0 \quad as \quad \epsilon \to 0 \quad if \quad \rho = t - \sigma_j(x), \ \ j \le m,$$

$$\sup_{0\le t\le T,|x|<\epsilon} |b(t,x)\rho_x| \to 0 \quad as \quad \epsilon \to 0 \quad if \quad \rho = t - s_m(x).$$

Noticing $\rho_t = 1$ it is clear that, taking T small,

$$\sigma(t,x) \ge CNb^2$$

with some $C > 0$. Since $\rho_{tt} = 0$ the second inequality is obvious.

We turn to the case $n_1 = 0$. Let $2t_h^*(x) \ge t_a^*(x)$. Recall that

$$b_t\sigma(t,x) = b^2\left[\frac{3N}{4}b_t^2 - 4Nb^2b_x^2 - Cb_t^2\right]$$

because $\rho = b$. By Lemma 3.11 we have $b_t \ge Kb > 0$ for any given K if taking T small in $\tilde{\omega}_h(T)$. Since b_x is bounded we get (Lemma 3.10)

$$b_t\sigma(t,x) \ge CNb^2b_t^2 \quad in \quad \tilde{\omega}_h(T).$$

Since $\omega(T) \subset \tilde{\omega}_h(T)$ it is clear that $\sigma(t,x) \ge CNb^2b_t$ in D. We turn to the second inequality. By Lemma 3.8 we see

$$\partial_t\left(\frac{b_t}{b}\right) \le C \quad in \quad \tilde{\omega}_h(T).$$

This shows that $b_{tt}b^{-1} \le C + b_t^2 b^{-2}$ in $\tilde{\omega}_h(T)$. From Lemma 3.11 again we have

$$b^{-2}b_t^2 \ge b^{-1}b_t \ge C \quad in \quad \tilde{\omega}_h(T)$$

taking T small and hence we get

$$b_{tt}b^{-1} \le 2b_t^2 b^{-2} \quad in \quad \tilde{\omega}_h(T).$$

Since $\omega(T) \subset \tilde{\omega}_h(T)$, we have the second inequality. Finally we study the case $n_1 = 0$ and $2t_a^*(x) \ge t_h^*(x)$. Recall

$$\tilde{b}_t \sigma(t,x) = b^2 \left[\frac{3N}{4} \tilde{b}_t^2 - 4Nb^2 \tilde{b}_x^2 - C\tilde{b}\tilde{b}_t b^{-1} b_t \right]$$

because $\rho = \tilde{b}$. Since $\tilde{b}_x$ is bounded (Lemma 3.10) and $\tilde{b} \ge b$ it follows from Lemma 3.11 that

$$b^2 \tilde{b}_x^2 \le C\tilde{b}^2 \le K^{-1}\tilde{b}_t^2 \quad in \quad \tilde{\omega}_a(T)$$

for any K taking T small. This shows that the second term can be canceled against the first term. On the other hand, since $2t_a^* \ge t_h^*$, from Lemma 3.16 we see that

$$b_t b^{-1} \le C\tilde{b}_t \tilde{b}^{-1} \quad in \quad \tilde{\omega}(T)$$

and hence $b_t b^{-1} \tilde{b}\tilde{b}_t \le C\tilde{b}_t^2$ in $\tilde{\omega}(T)$. This shows that

$$\sigma(t,x) \ge cNb^2 \tilde{b}_t^2 = cNb^2 \rho_t \quad in \quad \omega(T).$$

By Lemma 3.8 we see

$$\partial_t \left(\frac{\tilde{b}_t}{\tilde{b}} \right) \le C \quad in \quad \tilde{\omega}_a(T)$$

and hence $\tilde{b}_{tt}\tilde{b}^{-1} \le C + \tilde{b}_t^2 \tilde{b}^{-2}$ in $\tilde{\omega}_a(T)$. From Lemma 3.11 we get

$$\tilde{b}^{-2}\tilde{b}_t^2 \ge \tilde{b}^{-1}\tilde{b}_t \ge c \quad in \quad \tilde{\omega}_a(T)$$

with T small. Then one has

$$\tilde{b}_{tt}\tilde{b}^{-1} \le 2\tilde{b}_t^2 \tilde{b}^{-2} \quad in \quad \tilde{\omega}_a(T).$$

Noting $\rho = \tilde{b}$ and $\omega(T) \subset \tilde{\omega}_a(T)$ we have the desired assertion. □

We summarize what we have proved. Let us denote

$$\omega_j^+ = \{(t,x) \in \omega_j \mid t \ge \sigma_j(x)\}, \quad \omega_j^- = \{(t,x) \in \omega_j \mid t \le \sigma_j(x)\}.$$

Proposition 3.4. *We take ρ^{-N} with $\rho = \rho_{A,D}$ if $D = \omega(T)$, ρ^N with $\rho = \rho_{A,D}$ if $D = \omega_j^-$ and ρ^{-N} with $\rho = \rho_{A,D}$ if $D = \omega_j^+$. Then there is $c_1 > 0$ such that*

$$\begin{aligned}\frac{4}{N}\int_D |e^{-\theta t}\rho^{\pm N+1}\rho_t^{-1}||pu - h_x\partial_x u|^2 dxdt \ge &\int_{\partial D}\Gamma(u)\\ &+c_1 N\int_D |e^{-\theta t}\rho^{\pm N-1}\rho_t|E(u)dxdt\\ &+\theta\int_D e^{-\theta t}\rho^{\pm N}E(u)dxdt.\end{aligned}$$

Lemma 3.22. *Assume that the assumptions in Theorem 3.1 are verified, that is*

$$\Gamma(tY_\phi) \subset \frac{1}{2}\Gamma([h|a^\sharp|^2]_\phi), \quad \Gamma(tZ_\phi) \subset \frac{1}{2}\Gamma([h|a^\sharp|^2]_\phi), \quad \forall\phi \in \mathscr{C}^\pm(h|a^\sharp|^2).$$

Let $\rho = \rho_{A,D}$ and $D = \omega_j$ or $D = \omega(T)$. Then taking T small we have

$$\left|\rho(t,x)\frac{Y(t,x)}{\overline{a^\sharp}}\right|, \quad \left|\rho(t,x)\frac{Z(t,x)}{a^\sharp}\right| \le Cb(t,x)\rho_t(t,x)$$

in D.

Proof. We prove the assertion for Z because the proof for Y is just a repetition. From Proposition 3.2 with $F = h|a^\sharp|^2$ and $B = Z$ we have if $D = \omega_j$

$$|\rho(t,x)Z(t,x)| \le C|b(t,x)\tilde{b}(t,x)| \quad \text{in} \quad D.$$

On the other hand, since $|\tilde{b}(t,x)| = |a^\sharp(t,x)|$, $\rho_t = 1$ we get the desired assertion. Let $D = \omega(T)$ and $n_1 \ge 1$. Then the proof is the same. Let $D = \omega(T)$ and $n_1 = 0$. Proposition 3.2 gives

$$|Z(t,x)| \le C|\partial_t(b\tilde{b})|.$$

This shows that

$$\left|\frac{Z(t,x)}{a^\sharp(t,x)}\right| \le C\left|\frac{\partial_t(b\tilde{b})}{\tilde{b}}\right| = C\left(b_t + \frac{b\tilde{b}_t}{\tilde{b}}\right).$$

When $2t_h^*(x) \ge t_a^*(x)$ from Lemma 3.12 it follows that

$$\frac{\tilde{b}_t}{\tilde{b}} \le c\frac{b_t}{b} \quad in \quad \omega(T)$$

and hence we have

$$|\frac{Z}{a^\sharp}| \le c'(b_t + b_t) \le 2c'b_t.$$

Remarking that $\rho = b$ we get

$$|\rho\frac{Z}{a^\sharp}| \le c''b\rho_t \quad in \quad \omega(T).$$

We turn to the case $2t_a^*(x) \ge t_h^*(x)$. By Lemma 3.12 we have

$$\frac{b_t}{b} \le c\frac{\tilde{b}_t}{\tilde{b}} \quad in \quad \omega(T).$$

Hence we get

$$|\frac{Z}{a^\sharp}| \le C'(\frac{\tilde{b}+b}{\tilde{b}} + \frac{b\tilde{b}_t}{\tilde{b}}).$$

Since $\rho = \tilde{b}$ we see that

$$|\rho\frac{Z}{a^\sharp}| \le C''b\rho_t \quad in \quad \omega(T)$$

and hence the assertion. □

Lemma 3.23. *Let $D = \omega_j$ or $D = \omega(T)$ and $\rho = \rho_{A,D}$. Then we have*

$$\left|\frac{\partial_t a^\sharp}{a^\sharp}\right| \le C\frac{\rho_t}{\rho}, \quad \left|\partial_t\left(\frac{\partial_t a^\sharp}{a^\sharp}\right)\right| \le C\left(\frac{\rho_t}{\rho}\right)^2 \quad in \quad D.$$

Proof. Let $D = \omega_j$. Since

$$\frac{\partial_t a^\sharp}{a^\sharp} = \sum \frac{1}{t - t_j(x)} + \frac{\Psi_t}{\Psi}$$

and for $(t, x) \in \omega_j$ we have

$$|t - \sigma_j(x)| \le |t - \mathrm{Re}\, t_\mu(x)| \le |t - t_\mu(x)|$$

for all μ. It is clear that

$$|\rho(t,x)\frac{\partial_t a^\sharp}{a^\sharp}| \leq C \quad in \quad \omega_j$$

taking T small. This proves the assertion because $\rho_t = 1$. Similar arguments prove the second inequality when $D = \omega_j$ or $D = \omega(T)$, $n_1 \geq 1$. Let $D = \omega(T)$ and $n_1 = 0$. Assume that $2t_a^*(x) \geq t_h^*(x)$. Then from Lemma 3.13 it follows that

$$|\frac{\partial_t a^\sharp}{a^\sharp}| \leq C\frac{\tilde{b}_t}{\tilde{b}}, \quad |\partial_t(\frac{\partial_t a^\sharp}{a^\sharp})| \leq C\Big(\frac{\tilde{b}_t}{\tilde{b}}\Big)^2 \quad in \quad \tilde{\omega}_a(T)$$

and this proves the assertion because $\tilde{b} = \rho$. When $2t_h^*(x) \geq t_a^*(x)$ then using

$$\frac{\tilde{b}_t}{\tilde{b}} \leq C\frac{b_t}{b} \quad in \quad \tilde{\omega}(T)$$

(Lemma 3.12) we get the desired assertion. □

We pass to $ML^\sharp u = f$. Assume that u verifies $ML^\sharp u = f$. Recall that

$$\begin{aligned} ML^\sharp &= \begin{bmatrix} p + (Z/a^\sharp - h_x)\partial_x & 0 \\ 0 & p + (Y/\overline{a^\sharp} - h_x)\partial_x \end{bmatrix} + \tilde{R}\partial_t + \tilde{S} \\ &= (p - h_x\partial_x)I + \tilde{Q}\partial_x + \tilde{R}\partial_t + \tilde{S} \end{aligned}$$

where

$$\tilde{Q} = \begin{bmatrix} Z/a^\sharp & 0 \\ 0 & Y/\overline{a^\sharp} \end{bmatrix}, \quad \tilde{R} = \tilde{C} + A_x^\sharp + B^\sharp + {}^{co}B^\sharp, \quad \tilde{S} = MB^\sharp,$$

$$\tilde{C} = -\text{diag}\left(\frac{\partial_t \dot{a}^\sharp}{a^\sharp}, \frac{\partial_t \overline{a^\sharp}}{\overline{a^\sharp}}\right).$$

We assume that the hypothesis in Lemma 3.22 holds.

Lemma 3.24. *Let $D = \omega_j$ or $D = \omega(T)$ and $\rho = \rho_{A,D}$. Then we have*

$$\rho^2\rho_t^{-1}|\tilde{Q}|^2 \leq C(\tilde{Q})\rho_t b(t,x)^2 \quad in \quad D$$

with some $C(\tilde{Q})$.

Proof. It is clear from Lemma 3.22. □

Lemma 3.25. *Let $D = \omega_j$ or $D = \omega(T)$ and $\rho = \rho_{A,D}$. Then we have*

$$\rho^2\rho_t^{-1}|\tilde{R}|^2 \leq C(\tilde{R})\rho_t, \quad \rho^2\rho_t^{-1}|\tilde{S}|^2 \leq C(\tilde{S})\rho_t$$

with some $C(\tilde{R}) > 0$, $C(\tilde{S}) > 0$.

Proof. It is clear from Lemma 3.23. □

Note that

$$\begin{aligned}\rho^{\pm N+1}\rho_t^{-1}|pu-h_x\partial_x u|^2 &\le 2\rho^{\pm N+1}\rho_t^{-1}|ML^\sharp u|^2\\&\quad+\rho^{\pm N+1}\rho_t^{-1}\{C|\tilde{Q}|^2|\partial_x u|^2+C|\tilde{R}|^2|\partial_t u|^2+C|\tilde{S}|^2|u|^2\}\\&\le 2\rho^{\pm N+1}\rho_t^{-1}|ML^\sharp u|^2+\rho^{\pm N-1}\rho_t\{C(\tilde{Q})b^2|\partial_x u|^2\\&\quad+C(\tilde{R})|\partial_t u|^2+C(\tilde{S})|u|^2\}\end{aligned}$$

by Lemmas 3.24 and 3.25. Taking N and θ so that $c_1N^2\ge 8C(\tilde{Q}), 8C(\tilde{R})$ and $\theta\ge 8C(\tilde{S})$, it follows from Proposition 3.4 that

$$\begin{aligned}\frac{8}{N}\int_D e^{-\theta t}\rho^{\pm N+1}\rho_t^{-1}|ML^\sharp u|^2dxdt &\ge \int_{\partial D}\Gamma(u)\\&\quad+\frac{c_1}{2}\int_D e^{-\theta t}\rho^{\pm N-1}\rho_t E(u)dxdt\\&\quad+\frac{\theta}{2}\int_D e^{-\theta t}\rho^{\pm N}E(u)dxdt\end{aligned}$$

where $D=\omega_j^+$ or $D=\omega_j^-$ or $D=\omega(T)$.

3.9 Energy Estimates of Higher Order Derivatives

We start with

Lemma 3.26. *Let $D=\omega_j$ or $D=\omega(T)$ and $\rho=\rho_{A,D}$. Then we make $\rho\rho_t^{-1}$ as small as we please in D taking T small.*

Proof. Clear. □

Lemma 3.27. *Let $D=\omega_j^+$ or $D=\omega_j^-$ or $D=\omega(T)$ and $\rho=\rho_{A,D}$. Then we have*

$$\begin{aligned}&c_3N^{-1}\int_D e^{-\theta t}\rho^{\pm N+1}\rho_t^{-1}|(M+nA_x^\sharp)(L^\sharp-nA_x^\sharp)u|^2dxdt\\&\ge\int_{\partial D}\Gamma(u)+c_2\int_D e^{-\theta t}\rho^{\pm N-1}\rho_t E(u)dxdt+c_2\theta\int_D e^{-\theta t}\rho^{\pm N}E(u)dxdt\end{aligned}$$

for any $N\ge N_0(\tilde{Q},\tilde{R})+n$, $\theta\ge\theta_0(\tilde{S},n)$, $n\in\mathbb{N}$.

Proof. Note that

$$(M + nA_x^\sharp)(L^\sharp - nA_x^\sharp) = p - h_x\partial_x + \hat{Q}\partial_x + \tilde{R}\partial_t + \hat{S}$$

where $\hat{Q} = \tilde{Q} - nh_x I$, $\hat{S} = \tilde{S} + nA_x^\sharp B^\sharp - nMA_x^\sharp - n^2(A_x^\sharp)^2$ since $A^\sharp A_x^\sharp + A_x^\sharp A^\sharp = h_x$. Let $\epsilon > 0$ be given. Taking T small one can assume that

$$\rho^2\rho_t^{-1}|nh_x|^2 \leq \epsilon n^2 \rho_t b^2$$

since $h_x = 2bb_x$ and b_x is bounded by Lemma 3.26.

It is clear that $C(\hat{Q}) \leq 2(C(\tilde{Q}) + \epsilon n^2)$ and $C(\hat{S}) \leq 2(C(\tilde{S}) + cn^4)$ with some $c > 0$. Then taking $\epsilon > 0$, $N_0(\tilde{Q}, \tilde{R})$, $\theta_0(\tilde{S}, n)$ suitably we have

$$N \geq N_0(\tilde{Q}, \tilde{R}) + n,\ \theta \geq \theta_0(\tilde{S}, n) \Longrightarrow c_1 N^2 \geq 8C(\hat{Q}),\ 8C(\tilde{R}),\ \ \theta \geq 8C(\hat{S}).$$

Then we get the assertion applying the previous inequality. □

Proposition 3.5. *One can find $N_0 > 0$ such that for any $n \in \mathbb{N}$ there is $\theta_1(n)$ such that with $D = \omega_j^+$, $D = \omega_j^-$, $D = \omega(T)$ we have*

$$\sum_{k+l\leq n}\int_D |e^{-\theta t}\rho^{\pm N}||\partial_t^k\partial_x^l u|^2 dxdt + \sum_{l\leq n}\int_{\partial D}\Gamma(\partial_x^l u)$$
$$\leq C\sum_{k+l\leq n+1}\int_D |e^{-\theta t}\rho^{\pm N}||\partial_t^k\partial_x^l L^\sharp u|^2 dxdt$$
$$+C\sum_{k+l\leq n}\int_D |e^{-\theta t}\rho^{\pm N-1}||\partial_t^k\partial_x^l L^\sharp u|^2 dxdt$$

for any $N \geq N_0 + n$, $\theta \geq \theta_1(n)$ where $\rho = \rho_{A,D}$.

Proof. Take $N_0 = N_0(\tilde{Q}, \tilde{R})$. Then from Lemma 3.27 it follows that the integral

$$c_2\theta\int_D |e^{-\theta t}\rho^{\pm N}|E(\partial_x^q u)dxdt + \int_{\partial D}\Gamma(\partial_x^q u)$$

is estimated by

$$c_3 N^{-1}\int_D |e^{-\theta t}\rho^{\pm N+1}\rho_t^{-1}|\left|(M + qA_x^\sharp)(L^\sharp - qA_x^\sharp)\partial_x^q u\right|^2 dxdt.$$

Since $|\tilde{C}| \leq c(\rho_t/\rho)$ in D and

$$(L^\sharp - qA_x^\sharp)\partial_x^q u = \partial_x^q L^\sharp u - \sum_{j=0}^{q-1} B_j \partial_x^j u$$

this is bounded by constant (depends on q) times

$$\int_D |e^{-\theta t}\rho^{\pm N+1}\rho_t^{-1}|\big(|\partial_t\partial_x^q L^\sharp u|^2 + |\partial_x^{q+1}L^\sharp u|^2\big)dxdt$$

$$+\int_D |e^{-\theta t}\rho^{\pm N-1}\rho_t||\partial_x^q L^\sharp u|^2 dxdt$$

$$+\sum_{i+j\le q, i\le 1}\int_D |e^{-\theta t}\rho^{\pm N+1}\rho_t^{-1}||\partial_t^i\partial_x^j u|^2 dxdt$$

$$+\sum_{j\le q-1}\int_D |e^{-\theta t}\rho^{\pm N-1}\rho_t||\partial_x^j u|^2 dxdt.$$

The third and fourth terms are bounded by

$$C\sum_{j=0}^{q}\int_D |e^{-\theta t}\rho^{\pm N}|E(\partial_x^j u)dxdt.$$

Hence, taking θ large and summing up over $q = 0, 1, \dots, n$ we get

$$\frac{c_2}{2}\theta\sum_{j=0}^{n}\int_D |e^{-\theta t}\rho^{\pm N}||\partial_x^j u|^2 dxdt + \sum_{j=0}^{n}\int_{\partial D}\Gamma(\partial_x^j u)$$

$$\le C\sum_{j=0}^{n}\int_D |e^{-\theta t}\rho^{\pm N}|\big(|\partial_t\partial_x^j L^\sharp u|^2 + |\partial_x^{j+1}L^\sharp u|^2\big)dxdt$$

$$+C\sum_{j=0}^{n}\int_D |e^{-\theta t}\rho^{\pm N-1}||\partial_x^j L^\sharp u|^2 dxdt$$

where we have used $E(u) \ge c|u|^2$ with some $c > 0$. Noting that

$$\partial_t^k\partial_x^l u = \partial_t^{k-1}\partial_x^l L^\sharp u + \sum_{i\le k-1, j\le l+1} c_{ij}\partial_t^i\partial_x^j u$$

we consider

$$\sum_{k+l\le n, k\ge 1}\lambda^k\mu^l|\partial_t^k\partial_x^l u|^2$$

with $\lambda > 0$, $\mu > 0$ small and $\sum_l \mu^l < +\infty$. Since

$$\sum_{k+l\le n, k\ge 1} \lambda^k \mu^l \sum_{i\le k-1, j\le l+1} |\partial_t^i \partial_x^j u|^2$$
$$\le C \sum_{j=0}^{n} |\partial_x^j u|^2 + C\lambda\mu^{-1} \sum_{i+j\le n, i\ge 1} \lambda^i \mu^j |\partial_t^i \partial_x^j u|^2$$

taking $\lambda\mu^{-1}$ small enough so that the second term in the right-hand side cancels against the left-hand side we get

$$\sum_{k+l\le n, k\ge 1} \lambda^k \mu^l |\partial_t^k \partial_x^l u|^2 \le C \sum_{k+l\le n, k\ge 1} \lambda^k \mu^l |\partial_t^{k-1} \partial_x^l L^\sharp u|^2 + C \sum_{j=0}^{n} |\partial_x^j u|^2.$$

Now multiplying $|e^{-\theta t} \rho^{\pm N}|$ and integrating over D we have

$$\sum_{k+l\le n, k\ge 1} \int_D |e^{-\theta t} \rho^{\pm N}| |\partial_t^k \partial_x^l u|^2 dxdt$$
$$\le C \sum_{k+l\le n, k\ge 1} \int_D |e^{-\theta t} \rho^{\pm N}| |\partial_t^{k-1} \partial_x^l L^\sharp u|^2 dxdt$$
$$+ C \sum_{j=0}^{n} \int_D |e^{-\theta t} \rho^{\pm N}| |\partial_x^j u|^2 dxdt.$$

Since we have already estimated

$$\theta \sum_{j=0}^{n} \int_D |e^{-\theta t} \rho^{\pm N}| |\partial_x^j u|^2 dxdt$$

plugging this estimate into above inequality, we get the desired estimate. □

Remark. We will have the same estimates in $D = \tilde{\omega}(-T)$ with

$$\begin{cases} \rho(t,x) = \rho_0 = t - s_0(x) \text{ if } n_1 \ge 1, \\ \rho(t,x) = b(t,x) \text{ if } n_1 = 0 \text{ and } 2t_h^*(x) \ge t_a^*(x), \\ \rho(t,x) = \tilde{b}(t,x) \text{ if } n_1 = 0 \text{ and } 2t_a^*(x) \ge t_h^*(x) \end{cases}$$

assuming $b < 0$, $\tilde{b} < 0$ in $\tilde{\omega}_h(-T)$ and $\tilde{\omega}_a(-T)$ respectively.

3.10 Weighted Energy Estimates

In this section we collect weighted energy estimates in each subregion containing a pseudo-characteristic curve to get weighted energy estimates near the origin. Recall that $A^\sharp(0,0) = 0$ because $a^\sharp(0,0) = 0$.

Proposition 3.6. *Let $\varphi(x)$, $\varepsilon(x)$ be smooth apart from the origin such that $s_0(x) \le \varepsilon(x) \le \varphi(x) \le s_m(x)$. Put $r(t,x) = t - \varphi(x)$, $L^\sharp u = f$. Assume that*

$$\begin{cases} |A^\sharp(t,x)| \le C|x| \text{ in } |t| \le t^*(x), \\ |\varphi^{(k)}(x)|, |\varepsilon^{(k)}(x)| \le C_k|x|^{\delta-k} \text{ with some } \delta > 0 \text{ for } k = 0,1,\dots,Q, \\ \partial_t^k u(\varepsilon(x),x) = 0, \partial_t^k f(\varepsilon(x),x) = 0 \text{ for } k = 0,1,\dots,Q. \end{cases}$$

Then for any $q \in \mathbb{N}$ with $2q+1 \le Q$ there is a $w_q(t,x)$ verifying the followings

$$L^\sharp(u - w_q) = O(r^q), \quad u - w_q = O(r^{q+1})$$

where

$$|\partial_t^k \partial_x^l (u - w_q)|^2 \le C|x|^{-2l}|r|^{2(q+1-k-l)} t^*(x)^{2(Q-q-k-l-1)+1} \times \sum_{l_1+l_2 \le l} \int_{\varepsilon(x)}^{t^*(x)} |\partial_t^{Q+1+l_1}\partial_x^{l_2} u|^2 dt$$

for $|t| \le t^*(x)$, $k+l+q+1 \le Q$, $k+l \le q+1$,

$$\begin{aligned} &|\partial_t^k \partial_x^l (L^\sharp u - L^\sharp w_q)|^2 \\ &\le C|x|^{-2(l+1)}|r|^{2(q-k-l)} t^*(x)^{2(Q-q-l-1)+1} \sum_{l_1+l_2 \le l+1} \int_{\varepsilon(x)}^{\varphi(x)} |\partial_t^{Q+1+l_1}\partial_x^{l_2} u|^2 dt \\ &+ C|x|^{-2l} t^*(x)^{2(Q-q-l-1)+1} \sum_{l_1+l_2 \le l} \int_{\varepsilon(x)}^{\varphi(x)} |\partial_t^{Q+1+l_1}\partial_x^{l_2} f|^2 dxdt \\ &+ C|t-\varepsilon|^{2(Q-k)} \int_\varepsilon^t |\partial_t^{Q+1}\partial_x^l f|^2 dxdt \end{aligned}$$

for $q+l+1 \le Q$, $k+l \le q$ and

$$|\partial_t^k \partial_x^l w_q|^2 \le C|x|^{-2l} t^*(x)^{2(Q-q-l-1)+1} \sum_{l_1+l_2 \le l} \int_{\varepsilon(x)}^{\varphi(x)} |\partial_t^{Q+1+l_1}\partial_x^{l_2} u|^2 dt$$

for $q+l \le Q$.

We first show the following lemma.

Lemma 3.28. *Let $\psi(t,x) \in C^\infty$ be defined near the origin. Then one can write*

$$\psi(t,x) = \sum_{j=0}^{q} \psi_j(x) r^j + r^{q+1}\psi_q(t,x)$$

where $\psi_j(x)$, $\psi_q(t,x)$ verifies

$$|\partial_x^l \psi_j(x)| \le C_{jl}|x|^{-l}, \quad l = 0,1,\ldots,$$
$$|\partial_t^k \partial_x^l \psi_q(t,x)| \le C_{qkl}|x|^{-l}, \quad l = 0,1,\ldots.$$

Moreover if $\partial_t^\alpha \psi(\varepsilon(x),x) = 0$, $\alpha = 0,1,\ldots,Q$ then we have

$$|\partial_x^l \psi_j|^2 \le C \sum_{l_1+l_2\le l} |x|^{-2l}|\varphi-\varepsilon|^{2(Q-j-l)+1}\int_{\varepsilon(x)}^{\varphi(x)} |\partial_t^{Q+1+l_1}\partial_x^{l_2}\psi(\tau,x)|^2 d\tau$$

for $j + l \le Q$ and

$$|\partial_t^k \partial_x^l \psi_q|^2 \le C \sum_{l_1+l_2\le l} |x|^{-2l} t^*(x)^{2(Q-l-q-k-1)+1}\int_{\varepsilon(x)}^{t^*(x)} |\partial_t^{Q+1+l_1}\partial_x^{l_2}\psi|^2 d\tau$$

for $|t| \le t^*(x)$, $k+l+q+1 \le Q$.

Proof. Since it is clear that $\psi_j(x)$ and $\psi_q(t,x)$ are given by

$$\psi_j(x) = \frac{1}{j!}\partial_t^j \psi(\varphi(x),x),$$
$$\psi_q(t,x) = \frac{1}{q!}\int_0^1 (1-\tau)^q(\partial_t^{q+1}\psi)(\varphi(x)+\tau(t-\varphi(x)),x)d\tau$$

the first two inequalities are clear. Assume that $\partial_t^\alpha \psi(\varepsilon(x),x) = 0$, $\alpha = 0,1,\ldots,Q$, then we have

$$\partial_t^j \psi(t,x) = \frac{1}{(Q-j)!}\int_\epsilon^t (t-s)^{Q-j}\partial_t^{Q+1}\psi(s,x)ds.$$

This shows that

$$\psi_j(x) = \frac{(\varphi(x)-\varepsilon(x))^{Q-j+1}}{(Q-j)!j!}\int_0^1 (1-s)^{Q-j}\partial_t^{Q+1}\psi(s(\varphi-\varepsilon)+\varepsilon,x)ds.$$

Noting that

$$\left|\int_0^1 \partial_t^{Q+1+l_1}\partial_x^{l_2}\psi(s(\varphi-\varepsilon)+\varepsilon,x)ds\right|^2$$
$$\le C|\varphi-\varepsilon|^{-1}\int_\varepsilon^\varphi |\partial_t^{Q+1+l_1}\partial_x^{l_2}\psi(\tau,x)|^2 d\tau$$

we get the third inequality. Remarking that

$$\partial_t^k\psi_q(t,x)=\frac{1}{q!}\int_0^1 \tau^k(\partial_t^{q+k+1}\psi)(\varphi(x)+\tau(t-\varphi(x)),x)d\tau,$$
$$\partial_t^{q+k+1}\psi(t,x)=\frac{1}{(Q-q-k-1)!}\int_0^t (t-u)^{Q-k-q-1}\partial_t^{Q+1}\psi(u,x)du$$

we see

$$\partial_t^k\psi_q(t,x)=c\int_0^1 d\tau\int_0^1 \tau^k(\tau t+(1-\tau)\varphi-\varepsilon)^{Q-q-k}(1-u)^{Q-q-k-1}$$
$$\times\partial_t^{Q+1}\psi((\tau t+(1-\tau)\varphi-\varepsilon)u+\varepsilon,x)du.$$

Since $|\tau t+(1-\tau)\varphi-\varepsilon|\le|t^*-\varepsilon|\le 6t^*(x)$ for $|t|\le t^*(x)$ then

$$|\partial_t^k\partial_x^l\psi_q|\le C\sum_{l_1+l_2\le l}\int_0^1 d\tau\int_\varepsilon^{\tau t+(1-\tau)\varphi} t^*(x)^{Q-q-k-l-1}|x|^{-l}$$
$$\times|\partial_t^{Q+1+l_1}\partial_x^{l_2}\psi(u,x)|du$$

and hence

$$|\partial_t^k\partial_x^l\psi_q|^2\le C|x|^{-2l}t^*(x)^{2(Q-q-k-l-1)+1}$$
$$\times\sum_{l_1+l_2\le l}\int_\varepsilon^{t^*}|\partial_t^{Q+1+l_1}\partial_x^{l_2}\psi(u,x)|^2du$$

which is the desired inequality. □

Proof of Proposition 3.6. From Lemma 3.28 one can write

$$u(t,x)=\sum_{j=0}^q u_j(x)r^j+r^{q+1}V(t,x),\ f(t,x)=\sum_{j=0}^{q-1}f_j(x)r^j+r^qF_{q-1}(t,x).$$

Let us put

$$w_q(t,x)=\sum_{j=0}^q u_j(x)r(t,x)^j.$$

From Lemma 3.28 it follows that

$$|\partial_t^k\partial_x^l V|^2 \le C\sum_{l_1+l_2\le l}|x|^{-2l}t^*(x)^{2(Q-l-q-k-1)+1}\int_\varepsilon^{t^*}|\partial_t^{Q+1+l_1}\partial_x^{l_2}u|^2d\tau$$

for $|t|\le t^*(x)$, $k+l+q+1\le Q$. Hence we get

$$|\partial_t^k\partial_x^l(r^{q+1}V)|^2 \le C|x|^{-2l}|r|^{2(q+1-k-l)}t^*(x)^{2(Q-q-k-l-1)+1}$$
$$\times\sum_{l_1+l_2\le l}\int_\varepsilon^{t^*}|\partial_t^{Q+1+l_1}\partial_x^{l_2}u|^2dt$$

for $|t|\le t^*(x)$, $k+l+q+1\le Q$, $k+l\le q+1$. It is clear that one can write $L^\sharp(u-w_q)=r^qF$. We show the third estimate. From Lemma 3.28 we see

$$|u_j^{(l)}|^2\le C|x|^{-2l}|\varphi-\varepsilon|^{2(Q-j-l)+1}\sum_{l_1+l_2\le l}\int_\varepsilon^{\varphi}|\partial_t^{Q+1+l_1}\partial_x^{l_2}u|^2dt$$

for $j+l\le Q$ where $u_j^{(l)}=\partial_x^l u_j$. Since

$$|\partial_t^k\partial_x^l w_q|\le C\sum_{0\le j\le q,l_1+l_2=l}|u_j^{(l_1)}||x|^{-l_2}$$

then noting $|\varphi-\varepsilon|\le 6t^*(x)$ we have

$$|\partial_t^k\partial_x^l w_q|^2\le C|x|^{-2l}t^*(x)^{2(Q-l-q)+1}\sum_{l_1+l_2\le l}\int_\varepsilon^{\varphi}|\partial_t^{Q+1+l_1}\partial_x^{l_2}u|^2d\tau.$$

This is the third assertion. Finally we prove the second estimate. From $L^\sharp u=f$ and $L^\sharp(u-w_q)=r^qF$ we see $L^\sharp w_q=f-r^qF$. Hence we have

$$L^\sharp w_q=\sum_{j=0}^{q-1}f_j(x)r^j \mod O(r^q).$$

We now study $L^\sharp w_q$.

$$L^\sharp w_q=\sum_{j=0}^{q-1}f_jr^j+\sum_{\mu\ge q}\left(\sum_{i+j=\mu,i,j\le q}-A_i^\sharp u_j'+B_i^\sharp u_j+\sum_{i+j=\mu+1,i,j\le q}jA_i^\sharp u_j\varphi'\right)r^\mu$$
$$+r^{q+1}A_q\left(\sum_{j=0}^q-u_j'r^j+ju_jr^{j-1}\varphi'\right)+r^{q+1}B_q(\sum_{j=0}^q u_jr^j).$$

Note that

$$\left|\partial_t^k\partial_x^l \sum_{i+j=\mu,i,j\le q} -A_i^\sharp u_j' r^\mu + B_i^\sharp u_j r^\mu\right|^2$$

is bounded by

$$\sum_{l_1+l_2=l,k_1\le k} r^{2(\mu-k_1-l_1)}|x|^{-2(l+1)}|\varphi-\varepsilon|^{2(Q-j-l_2-1)+1}$$
$$\times \sum_{j_1+j_2\le l_2+1}\int_\varepsilon^\varphi |\partial_t^{Q+1+j_1}\partial_x^{j_2}u|^2dt$$

and hence by

$$r^{2(\mu-k-l)}|x|^{-2(l+1)}t^*(x)^{2(Q-q-l-1)+1}\sum_{l_1+l_2\le l+1}\int_\varepsilon^\varphi |\partial_t^{Q+1+l_1}\partial_x^{l_2}u|^2dt. \tag{3.17}$$

Similarly the term

$$\left|\partial_t^k\partial_x^l \sum_{i+j=\mu+1,i,j\le q} A_i^\sharp u_j\varphi' r^\mu\right|^2$$

is estimated by

$$r^{2(\mu-k-l)}|x|^{-2(l+1)}t^*(x)^{2(Q-q-l-1)+1}\sum_{l_1+l_2\le l+1}\int_\varepsilon^\varphi |\partial_t^{Q+1+l_1}\partial_x^{l_2}u|^2dt$$

and

$$|\partial_t^k\partial_x^j\sum_{j=0}^q r^{q+1}A_q(u_j'r^j - ju_j\varphi' r^{j-1})|^2$$

is bounded by

$$r^{2(q+1-k-l)}|x|^{-2(l+1)}t^*(x)^{2(Q-q-l-1)+1}\sum_{l_1+l_2\le l+1}\int_\varepsilon^\varphi |\partial_t^{Q+1+l_1}\partial_x^{l_2}u|^2dt. \tag{3.18}$$

One can estimate the term

$$|\partial_t^k\partial_x^l\sum_{j=0}^q r^{q+1}B_q u_j r^j|^2$$

by the same argument. Thus we get

$$|\partial_t^k \partial_x^l (L^\sharp w_q - \sum_{j=0}^{q-1} f_j r^j)|^2 \le C r^{2(q-k-l)} |x|^{-2(l+1)} t^*(x)^{2(Q-q-l-1)+1}$$
$$\times \sum_{l_1+l_2 \le l+1} \int_\varepsilon^\varphi |\partial_t^{Q+1+l_1} \partial_x^{l_2} u|^2 dt.$$

Since

$$L^\sharp u - L^\sharp w_q = -(L^\sharp w_q - \sum_{j=0}^{q-1} f_j r^j) + r^q F_{q-1}$$

it remains to estimate $|\partial_t^k \partial_x^l r^q F_{q-1}|$. From Lemma 3.28 it follows that

$$|\partial_t^k \partial_x^l r^q F_{q-1}|^2 \le C|\partial_t^k \partial_x^l f|^2 + C|\partial_t^k \partial_x^l \sum_{j=0}^{q-1} f_j r^j|^2$$
$$\le C|\partial_t^k \partial_x^l f|^2 + C|x|^{-2l} t^*(x)^{2(Q-q-l)+1} \sum_{l_1+l_2 \le l} \int_\varepsilon^\varphi |\partial_t^{Q+1+l_1} \partial_x^{l_2} f|^2 dt.$$

Noting

$$|\partial_t^k \partial_x^l f|^2 \le C|t-\varepsilon|^{2(Q-k)} \int_\varepsilon^t |\partial_t^{Q+1} \partial_x^l f|^2 dt$$

we conclude the proof. □

We introduce several notations.

$$\Omega_\nu = \{(t,x) \mid |x| \le \delta(T-t), s_0(x) \le t \le s_\nu(x)\}, \quad \nu = 0, \ldots, m,$$
$$\tilde{\Omega}_{\nu+1} = \{(t,x) \mid |x| \le \delta(T-t), s_0(x) \le t \le \sigma_{\nu+1}(x)\}, \quad \nu = 0, \ldots, m-1$$

and

$$\omega_\nu^- = \{(t,x) \mid |x| \le \delta(T-t), s_{\nu-1}(x) \le t \le \sigma_\nu(x)\}, \quad \nu = 1, \ldots, m,$$
$$\omega_\nu^+ = \{(t,x) \mid |x| \le \delta(T-t), \sigma_\nu(x) \le t \le s_\nu(x)\}, \quad \nu = 1, \ldots, m.$$

Here we note that

$$\tilde{\Omega}_{\nu+1} = \Omega_\nu \cup \omega_{\nu+1}^-, \quad \Omega_{\nu+1} = \tilde{\Omega}_{\nu+1} \cup \omega_{\nu+1}^+.$$

Now we introduce the inductive hypothesis.

INDUCTIVE HYPOTHESIS: For any $n \in \mathbb{N}$ there are $Q_\nu = Q_\nu(n) \geq n$ and $q_\nu = q_\nu(n) \geq n$ such that

$$(H_\nu) \quad \begin{cases} L^\sharp u = f, \;\; \partial_t^\alpha u(s_0(x), x) = 0, \;\; \partial_t^\alpha f(s_0(x), x) = 0, \;\; \alpha = 0, 1, \ldots, Q_\nu \\ \Longrightarrow \sum_{k+l \leq n} \int_{\Omega_\nu} |\partial_t^k \partial_x^l u|^2 dxdt \leq C \sum_{k+l \leq q_\nu(n)} \int_{\Omega_m} |\partial_t^k \partial_x^l f|^2 dxdt. \end{cases}$$

Let $\kappa > 0$ be a positive number such that

$$t^*(x) = O(|x|^\kappa).$$

In Proposition 3.6 we take $\varphi = s_\nu$ and construct w_q and study the equation

$$L^\sharp(u - w_q) = f - L^\sharp w_q$$

in $\Omega = \omega_{\nu+1}^-$. In Proposition 3.4, taking $N = 2(N_0 + n)$, $\theta = \theta_1(n + N/2)$ we get

$$\begin{aligned} &\sum_{l \leq n+N/2} \int_{\partial\omega_{\nu+1}^-} \Gamma(\partial_x^l(u - w_q)) \\ &\quad + \sum_{k+l \leq n+N/2} \int_{\omega_{\nu+1}^-} e^{-\theta t} |\rho_{\nu+1}^N| |\partial_t^k \partial_x^l (u - w_q)|^2 dxdt \\ &\quad \leq C \sum_{k+l \leq n+1+N/2} \int_{\omega_{\nu+1}^-} e^{-\theta t} |\rho_{\nu+1}^N| |\partial_t^k \partial_x^l L^\sharp(u - w_q)|^2 dxdt \\ &\quad + C \sum_{k+l \leq n+N/2} \int_{\omega_{\nu+1}^-} e^{-\theta t} |\rho_{\nu+1}^{N-1}| |\partial_t^k \partial_x^l L^\sharp(u - w_q)|^2 dxdt. \end{aligned}$$

Taking q, Q so that

$$2(q - n) \geq N + 1, \quad 2\kappa(Q - q - n - 1 - N/2) \geq 2n \tag{3.19}$$

then Proposition 3.6 shows

$$\partial_t^\alpha \partial_t^k \partial_x^l (u - w_q)(s_\nu(x), x) = 0, \;\; k + l \leq n, \; \alpha \leq N/2 + 1$$

because we have $q+1-(k+\alpha)-l \geq q+1-(N/2+n+1) = q-(N/2+n) > 0$ and $2\kappa(Q-q-k-\alpha-l-1)-2l \geq 2\kappa(Q-q-n-N/2-2)-2l \geq 2n-2l \geq 0$.

Lemma 3.29. *Assume that $(\partial_t^\alpha u)(s_\nu(x), x) = 0$, $\alpha = 0, 1, \ldots, p + N/2 + 1$. Then there is $C(N) > 0$ such that*

$$\int_{\omega_{\nu+1}^-} e^{-\theta t} |\rho_{\nu+1}^N| |\partial_t^{p+N/2} u|^2 dxdt \geq C(N) \int_{\omega_{\nu+1}^-} e^{-\theta t} |\partial_t^p u|^2 dxdt.$$

Proof. Note that

$$\partial_t|\partial_t^p u|^2 = \partial_t^{p+1}u\cdot\overline{\partial_t^p u} + \overline{\partial_t^{p+1}u}\cdot\partial_t^p u.$$

Multiply $-\rho^{2q+1}(\rho = \rho_{\nu+1})$ on the equation to get

$$\begin{aligned}&-\partial_t(\rho^{2q+1}|\partial_t^p u|^2) + (2q+1)\rho^{2q}|\partial_t^p u|^2\\ &= -\rho^{2q+1}(\partial_t^{p+1}u\cdot\overline{\partial_t^p u} + \overline{\partial_t^{p+1}u}\cdot\partial_t^p u).\end{aligned}$$

Integrating over $\omega_{\nu+1}^-$ we get

$$\begin{aligned}&-\int_{\omega_{\nu+1}^-}\partial_t(\rho^{2q+1}|\partial_t^p u|^2)dxdt + (2q+1)\int_{\omega_{\nu+1}^-}\rho^{2q}|\partial_t^p u|^2 dxdt\\ &\leq 2\int_{\omega_{\nu+1}^-}\rho^{2q+2}|\partial_t^{p+1}u|^2 dxdt + \frac{1}{2}\int_{\omega_{\nu+1}^-}\rho^{2q}|\partial_t^p u|^2 dxdt\end{aligned}$$

so that

$$\begin{aligned}&(2q+\frac{1}{2})\int_{\omega_{\nu+1}^-}\rho^{2q}|\partial_t^p u|^2 dxdt - \int_{\partial\omega_{\nu+1}^-}(\rho^{2q+1}|\partial_t^p u|^2)dx\\ &\leq 2\int_{\omega_{\nu+1}^-}\rho^{2q+2}|\partial_t^{p+1}u|^2 dxdt.\end{aligned}$$

Since $\partial_t^p u = 0$ on $t = s_\nu(x)$ and $t = \sigma_{\nu+1}(x)$, $|x| = \delta(T-t)$ are space-like curves we get

$$\int_{\partial\omega_{\nu+1}^-}(\rho^{2q+1}|\partial_t^p u|^2)dx \leq 0.$$

Hence we have

$$(2q+\frac{1}{2})\int_{\omega_{\nu+1}^-}\rho^{2q}|\partial_t^p u|^2 dxdt \leq 2\int_{\omega_{\nu+1}^-}\rho^{2q+2}|\partial_t^{p+1}u|^2 dxdt.$$

Inductively we get the assertion starting with $q = 0$. □

Applying Lemma 3.29 we get

$$\begin{aligned}&\sum_{k+l\leq n}\int_{\omega_{\nu+1}^-}e^{-\theta t}|\partial_t^k\partial_x^l(u-w_q)|^2 dxdt\\ &\leq C\sum_{k+l\leq n+1+N/2}\int_{\omega_{\nu+1}^-}e^{-\theta t}|\rho_{\nu+1}^N||\partial_t^k\partial_x^l L^\sharp(u-w_q)|^2 dxdt\\ &\quad + C\sum_{k+l\leq n+N/2}\int_{\omega_{\nu+1}^-}e^{-\theta t}|\rho_{\nu+1}^{N-1}||\partial_t^k\partial_x^l L^\sharp(u-w_q)|^2 dxdt.\end{aligned}$$

Assuming that q, Q verify

$$2\kappa(Q-q-l-1) \geq 2(n+2+\frac{N}{2}) \tag{3.20}$$

we have from Proposition 3.6 that

$$\sum_{k+l\leq n+1+N/2} \int_{\omega^-_{\nu+1}} e^{-\theta t}|\rho^N_{\nu+1}||\partial_t^k\partial_x^l L^\sharp(u-w_q)|^2 dxdt$$
$$\leq C \sum_{k+l\leq Q+n+N/2+3} \int_{\Omega_\nu} e^{-\theta t}|\partial_t^k\partial_x^l u|^2 dxdt$$
$$+C \sum_{k+l\leq Q+3+n+N/2} \int_{\Omega_m} e^{-\theta t}|\partial_t^k\partial_x^l f|^2 dxdt.$$

We choose q, Q so that (recall $N=2(N_0+n)$)

$$q \geq N_0+2n+1, \quad \kappa Q \geq (\kappa+1)(N_0+2n+q+2) \tag{3.21}$$

then it is easy to check that these q, Q verify (3.19) and (3.20). Noticing $|\partial_t^k\partial_x^l u| \leq |\partial_t^k\partial_x^l(u-w_q)| + |\partial_t^k\partial_x^l w_q|$ we summarize what we have proved;
Estimates in $\tilde{\Omega}_\nu$: if $\partial_t^\alpha u(0,x)=0$, $\partial_t^\alpha f(0,x)=0$ for $\alpha=0,1,\ldots,\tilde{Q}_\nu(n)$ then we have

$$\sum_{k+l\leq n} \int_{\tilde{\Omega}_{\nu+1}} |\partial_t^k\partial_x^l u|^2 dxdt \leq C \sum_{k+l\leq \tilde{q}_\nu(n)} \int_{\Omega_m} |\partial_t^k\partial_x^l f|^2 dxdt$$

where $\tilde{q}_\nu(n)=q_\nu(Q+2n+N_0+3)$, $\tilde{Q}_\nu(n)=Q_\nu(Q+2n+N_0+3)$.

We go to the next step. Let $\varphi=\sigma_{\nu+1}$ we consider $L^\sharp(u-w_q)=f-L^\sharp w_q$ in the region $\omega^+_{\nu+1}$. From Proposition 3.6 it follows

$$\sum_{l\leq n} \int_{\partial\omega^+_{\nu+1}} \Gamma(\partial_x^l(u-w_q)) + \sum_{k+l\leq n} \int_{\omega^+_{\nu+1}} e^{-\theta t}|\rho^{-N}_{\nu+1}||\partial_t^k\partial_x^l(u-w_q)|^2 dxdt$$
$$\leq C \sum_{k+l\leq n+1} \int_{\omega^+_{\nu+1}} e^{-\theta t}|\rho^{-N}_{\nu+1}||\partial_t^k\partial_x^l L^\sharp(u-w_q)|^2 dxdt$$
$$+C \sum_{k+l\leq n} \int_{\omega^+_{\nu+1}} e^{-\theta t}|\rho^{-N-1}_{\nu+1}||\partial_t^k\partial_x^l L^\sharp(u-w_q)|^2 dxdt.$$

From Proposition 3.6 we have

$$(\rho^{-N}_{\nu+1}\partial_x^l(u-w_q))(\sigma_{\nu+1}(x),x)=0, \quad l\leq n$$

if

$$(q-n) \geq N+1, \quad 2\kappa(Q-q-n-1) \geq 2(n+1).$$

Since $t = s_{\nu+1}(x)$ and $|x| = \delta(T-t)$ are space-like curves, thanks to Proposition 3.6, the above inequality yields

$$\sum_{k+l\leq n} \int_{\omega_{\nu+1}^+} |\partial_t^k \partial_x^l (u-w_q)|^2 dxdt$$
$$\leq C \sum_{k+l\leq Q+n+2} \left(\int_{\tilde{\Omega}_{\nu+1}} |\partial_t^k \partial_x^l u|^2 dxdt + \int_{\Omega_m} |\partial_t^k \partial_x^l f|^2 dxdt \right).$$

Then by induction hypothesis one has

$$\sum_{k+l\leq n} \int_{\Omega_{\nu+1}} |\partial_t^k \partial_x^l u|^2 dxdt \leq C \sum_{k+l\leq q_{\nu+1}(n)} \int_{\Omega_m} |\partial_t^k \partial_x^l f|^2 dxdt$$

for any u, f with

$$\partial_t^\alpha u(s_0(x), x) = 0, \quad \partial_t^\alpha f(s_0(x), x) = 0, \quad \alpha = 0, 1, \ldots, Q_{\nu+1}(n)$$

where $q_{\nu+1}(n) = \tilde{q}_\nu(Q+n+2)$, $Q_{\nu+1}(n) = \tilde{Q}_\nu(Q+n+2)$. This proves $(H_{\nu+1})$.

Finally we derive energy estimates in $\omega(T)$. We remark that

$$C\rho \geq (|x|^{c_1} + |t|^{c_2}) \quad \text{in} \quad \omega(T)$$

with some $c_i > 0$ when $n_1 = 0$ because we have

$$C\rho \geq \prod |t - t_j(x)| \geq \prod (t - |t_j(x)|) \geq \prod \frac{2}{3} t$$
$$\geq \prod \frac{1}{3}(t + t^*(x)) \geq \prod \frac{1}{3}(|t| + |x|^\kappa).$$

When $n_1 \geq 1$ we see

$$\rho = \rho_{m+1} = t - s_m(x) \geq \frac{2}{3}t + \frac{t}{3} - s_m \geq \frac{1}{2}(t + t^*(x)) \geq \frac{1}{2}(|t| + |x|^\kappa).$$

Take $\varphi = s_m(x)$ and q, Q are large in Proposition 3.6, then one gets

$$\sum_{k+l\leq n+1} \int_{\omega(T)} e^{-\theta t} \rho^{-N} |\partial_t^k \partial_x^l L^\sharp (u-w_q)|^2 dxdt$$
$$\leq C \sum_{k+l\leq Q+n+3} \left(\int_{\Omega_m} |\partial_t^k \partial_x^l u|^2 dxdt + \int_S |\partial_t^k \partial_x^l f|^2 dxdt \right)$$

where

$$S = \{(t,x) \mid |x| \le \delta(T-t), s_0 \le t \le s_m(\bar{x})\}.$$

Hence we have

$$\sum_{k+l\le n} \int_{\omega(T)} |\partial_t^k \partial_x^l u|^2 dxdt$$

$$\le C \sum_{k+l\le Q+n+3} \left(\int_{\Omega_m} |\partial_t^k \partial_x^l u|^2 dxdt + \int_S |\partial_t^k \partial_x^l f|^2 dxdt \right).$$

Thus we have proved

Proposition 3.7. *Let W be an open neighborhood of the origin and assume that the assumptions in Theorem 3.1 are verified. Then there are δ, T such that for any $n \in \mathbb{N}$ one can find $q(n)$, $Q(n)$ so that we have*

$$\sum_{k+l\le n} \int_S |\partial_t^k \partial_x^l u|^2 dxdt \le C_n \sum_{k+l\le q(n)} \int_S |\partial_t^k \partial_x^l L^\sharp u|^2 dxdt$$

for any $u \in C^\infty(W)$ with $\partial_t^\alpha u(s_0(x),x) = 0$, $\alpha = 0, 1, \ldots, Q(n)$.

Remark. If $u(t,x) = 0$ in $t \le c$ with some $c > 0$ then it is clear that $\partial_t^\alpha u(s_0(x),0) = 0$ for any α. If $u(t,x) = 0$ in $t \le -c$ with some $c > 0$ we apply Proposition 3.6 with $\varepsilon = -c$ and $\varphi = s_0(x)$ to get $u - w_q = O(|t - s_0(x)|^{q+1})$. On the other hand from the Remark after Proposition 3.5 and Lemma 3.29 we have

$$\sum_{k+l\le n} \int_{\tilde{\omega}(-c)} |\partial_t^k \partial_x^l u|^2 dxdt \le C \sum_{k+l\le n'} \int_{\tilde{\omega}(-c)} |\partial_t^k \partial_x^l f|^2 dxdt$$

then repeating similar arguments we conclude that Proposition 3.7 holds in this case also.

Theorem 3.3. *Assume that the assumptions in Theorem 3.1 are verified. Then the Cauchy problem* (CP) *is C^∞ well posed.*

Proof. Let f and u be C^∞ near the origin such that $L^\sharp u = f$ and $u(\tau,x) = 0$. It is clear that one can compute

$$\partial_t^j u(\tau,x) = w_j(f)$$

from $L^\sharp u = f$ which is a linear combination of $\partial_t^k \partial_x^l f(\tau,x)$ with $k + l \le j - 1$. Let $n \in \mathbb{N}$ be fixed and let us set

$$\rho_n(f) = \sum_{j=1}^{n} w_j(f)\frac{t^j}{j!}$$

then it is clear that $u - \rho_n(f) = O(|t-\tau|^{n+1})$ and $L^\sharp(u - \rho_n(f)) = O(|t-\tau|^n)$.

Let f be C^∞ such that $f = 0$ in $t \le \tau$. We choose a sequence of polynomials $\{f_p(t,x)\}$ so that

$$\sup_{k+l\le Q,(t,x)\in V} |\partial_t^k \partial_x^l (f(t,x) - f_p(t,x))| \to 0 \quad p \to \infty$$

where V is a neighborhood of the origin and $Q > n + q(n)$. By the Cauchy–Kowalevsky theorem the Cauchy problem

$$\begin{cases} L^\sharp u_p = f_p, \\ u_p(\tau, x) = 0 \end{cases}$$

has a solution u_p in a fixed domain W (independent of p). One can assume $V \subset W$. Let us define

$$\tilde{u}_p(t,x) = \begin{cases} u_p(t,x) - \rho_n(f_p) & t \ge \tau, \\ 0 & t < \tau \end{cases}$$

which is C^n so that we have

$$L^\sharp \tilde{u}_p = \begin{cases} f_p - L^\sharp(\rho_n(f_p)) & t \ge \tau, \\ 0 & t < \tau. \end{cases}$$

Choosing $S \subset V$ and applying Proposition 3.7 we get

$$\sum_{k+l\le n} \int_S |\partial_t^k \partial_x^l \tilde{u}_p|^2 dxdt \le C \sum_{k+l\le q(n)} \int_{S\cap\{t\ge\tau\}} |\partial_t^k \partial_x^l (f_p - L^\sharp(\rho_n(f_p)))|^2 dxdt.$$

Since $\sup_S |\partial_t^k \partial_x^l L^\sharp(\rho_n(f_p) - \rho_n(f))| \to 0$ as $p \to \infty$ if $k+l+n \le q(n)+n < Q$ then $\{\tilde{u}_p\}$ is a Cauchy sequence in $H^n(S)$ and hence there exists $u \in H^n(S)$ such that

$$\sum_{k+l\le n} \int_S |\partial_t^k \partial_x^l (\tilde{u}_p - u)|^2 dxdt \to 0, \quad p \to \infty.$$

Thus u is a desired solution to our Cauchy problem. □

3.11 Conditions for Well-Posedness

In this section we introduce the conditions $(C^\pm)$ which are equivalent to those in Theorem 3.1 and are more convenient to handle. The equivalence will be proved in Sect. 3.14. We start with

Definition 3.9. Let $\gamma \in \mathbb{Q}_+$. We say $\phi(x) \in \mathscr{G}^{\pm}(\gamma)$ if $\phi(x)$ is defined in $0 < \pm x < r(\phi)$ with some $r(\phi) > 0$ and expressed by convergent Puiseux series

$$\phi(x) = \sum_{j=0} C_j(\pm x)^{j/p}, \quad C_j \in \mathbb{R}, \ \ 0 < \pm x < r(\phi)$$

with some $p \in \mathbb{N}$ satisfying $|\phi(x)| \le C(\pm x)^{\gamma}$ in $0 < \pm x < r(\phi)$ with some $C > 0$. Hence it is clear that

$$\mathscr{C}^{\pm}(h|a^{\sharp}|^2) \subset \mathscr{G}^{\pm}(\gamma)$$

if $t^*(x) \sim |x|^{\gamma}$. We also define $\sigma(\phi)$ for $\phi \in \mathscr{G}^{\pm}(\gamma)$ by

$$C^{-1}(\pm x)^{\sigma(\phi)} \le |\phi(x)| \le C(\pm x)^{\sigma(\phi)}$$

with some $C > 0$.

Definition 3.10. Let $f(t,x)$ be real analytic near the origin and $f(0,0) = 0$. Let $p, q \in \mathbb{Q}_+$ and $\phi \in \mathscr{G}^{\pm}(\gamma)$. We define $\mu(f_\phi; p, q)$ by

$$f_\phi(s^p t, s^q x) = s^{\mu}(f^0(t,x) + o(s)), \quad s \to 0$$

where $f^0(t,x)$ does not vanish identically. Let $f(t,x)$, $g(t,x)$ be real analytic near the origin. We define

$$\mu([\frac{f}{g}]_\phi; p, q) = \mu(f_\phi; p, q) - \mu(g_\phi; p, q).$$

Remark. Note that $\mu(f_\phi; p, q)$ is uniquely determined by $\Gamma(f_\phi)$. Indeed write

$$f_\phi(s^p t, s^q x) = \sum C_{ij}(s^p t)^i (s^q x)^{j/\alpha} = \sum C_{ij} s^{pi+qj/\alpha} t^i x^{j/\alpha}$$

then we see

$$\mu = \min_{C_{ij} \neq 0} \{pi + \frac{qj}{\alpha}\}.$$

This means that the straight line $pt + qx/\alpha = \mu$ is tangent to $\Gamma(f_\phi)$. It is obvious that $\mu([fg]_\phi; p, q) = \mu(f_\phi; p, q) + \mu(g_\phi; p, q)$.

We introduce the following condition.

CONDITION $(C^{\pm})$: Let γ be a positive number such that $t^*(x) \sim |x|^{\gamma}$.

$$(C^{\pm}; Y) \qquad \begin{cases} \text{For any } p, q \in \mathbb{Q}_+ \text{ and } \phi \in \mathscr{G}^{\pm}(\gamma) \text{ with} \\ p \ge \sigma(\phi) q, \quad \mu(h_\phi; p, q) > 2q(1 - \sigma(\phi)) \ \text{ we have} \\ 2p + 2\mu([\frac{Y}{a^{\sharp}}]_\phi; p, q) \ge \mu(h_\phi; p, q). \end{cases}$$

Here $\sigma(\phi)q$ should be read as p if $\phi \equiv 0$. The condition $(C^{\pm}; Z)$ is similarly defined replacing $2\mu([\frac{Y}{a^\sharp}]_\phi; p, q)$ by $2\mu([\frac{Z}{a^\sharp}]_\phi; p, q)$. We say that the condition $(C^{\pm})$ is satisfied if and only if both $(C^{\pm}; Y)$ and $(C^{\pm}; Z)$ are verified.

Lemma 3.30. *Let $f(t,x)$ be real analytic near the origin and $f(0,0) = 0$. Then*

$$\mu([\frac{\partial_t f}{f}]_\phi; p, q) \geq -p, \quad \mu([\frac{\partial_t^2 f}{f}]_\phi; p, q) \geq -2p.$$

Moreover

$$\mu([\partial_t(\frac{\partial_t f}{f})]_\phi; p, q) \geq -2p.$$

Proof. Let $f_\phi(s^p t, s^q x) = s^{\mu(f_\phi; p,q)}(f^0(t,x) + o(1))$. On the other hand, writing

$$f_\phi(t,x) = x^{\tilde{n}}(t^{\tilde{m}} + f_1(x)t^{\tilde{m}-1} + \cdots + f_{\tilde{m}}(x))\Phi_\phi(t,x)$$

we have

$$t\partial_t f_\phi(t,x) = x^{\tilde{n}}(\tilde{m}t^{\tilde{m}} + (\tilde{m}-1)f_1(x)t^{\tilde{m}-1} + \cdots + f_{\tilde{m}-1}(x)t)\Phi_\phi(t,x)$$
$$+x^{\tilde{n}}(t^{\tilde{m}} + \cdots + f_{\tilde{m}}(x))t(\partial_t\Phi)_\phi(t,x).$$

It is clear that $\Gamma(t\partial_t f_\phi) \subset \Gamma(f_\phi)$ by definition. This gives

$$(t\partial_t f_\phi)(s^p t, s^q x) = s^{\mu^*}(c^0(t,x) + o(1)), \quad \mu^* \geq \mu(f_\phi; p, q).$$

Since $\partial_t f_\phi = (\partial_t f)_\phi$ we see

$$s^p t(\partial_t f)_\phi(s^p t, s^q x) = s^{\mu^*}(c^0(t,x) + o(1))$$

and hence $(\partial_t f)_\phi(s^p t, s^q x) = s^{\mu^* - p}(c^0(t,x)/t + o(1))$. This proves that

$$\mu([\partial_t f]_\phi; p, q) = \mu^* - p$$

and hence $\mu([\partial_t f]_\phi; p, q) - \mu(f_\phi; p, q) = \mu^* - \mu - p \geq -p$.

The second inequality is proved similarly because $\Gamma(t^2\partial_t^2 f_\phi) \subset \Gamma(f_\phi)$. We turn to the last inequality. Note that

$$\partial_t(\frac{\partial_t f}{f}) = \frac{\partial_t^2 f}{f} - (\frac{\partial_t f}{f})^2, \quad \mu([(\frac{\partial_t f}{f})^2]_\phi; p, q) = 2\mu([\frac{\partial_t f}{f}]_\phi; p, q).$$

Then we conclude that

$$\mu([\partial_t(\frac{\partial_t f}{f})]_\phi; p, q) \geq \min\{\mu([\frac{\partial_t^2 f}{f}]_\phi; p, q), \mu([(\frac{\partial_t f}{f})^2]_\phi; p, q)\} \geq -2p$$

which is the desired assertion. □

Recall that $L^\sharp M = p + Q\partial_x + R\partial_t + S$ and

$$Q = \mathrm{diag}(\frac{Y}{\overline{a^\sharp}}, \frac{Z}{a^\sharp}), \quad R = C - A_x^\sharp + B^\sharp + {}^{co}B^\sharp,$$

$$S = L^\sharp C + L^\sharp({}^{co}B^\sharp - A_x^\sharp), \quad C = \mathrm{diag}(\frac{\partial_t \overline{a^\sharp}}{\overline{a^\sharp}}, \frac{\partial_t a^\sharp}{a^\sharp}).$$

Lemma 3.31. *Let $S = (s_{ij})$, $R = (r_{ij})$. Then we have*

$$\mu([s_{ij}]_\phi; p, q) \geq -2p, \quad \mu([r_{ij}]_\phi; p, q) \geq -p.$$

Proof. It suffices to study $L^\sharp C = \partial_t C - A^\sharp \partial_x C$. Since

$$\partial_t C = \mathrm{diag}\Big(\partial_t(\frac{\partial_t \overline{a^\sharp}}{\overline{a^\sharp}}), \partial_t(\frac{\partial_t a^\sharp}{a^\sharp})\Big)$$

the assertion $\mu([\partial_t C]_\phi; p, q) \geq -2p$ follows from Lemma 3.30. Note

$$A^\sharp \partial_x C = \begin{bmatrix} c^\sharp \partial_x(\partial_t \overline{a^\sharp}/\overline{a^\sharp}) & a^\sharp \partial_x(\partial_t a^\sharp/a^\sharp) \\ \overline{a^\sharp} \partial_x(\partial_t \overline{a^\sharp}/\overline{a^\sharp}) & \overline{c^\sharp} \partial_x(\partial_t a^\sharp/a^\sharp) \end{bmatrix}.$$

We study the (1,1)-th entry

$$c^\sharp \partial_x(\frac{\partial_t \overline{a^\sharp}}{\overline{a^\sharp}}) = \frac{c^\sharp}{\overline{a^\sharp}} \partial_x \partial_t \overline{a^\sharp} - \partial_x \overline{a^\sharp} \frac{c^\sharp}{\overline{a^\sharp}} \frac{\partial_t \overline{a^\sharp}}{\overline{a^\sharp}}.$$

Since $|c^\sharp/\overline{a^\sharp}| \leq 1$ we see that

$$\mu([c^\sharp \partial_x(\frac{\partial_t \overline{a^\sharp}}{\overline{a^\sharp}})]_\phi; p, q) \geq -2p.$$

Similarly we get $\mu([A^\sharp \partial_x C]_\phi; p, q) \geq -p$. We turn to R. From Lemma 3.30 it follows immediately that

$$\mu([R]_\phi; p, q) \geq -p$$

which proves the assertion. □

Lemma 3.32. *We have*

$$\mu([\frac{Y}{\overline{a^\sharp}}]_\phi; p, q) \geq 0, \quad \mu([\frac{Z}{a^\sharp}]_\phi; p, q) \geq 0.$$

Proof. We consider $Y/\overline{a^\sharp}$. The argument for $Z/a^\sharp$ is similar. Noting that

$$\frac{Y}{\overline{a^\sharp}} = \partial_t c^\sharp - (\frac{c^\sharp}{\overline{a^\sharp}})\partial_t \overline{a^\sharp} + \mathrm{Tr}\,(AB)$$

the assertion follows immediately because $\mu([c^\sharp/\overline{a^\sharp}]_\phi; p, q) \geq 0$. □

In what follows we assume that (C^+) fails, that is there are $\phi \in \mathscr{G}^+(\gamma)$, p, $q \in \mathbb{Q}_+$ with $p \geq \sigma(\phi)q$, $\mu(h_\phi; p, q) > 2q(1-\sigma(\phi))$ ($q\sigma(\phi) = p$ if $\phi \equiv 0$) such that we have

$$\text{either } 2p+2\mu([\frac{Y}{\overline{a^\sharp}}]_\phi; p, q) < \mu(h_\phi; p, q) \text{ or } 2p+2\mu([\frac{Z}{a^\sharp}]_\phi; p, q) < \mu(h_\phi; p, q).$$

Without restrictions one may assume that

$$2p + 2\mu([\frac{Y}{\overline{a^\sharp}}]_\phi; p, q) < \mu(h_\phi; p, q), \quad \mu([\frac{Y}{\overline{a^\sharp}}]_\phi; p, q) \leq \mu([\frac{Z}{a^\sharp}]_\phi; p, q) \tag{3.22}$$

because the arguments below are parallel both for $Y/\overline{a^\sharp}$ and $Z/a^\sharp$.

Proposition 3.8. *Assume that (C^+) fails. Then there are p, $q \in \mathbb{Q}_+$, $\phi \in \mathscr{G}^+(\gamma)$ with $p \geq \sigma(\phi)q$, $1 > q(1-\sigma(\phi))$, $\mu(h_\phi; p, q) \geq 2$ such that*

$$\mu([\frac{Y}{\overline{a^\sharp}}]_\phi; p, q) + p < 1, \quad 2q(1-\sigma(\phi)) - 1 - p - \mu([\frac{Y}{\overline{a^\sharp}}]_\phi; p, q) < 0$$

where $q\sigma(\phi)$ should read as p if $\phi \equiv 0$.

Proof. Let $\phi \not\equiv 0$. Then we replace p, q in (3.22) by

$$\frac{2p}{\mu(h_\phi; p, q)}, \quad \frac{2q}{\mu(h_\phi; p, q)}.$$

Then remarking that $\mu(h_\phi; \kappa p, \kappa q) = \kappa\mu(h_\phi; p, q)$ we may suppose that in (3.22)

$$p \geq \sigma(\phi)q, \quad 1 > q(1-\sigma(\phi)), \quad \mu(h_\phi; p, q) = 2,$$
$$p + \mu([\frac{Y}{\overline{a^\sharp}}]_\phi; p, q) < 1.$$

In the case $\phi \equiv 0$ we make the same convention.

Let us put

$$f(p) = 1 - p - \mu([\frac{Y}{a^\sharp}]_\phi; p, q),$$
$$g(p) = 2q(1 - \sigma(\phi)) - 1 - p - \mu([\frac{Y}{a^\sharp}]_\phi; p, q).$$

Suppose that $g(p) \geq 0$. Otherwise nothing to be proved. We note that $p < 1$ because

$$p + \mu([\frac{Y}{a^\sharp}]_\phi; p, q) < 1, \ \ \mu([\frac{Y}{a^\sharp}]_\phi; p, q) \geq 0.$$

Remark that

$$f(p) - g(p) = 2(1 - q(1 - \sigma(\phi))) > 0.$$

On the other hand we see $f(1) \leq 0$ and $g(1) < 0$ since $\mu([Y/a^\sharp]_\phi; p, q) \geq 0$. Write

$$\mu([\frac{Y}{a^\sharp}]_\phi; p, q) = \mu(Y_\phi; p, q) - \mu([a^\sharp]_\phi; p, q)$$

then we see that $\mu([Y/a^\sharp]_\phi; p, q)$ is continuous with respect to p. Then there exists $p \leq p^* < 1$ such that

$$g(p^*) = 0, \quad g(p) < 0, \quad p^* < p < 1.$$

Since $f(p^*) > g(p^*) = 0$ one can take $\hat{p}$ so close to p^* $(p^* < \hat{p})$ so that $f(\hat{p}) > 0$ and $g(\hat{p}) < 0$. This $\hat{p}$ is a desired one. □

Remark. Since $p \geq \sigma q$, $1 > q(1 - \sigma)$ this shows that $1 + p > q$.

Lemma 3.33. *Assume that* $p \geq \sigma(p)q$, $\mu + p < 1$, $2q(1 - \sigma(\phi)) - 1 - p - \mu < 0$. *Set* $\delta = (1 + p - q)^{-1}$ *and* $2\sigma_1 = 1 - \delta\mu + \delta q - 2\delta p$. *Then we have*

$$\sigma_1 - \delta q \sigma(\phi) - 1 + \delta p < 0.$$

In particular $\sigma_1 < 1 - \delta(p - \sigma(\phi)q) \leq 1$.

Proof. We plug $1 = \delta(1 + p - q)$ into $1 - \delta\mu + \delta q - 2\delta p$ then we get

$$2\sigma_1 = \delta(1 + p - q) - \delta\mu + \delta q - 2\delta p = \delta - \delta\mu - \delta p = \delta(1 - \mu - p).$$

We compute $\delta(2q(1 - \sigma) - 1 - p - \mu) < 0$ which is

$$\begin{aligned} 2q\delta - 2q\delta\sigma - \delta - \delta p - \delta\mu &= \delta(1 - p - \mu) - 2q\delta\sigma - 2\delta(1 - q) \\ &= 2\sigma_1 - 2q\delta\sigma - 2 + 2\delta p \end{aligned}$$

because $\delta(1 - q) = 1 - \delta p = 2(\sigma_1 - \delta q \sigma - 1 + \delta p)$. This proves the assertion. □

3.12 Construction of Asymptotic Solutions

Our strategy for proving the necessity of the condition (C^+) is as follows; assume that (C^+) fails then from Proposition 3.8 there exist $p, q \in \mathbb{Q}_+$ and $\phi \in \mathcal{G}^+(\gamma)$ such that $\mu = \mu([Y/\overline{a^\sharp}]_\phi; p, q)$ verifies

$$\begin{aligned} &\mu \geq 2, \quad p \geq \sigma q, \ 1 > q(1-\sigma), \\ &\mu + p < 1, \quad 2q(1-\sigma) - 1 - p - \mu < 0 \end{aligned} \tag{3.23}$$

where if $\phi \equiv 0$ then $q\sigma$ should be read as p. Let U be a neighborhood of the origin and $\phi \in \mathcal{G}^+(\gamma)$. We introduce a system of local coordinates $x = (x_1, x_2)$ in $U \cap \{x_2 > 0\}$ so that

$$x_1 = t - \phi(x), \quad x_2 = x.$$

Definition 3.11. Let L be a differential operator defined near the origin. Then we denote by L_ϕ the representation of L in a system of local coordinates (x_1, x_2). Let f be a smooth function near the origin. We denote by $f_\phi(x) = f(x_1 + \phi(x_2), x_2)$ the representation of f in a system of local coordinates (x_1, x_2).

Then from Corollary 1.1, or rather its proof, we have a priori estimates for $L^\sharp_\phi$ while one can construct asymptotic solutions contradicting thus obtained a priori estimates.

In this section we construct desired asymptotic solutions. Let

$$(L^\sharp M)_\phi = \sum_{i,j=1}^{2} h^{(ij)}(x)\partial_i\partial_j + \sum_{i=1}^{2} B^{(i)}(x)\partial_i + F(x)$$

where $\partial_i = \partial/\partial x_i$ and $h^{(ij)}$ has the form

$$\begin{aligned} h^{(11)}(x) &= 1 - h_\phi(x)\phi'(x_2)^2, \ h^{(12)}(x) = 2h_\phi(x)\phi'(x_2), \ h^{(22)}(x) = -h_\phi(x), \\ B^{(2)}(x) &= Q_\phi(x), \ B^{(1)}(x) = h_\phi(x)\phi''(x_2) - \phi'(x_2)Q_\phi(x) + R_\phi(x), \\ F(x) &= S_\phi(x). \end{aligned}$$

Recall that $L^\sharp M = \partial_t^2 - h\partial_x^2 + Q\partial_x + R\partial_t + S$ and

$$h(t,x) = x^{2n_1}(t^{2m_1} + h_1(x)t^{2m_1-1} + \cdots + h_{2m_1}(x))e(t,x)^2.$$

Then it is clear that one can write

$$h_\phi(x) = \sum_{(\alpha,\beta)\in M(\phi)} h_{\alpha\beta}(x)x_1^\alpha x_2^\beta(1 + O(x_2^{1/\theta}))$$

with some $\theta = \theta(\phi) \in \mathbb{N}$ where $M(\phi)$ is defined as

$$M(\phi) = \{(\alpha, \beta) \in \mathbb{N} \times (\mathbb{N}/\theta) \mid \lim_{x_1 \to 0, x_2 \downarrow 0} h_{\alpha\beta}(x) = h^*_{\alpha\beta} \neq 0\}.$$

Recall that the Newton polygon $\Gamma(h_\phi)$ is defined by $\{(\alpha, \beta) \mid (\alpha, \beta) \in M(\phi)\}$. Note that

$$\mu(h_\phi; p, q) \geq 2 \Longrightarrow \alpha p + \beta q \geq 2, \quad \forall(\alpha, \beta) \in M(\phi).$$

Then we get with some real c

$$h^{(22)}(x) = -\sum_{(\alpha,\beta)\in M(\phi)} h_{\alpha\beta}(x) x_1^\alpha x_2^\beta (1 + O(x_2^{1/\theta})),$$

$$h^{(12)}(x) = 2\sum_{(\alpha,\beta)\in M(\phi)} c h_{\alpha\beta}(x) x_1^\alpha x_2^{\beta+(\sigma-1)} (1 + O(x_2^{1/\theta})),$$

$$h^{(11)}(x) = 1 - \sum_{(\alpha,\beta)\in M(\phi)} c^2 h_{\alpha\beta} x_1^\alpha x_2^{\beta+2(\sigma-1)} (1 + O(x_2^{1/\theta})).$$

We make a dilation of coordinates: $x_1 = \lambda^{-\delta p} y_1$, $x_2 = \lambda^{-\delta q} y_2$. Let P_λ denote the representation of P in the coordinates $y = (y_1, y_2)$ then we have

$$\begin{aligned}
\lambda^{-2\delta p}(L^\sharp M)_{\phi,\lambda} = {} & h_\lambda^{(11)}(y)\partial_1^2 + h_\lambda^{(12)}(y)\lambda^{\delta q - \delta p}\partial_1\partial_2 \\
& + h_\lambda^{(22)}(y)\lambda^{2\delta q - 2\delta p}\partial_2^2 + B_\lambda^{(1)}(y)\lambda^{-\delta p}\partial_1 \\
& + B_\lambda^{(2)}(y)\lambda^{\delta q - 2\delta p}\partial_2 + F_\lambda(y)\lambda^{-2\delta p}
\end{aligned}$$

where $f_\lambda(y) = f(\lambda^{-\delta p} y_1, \lambda^{-\delta q} y_2)$. Let us take τ as the least common denominator of δ, p, q, σ, σ_1, $1/\theta$.

Lemma 3.34. *We have*

$$\begin{aligned}
\lambda^{2\sigma_1} h_\lambda^{(11)}(y) &= \lambda^{2\sigma_1}(1 + O(\lambda^{-1/\tau})), \\
\lambda^{\delta q - \delta p + \sigma_1 + 1} h_\lambda^{(12)}(y) &= O(\lambda^{-1/\tau}), \\
\lambda^{2\delta q - 2\delta p + 2} h_\lambda^{(22)}(y) &= O(1).
\end{aligned}$$

Proof. Note that $-\delta\alpha p - \delta\beta q - 2\delta q(\sigma - 1) = -\delta(\alpha p + \beta q - 2q(1 - \sigma))$. From $\mu(h_\phi; p, q) \geq 2$ we see $\alpha p + \beta q \geq 2$ if $(\alpha, \beta) \in M(\phi)$ and hence it follows that $\alpha p + \beta q - 2q(1 - \sigma) > 2q(1 - \sigma)$. That is

$$\lambda^{2\sigma_1} h_\lambda^{(11)}(y) = \lambda^{2\sigma_1}(1 + O(\lambda^{-1/\tau})).$$

We next study $\lambda^{\delta q-\delta p+\sigma_1+1}h_\lambda^{(12)}(y)$. Recall that

$$\begin{aligned}
&-\alpha\delta p-\beta\delta q-\delta q(\sigma-1)+\delta q-\delta p+\sigma_1+1\\
&=\delta(-\alpha p-\beta q)-\delta q\sigma+2\delta q-\delta p+\sigma_1+1\\
&=\delta(-\alpha p-\beta q)-\delta q\sigma-2+2\delta(1+p)-\delta p+\sigma_1+1\\
&=\delta(2-\alpha p-\beta q)+(\sigma_1-\delta q\sigma-1+\delta p)<0
\end{aligned}$$

by Lemma 3.33 and the fact $\alpha p+\beta q\geq 2$ for $(\alpha,\beta)\in M(\phi)$. This proves that

$$\lambda^{\delta q-\delta p+\sigma_1+1}h_\lambda^{(12)}(y)=O(\lambda^{-1/\tau}).$$

Finally we study $\lambda^{2\delta q-2\delta p+2}h_\lambda^{(22)}(y)$. Then we see

$$\begin{aligned}
&2\delta q-2\delta p+2-\alpha\delta p-\beta\delta q\\
&\quad=2\delta q-2\delta p+2\delta(1+p-q)-\alpha\delta p-\beta\delta q=\delta(2-\alpha p-\beta q)\leq 0
\end{aligned}$$

because $(\alpha,\beta)\in M(\phi)$ and hence the assertion. □

Lemma 3.35. *We have*

$$\begin{aligned}
\lambda^{\delta q-2\delta p+1}B_\lambda^{(2)}(y)&=\lambda^{\delta q-2\delta p+1}Q_{\phi,\lambda}(y)=\lambda^{2\sigma_1}[Q_\phi^0(y)+O(\lambda^{-1/\tau})],\\
\text{diagonal entry of }\lambda^{-\delta p+\sigma_1}B_\lambda^{(1)}&=O(\lambda^{2\sigma_1-1/\tau}),\\
\text{off diagonal entry of }\lambda^{-\delta p+\sigma_1}B_\lambda^{(1)}&=O(\lambda^{-\delta p+\sigma_1}),\\
\lambda^{-2\delta p}F_\lambda&=O(1).
\end{aligned}$$

Proof. By definition $\delta q-2\delta p+1=2\sigma_1+\delta\mu([Y/\overline{a^\sharp}]_\phi;p,q)$. Noting that the fact $\mu([Y/\overline{a^\sharp}]_\phi;-\delta p,-\delta q)=-\delta\mu([Y/\overline{a^\sharp}]_\phi;p,q)$ we get the first assertion. We next study $\lambda^{-\delta p+\sigma_1}B_\lambda^{(1)}(y)$. Recall

$$B^{(1)}(x)=h_\phi(x)\phi''(x_2)-\phi'(x_2)Q_\phi(x)+R_\phi(x).$$

Note that $\lambda^{-\delta p+\sigma_1}(h_\phi\phi'')_\lambda$ yields a term with the power $-\delta p+\sigma_1-\delta\alpha p-\delta\beta q-\delta q(\sigma-2)$. We plug $2\delta q=2\delta(1+p)-2$ and hence this gives the power

$$-\delta(\alpha p+\beta q-2)+(\sigma_1-\delta q\sigma-1+\delta p)-1<-1$$

by Lemma 3.33. This shows $\lambda^{-\delta p+\sigma_1}(h_\phi\phi'')_\lambda=O(\lambda^{-1})$. We turn to the term $\lambda^{-\delta p+\sigma_1}(\phi' Q_\phi)_\lambda$. Note that

$$\begin{aligned}
&-\delta p+\sigma_1-\delta q(\sigma-1)+\mu([Y/\overline{a^\sharp}]_\phi;-\delta p,-\delta q)\\
&\quad=-\delta\mu+\delta(1+p)-1-\delta q\sigma-\delta p+\sigma_1
\end{aligned}$$

$$= \delta(1 - p - \mu) - 1 - \delta q\sigma + \delta p + \sigma_1$$
$$= 2\sigma_1 + (\sigma_1 - \delta q\sigma - 1 + \delta p) < 2\sigma_1$$

which follows from Lemma 3.33. This gives $\lambda^{-\delta p+\sigma_1}(\phi' Q_\phi)_\lambda = O(\lambda^{2\sigma_1 - 1/\tau})$. Recall $R = C + G$ with smooth G. From Lemma 3.30 it follows that $C_{\phi,\lambda} = O(\lambda^{\delta p})$ and hence $\lambda^{-\delta p+\sigma_1} R_{\phi,\lambda}(y) = O(\lambda^{\sigma_1})$. Finally we consider $\lambda^{-2\delta p} F_\lambda$. Since $F = S_\phi$ and $S = L^\sharp C$+smooth term, it is enough to consider $L^\sharp C$. From Lemma 3.31 it follows that

$$S_{\phi,\lambda} = O(\lambda^{2\delta p})$$

and hence the desired result. □

Let us define ν by $\nu = \sigma_1 \tau$.

Proposition 3.9. *Assume that* (C^+) *fails. Then for any given* $\hat{y} = (\hat{y}_1, \hat{y}_2)$, $\hat{y}_2 > 0$, *any given neighborhood* $U(\hat{y})$ *of* $\hat{y}$ *and any given* $N \in \mathbb{N}$ *there is* $\bar{y} \in U(\hat{y})$, *a neighborhood* W *of* $\bar{y}$ ($W \subset U(\hat{y})$) *and analytic functions* $l^j(y)$, $1 \le j \le \nu$ *and* $u_n(y)$, $0 \le n \le N$ *defined in* W *such that*

$$E(y,\lambda)^{-1}\lambda^{-2\delta p} L^\sharp_{\phi,\lambda} U_\lambda = O(\lambda^{2\sigma_1 - (\nu+N+1)/\tau}) \ \text{in } W$$

where

$$E(y,\lambda) = \exp\{i(\mu y_2 \lambda + \sum_{j=1}^{\nu} l^j(y)\lambda^{\sigma_j})\},$$

$$U_\lambda = E(y,\lambda)\lambda^\kappa \sum_{n=0}^{N} \lambda^{-n/\tau} u_n(y)$$

and $\sigma_j = (\nu + 1 - j)/\tau$, $\kappa = \kappa(p,q)$. *Moreover*

$$\operatorname{Im} l^1(y) \ge (y_2 - \bar{y}_2)^2 + \delta_0(\bar{y}_1 - y_1)$$

in $W \cap \{y_1 \le \bar{y}_1\}$ *with some* $\delta_0 > 0$ *and* $u_0(\bar{y}) \ne 0$.

Proof. Recall that

$$\lambda^{\delta q - 2\delta p + 1} B^{(2)}_\lambda(y) = \lambda^{\delta q - 2\delta p + 1} Q_{\phi,\lambda}(y)$$
$$= \lambda^{\delta q - 2\delta p + 1}\begin{bmatrix} [Y/\overline{a^\sharp}]_{\phi,\lambda} & 0 \\ 0 & [Z/a^\sharp]_{\phi,\lambda} \end{bmatrix}$$
$$= \lambda^{2\sigma_1}\begin{bmatrix} \sum_{j=0} C^1_j(y)\lambda^{-j/\tau} & 0 \\ 0 & \sum_{j=0} C^2_j(y)\lambda^{-j/\tau} \end{bmatrix}.$$

From the assumption we have $C_0^1(y) \neq 0$. We look for U_λ in the form

$$U_\lambda = M_{\phi,\lambda} u_\lambda, \quad u_\lambda = E(y,\lambda) \sum_{n=0}^{N} \lambda^{-n/\tau} v_n(y)$$

so that the problem is reduced to solve

$$E(y,\lambda)^{-1} (L^\sharp M)_{\phi,\lambda} E(y,\lambda) \sum_{n=0}^{N} v_n(y) \lambda^{-n/\tau} = O(\lambda^{-N_1})$$

with $N_1 = -2\delta p - 2\sigma_1 + (\nu + N + 1)/\tau$. This turns out to be

$$\begin{aligned} &\lambda^{2\sigma_1+2\delta p} \Big\{ \sum_{j=1}^{\nu} \mathscr{L}_j(l^1,\dots,l^j) \lambda^{-(j-1)/\tau} \sum_{n=1} v_n \lambda^{-n/\tau} \\ &+ \sum_{n=0}^{N} (2\sqrt{-1}\, l^1_{y_1} \frac{\partial}{\partial y_1} v_n + R_n(l^1,\dots,l^\nu, v_0,\dots,v_{n-1})) \lambda^{-(n+\nu)/\tau} \Big\} \\ &+ O(\lambda^{2\sigma_1+2\delta p-(\nu+N+1)/\tau}) = O(\lambda^{-N_1}) \end{aligned} \tag{3.24}$$

where

$$\mathscr{L}_j = \begin{bmatrix} \mathscr{L}_j^1 & 0 \\ 0 & \mathscr{L}_j^2 \end{bmatrix}, \quad \mathscr{L}_1^i(l^1) = -(l^1_{y_1})^2 + \sqrt{-1} C_0^i(y), \quad v_n = \begin{bmatrix} v_n^1 \\ v_n^2 \end{bmatrix},$$

$$\mathscr{L}_j^i(l^1,\dots,l^j) = -2 l^1_{y_1} l^j_{y_1} + K_j^i(l^1,\dots,l^{j-1}), \ j \geq 2$$

and K_j^i, R_n are non linear differential operators with real analytic coefficients. More precisely

$$\mathscr{L}_j^i = \Phi_j(C_0^i,\dots,C_{j-2}^i, l^1,\dots,l^j) + \sqrt{-1} C_{j-1}^i(y), \quad 1 \leq j \leq \nu$$

where Φ_j are independent of i. To see this it is enough to note that off diagonal part of the coefficients does not enter to the determination.

Let $U(\hat{y})$ be given. We divide the cases into two:

(a) $C_j^1(y) = C_j^2(y)$ in U for $0 \leq j \leq \nu - 1$,
(b) there exists $k \leq \nu - 1$ and $\bar{y} \in U$ such that

$$C_j^1(y) = C_j^2(y) \quad \text{in} \quad U, \quad 0 \leq j \leq k-1, \quad C_k^1(\bar{y}) \neq C_k^2(\bar{y}).$$

In case (b) we choose $W_1 = W_1(\bar{y}) \subset U$ so that

$$|C_k^1(y) - C_k^2(y)| \geq c > 0 \quad \text{in} \quad W_1.$$

We first define $l^j(y)$, $1 \le j \le \nu$. Take $\mu \in \mathbb{R}$ and $W_2 \subset W_1$ so that

$$-\zeta^2 + \sqrt{-1}\mu C_0^1(y) = 0$$

has a root $F(y)$ with $\mathrm{Im}\, F(y) < -\delta_0 < 0$ in W_2. Note that $|F(y)| \sim \sqrt{|\mu|}$. We solve the Cauchy problem

$$l^1_{y_1} = F(y), \quad l^1|_{y_1=\bar{y}_1} = \sqrt{-1}(y_2 - \bar{y}_2)^2.$$

This gives that $|l^1_{y_1}| \sim \sqrt{|\mu|}$. We define $l^j(y)$ successively by solving

$$\begin{cases} \mathscr{L}^1_j(l^1,\ldots,l^j) = -2l^1_{y_1} l^j_{y_1} + K^1_j(l^1,\ldots,l^{j-1}) = 0, \\ l^j|_{y_1=\bar{y}_1} = 0 \end{cases}$$

for $2 \le j \le \nu$. In the case (a) we have clearly that

$$\mathscr{L}^2_j(l^1,\ldots,l^j) = 0 \quad \text{in} \quad W_2 \quad \text{for} \quad j = 1,\ldots,\nu$$

and in the case (b) we have

$$\mathscr{L}^2_j(l^1,\ldots,l^j) = 0 \text{ in } W_2 \text{ for } j = 1,\ldots,k, \ |\mathscr{L}^2_{k+1}(l^1,\ldots,l^{k+1})| \ge c' > 0 \text{ in } W_2.$$

We first study the case (b). We observe the second equation of (3.24) which is equal to, up to the factor $\lambda^{2\sigma_1+2\delta p}$

$$\sum_{n=0}^{N+\nu-k-1} \{\mathscr{L}^2_{k+1}(l^1,\ldots,l^{k+1})v^2_n + R^2_n(l^1,\ldots,l^\nu,v_0,\ldots,v_{n-1})\}\lambda^{-(n+k)/\tau}$$
$$+O(\lambda^{-(\nu+N-k)/\tau}) = O(\lambda^{-N_1}).$$

Hence the second equation is reduced to

$$\mathscr{L}^2_{k+1}(l^1,\ldots,l^{k+1})v^2_n + R^2_n(l^1,\ldots,l^\nu,v_0,\ldots,v_{n-1}) = 0.$$

Here we remark that

$$R^2_n(l^1,\ldots,l^\nu,v_0,\ldots,v_{n-1})|_{v_0^2=\cdots=v_{n-1}^2=0} = 0 \text{ for } n \le \nu - k - 1.$$

On the other hand the first equation is

$$\sum_{n=0}^{N}(2\sqrt{-1}\, l^1_{y_1}\frac{\partial}{\partial y_1}v^1_n + R^1_n(l^0,\ldots,l^\nu,v_1,\ldots,v_{n-1}))\lambda^{-(n+\nu))/\tau}$$
$$+O(\lambda^{-(\nu+N+1)/\tau})$$

and hence we are led to the equation

$$2\sqrt{-1}\,l^1_{y_1}\frac{\partial}{\partial y_1}v^1_n + R^1_n(l^1,\ldots,l^\nu,v_0,\ldots,v_{n-1}) = 0.$$

Thus (3.24) is reduced to

$$\begin{cases}\mathcal{L}^2_{k+1}(l^1,\ldots,l^{k+1})v^2_n + R^2_n(l^1,\ldots,l^\nu,v_0,\ldots,v_{n-1}) = 0,\\ 2\sqrt{-1}\,l^1_{y_1}\dfrac{\partial}{\partial y_1}v^1_n + R^1_n(l^1,\ldots,l^\nu,v_0,\ldots,v_{n-1}) = 0.\end{cases}$$

Starting with $v^2_0 = 0$ one can solve this system successively with initial conditions

$$v^1_0|_{y_1=\bar{y}_1} \neq 0,\;\; v^1_n|_{y_1=\bar{y}_1} = 0,\;\; n = 1,2,\ldots,N$$

because $v^2_0 = 0$ verifies the first equation.

We turn to the case (a). Up to the factor $\lambda^{2\sigma_1+2\delta p}$ (3.24) is

$$\sum_{n=0}^{N}\{2\sqrt{-1}\,l^1_{y_1}\frac{\partial}{\partial y_1}v_n + R_n(l^1,\ldots,l^\nu,v_0,\ldots,v_{n-1})\}\lambda^{-(n+\nu)/\tau}$$
$$+O(\lambda^{-(N+\nu+1)/\tau}) = O(\lambda^{-N_1}).$$

Hence we are led to

$$\begin{cases}2\sqrt{-1}\,l^1_{y_1}\dfrac{\partial}{\partial y_1}v_n + R_n(l^1,\ldots,l^\nu,v_0,\ldots,v_{n-1}) = 0,\\ v_0|_{y_1=\bar{y}_1} = \begin{pmatrix}\neq 0\\ 0\end{pmatrix},\;\; v_n|_{y_1=\bar{y}_1} = 0,\;\; n = 1,2,\ldots,N\end{cases}$$

from which one can solve v_n successively.

To finish the proof it suffices to prove the following lemma.

Lemma 3.36. *We can choose v_0 so that U_λ is non trivial, that is there is a $\tilde{\kappa} \in \mathbb{Q}_+$ independent of N such that one can write*

$$U_\lambda = M_{\phi,\lambda}E(y,\lambda)\sum_{n=0} v_n\lambda^{-n/\tau} = E(y,\lambda)\lambda^{\tilde{\kappa}}\sum_{n=0} u_n\lambda^{-n/\tau}$$

where $u_0(\bar{y}) \neq 0$.

Proof. Recall $M = \partial_t + A^\sharp\partial_x$ where

$$A^\sharp_{\phi,\lambda} = \begin{bmatrix}\lambda^{-\alpha}(a(y) + O(\lambda^{-1/\tau}) & \lambda^{-\beta}(b(y) + O(\lambda^{-1/\tau})\\ \lambda^{-\beta}(\overline{b(y)} + O(\lambda^{-1/\tau}) & -\lambda^{-\alpha}(a(y) + O(\lambda^{-1/\tau})\end{bmatrix}$$

with $\beta \le \alpha$ and it can be assumed $b(\bar{y}) \neq 0$ thanks to Lemma 3.2. Recall also

$$\lambda^{-2\delta p} M_{\phi,\lambda} = \lambda^{-\delta p}(I - \phi'(\lambda^{-\delta q} y_2) A^{\sharp}_{\phi,\lambda})\partial_1 + \lambda^{\delta q - 2\delta p} A^{\sharp}_{\phi,\lambda}\partial_2 + \lambda^{-2\delta p}\tilde{C}_{\phi,\lambda}.$$

We observe

$$\begin{aligned} -\delta q(\sigma - 1) - \delta p + \sigma_1 &= -\delta q\sigma + \delta q - \delta p + \sigma_1 = -\delta q\sigma + \delta - 1 + \sigma_1 \\ &= \delta(1 - p - \mu) + \delta p + \delta\mu - \delta q\sigma - 1 + \sigma_1 \\ &< \delta(1 - p - \mu) + \delta\mu = 1 + \delta q - 2\delta q \end{aligned}$$

by Lemma 3.33. This proves that

$$\lambda^{-\delta p + \sigma_1}\phi'(\lambda^{-\delta q} y_2) = o(\lambda^{1 + \delta q - 2\delta p}).$$

Since $\tilde{C}_{\phi,\lambda} = O(\lambda^{\delta p})$ we get $\lambda^{-2\delta p}\tilde{C}_{\phi,\lambda} = O(\lambda^{-\delta p})$ and hence

$$\lambda^{-2\delta p}\tilde{C}_{\phi,\lambda} = \lambda^{\delta p}\left(\begin{bmatrix} c(y) & 0 \\ 0 & \overline{c(y)} \end{bmatrix} + O(\lambda^{-1/\tau})\right) = \lambda^{\delta p}(c^0(y) + O(\lambda^{-1/\tau})).$$

We note that $\delta q - 2\delta p + 1 = 2\sigma_1 + \delta\mu$ by Lemma 3.33. Let us set

$$\kappa = \max\{2\sigma_1 + \delta\mu - \beta, -\delta p\}.$$

Then we conclude that

$$\lambda^{-2\delta p} E(y,\lambda)^{-1} M_{\phi,\lambda} E(y,\lambda) = \lambda^{\kappa}\{\mu \begin{bmatrix} \lambda^{\beta-\alpha} a(y) & b(y) \\ \overline{b(y)} & -\lambda^{\beta-\alpha} a(y) \end{bmatrix} + O(\lambda^{-1/\tau})\}$$

when $\kappa = 2\sigma_1 + \delta\mu - \beta > -\delta p$. Since

$$v_0(\bar{y}) = \begin{bmatrix} v_0^1(\bar{y}) \\ 0 \end{bmatrix} \quad \text{(case (b))} \quad \text{or} \quad v_0(\bar{y}) = \begin{bmatrix} v_0^1(\bar{y}) \\ v_0^2(\bar{y}) \end{bmatrix}$$

choosing $v_0^2(y)$ suitably we get the assertion because $b(\bar{y}) \neq 0$. If $\kappa = 2\sigma_1 + \delta\mu - \beta = -\delta p$ then we see

$$\begin{aligned} \lambda^{-2\delta p} E(y,\lambda)^{-1} M_{\phi,\lambda} E(y,\lambda) = \lambda^{\kappa}\Big\{\mu &\begin{bmatrix} \lambda^{\beta-\alpha} a(y) & b(y) \\ \overline{b(y)} & -\lambda^{\beta-\alpha} a(y) \end{bmatrix} \\ &+ \begin{bmatrix} l^1_{y_1} + c(y) & 0 \\ 0 & l^1_{y_1} + \overline{c(y)} \end{bmatrix} + O(\lambda^{-1/\tau})\Big\}. \end{aligned}$$

The choice

$$v_0(\bar{y}) = \begin{pmatrix} v_0^1(\bar{y}) \\ 0 \end{pmatrix}$$

proves the assertion clearly. Finally if $\kappa = -\delta p > 2\sigma_1 + \delta\mu - \beta$ then

$$\lambda^{-2\delta p} E(y,\lambda)^{-1} M_{\phi,\lambda} E(y,\lambda)$$
$$= \lambda^{\kappa}\Big\{ \begin{bmatrix} l^1_{y_1} + c(y) & 0 \\ 0 & l^1_{y_1} + \overline{c(y)} \end{bmatrix} + O(\lambda^{-1/\tau})\Big\}.$$

Since $l^1_{y_1} = F(y) = \{\sqrt{-1}\mu C_0^1(y)\}^{1/2}$ it is clear that one can choose μ so that $l^1_{y_1} + c(y) \neq 0$ and hence the result. □

3.13 Proof of Necessity

In this section we show

Theorem 3.4. *Assume that the Cauchy problem* (CP) *is* C^∞ *well posed near the origin. Then* $(C^\pm)$ *are verified.*

We only show the necessity of (C^+) since the necessity of (C^-) is proved with obvious modifications. Let $t^*(x) \sim |x|^\gamma$ and denote

$$D(r,M) = \{(t,x) \mid 0 < x < r,\ 0 < t < Mx^\gamma\},$$
$$\Delta(\hat{t},\hat{x};c) = \{(t,x) \mid (t-\hat{t}) + c^{-1}|x-\hat{x}| \le 0, 0 \le t \le \hat{t}\}.$$

From Lemma 3.14 it follows that

$$|h(t,x)| \le C(M)^2 r^2 \quad \text{in} \quad (t,x) \in D(r,M). \tag{3.25}$$

Lemma 3.15 implies that every curve $t = \phi(x)$, $\phi(x) \in \mathscr{G}^+(\gamma)$ is a space-like curve

$$h(t,x)\phi'(x)^2 < 1.$$

We state this in a more precise way. Let us put

$$\mu(M,\gamma,\hat{x}) = \begin{cases} C(M)^{-1}(2M)^{-1} & \text{if } \gamma \ge 1, \\ C(M)^{-1}(2M)^{-1}\hat{x}^{1-\gamma} & \text{if } \gamma < 1. \end{cases}$$

Then we have

Lemma 3.37. *There is a* $X = X(M,\gamma) > 0$ *such that*

$$(\hat{t},\hat{x}) \in D(\mu,M),\ 0 < \hat{x} < X \Longrightarrow \Delta(\hat{t},\hat{x}; C(M)\mu) \subset D(\mu,M)$$

where $\mu = \mu(M,\gamma,\hat{x})$.

Proof. When $\gamma \geq 1$ we choose X so that

$$0 < \hat{x} < X \Longrightarrow \gamma\hat{x}^{\gamma-1} < 2.$$

With this choice of X we have

$$\gamma M\hat{x}^{\gamma-1} < (C(M)\mu)^{-1} = \begin{cases} 2M & \text{if} \quad \gamma \geq 1, \\ 2M\hat{x}^{\gamma-1} & \text{if} \quad \gamma < 1 \end{cases}$$

if $0 < \hat{x} < X$. On the other hand it is clear that

$$0 < \hat{x} < X \Longrightarrow \hat{x} + C(M)\mu\hat{x}^{\gamma} < \mu$$

taking X small. Thus we get the assertion. □

From (3.25) we see $|h(t,x)| \leq C(M)^2\mu(M,\gamma,\hat{x})^2$ for any $(t,x) \in D(\mu(M,\gamma,\hat{x}),M)$ then Lemma 3.37 implies that $\Delta(\hat{t},\hat{x}; C(M)\mu)$ is a dependence domain of $(\hat{t},\hat{x})$ provided that $0 < \hat{x} < X, 0 < \hat{t} < M\hat{x}^{\gamma}$. That is (see for example [44, 63])

$$Lu = 0 \ \text{ in } \ \Delta(\hat{t},\hat{x}; C(M)\mu), \ \ u(t,x) = 0 \ \text{ in } \ t \leq 0 \Longrightarrow u(\hat{t},\hat{x}) = 0.$$

Now let $\phi(x) \in \mathscr{G}^{\pm}(\gamma)$ and consider the change of systems of local coordinates

$$T_{\phi} : U \cap \{x > 0\} \ni (t,x) \mapsto (x_1,x_2) = (t - \phi(x), x) \in W \cap \{x_2 > 0\}.$$

With small $\delta > 0$ we put

$$E = E(M,\gamma,\phi) = \{(x_1,x_2) \mid 0 < x_2 < \delta, 0 < x_1 < Mx_2^{\gamma} - \phi(x_2)\}$$

and denote by $\bar{E}$ the closure of E. Let $\tau > 0$ and denote

$$K(\tau) = T_{\phi}^{-1}(\bar{E}^{\tau}) = \{(t,x) \mid 0 \leq x \leq \delta, \phi(x) \leq t \leq Mx^{\gamma}, t \leq \phi(x) + \tau\},$$

$$K = T_{\phi}^{-1}(\bar{E}) = \{(t,x) \mid 0 \leq x \leq \delta, \phi(x) \leq t \leq Mx^{\gamma}\}.$$

Since $\sigma(\phi) \leq \gamma$ from the observations made above we conclude that

$$Lw = 0 \ \text{ in } \ K(\tau), \ \ w \in C_0^{\infty}(K) \Longrightarrow w = 0 \ \text{ in } \ K(\tau).$$

Then from Corollary 1.1, or rather from its proof we have

$$|u|_{C^0(K(\tau))} \le C|Lu|_{C^r(K(\tau))}$$

with some $r \in \mathbb{N}$ for any $u \in C_0^\infty(K)$ and any $|\tau| < \epsilon$.

Proposition 3.10. *Assume that the Cauchy problem* (CP) *for L is C^∞ well posed near the origin. Then there are M, a neighborhood $\tilde{W}$ of the origin, $C > 0$ and $r \in \mathbb{N}$ such that*

$$\sup_{0\le x_1\le\tau} |u| \le C \sup_{0\le x_1\le\tau, k+l+m\le r} |x_2^{-k(1-\sigma)-m}\partial_{x_1}^k \partial_{x_2}^l L_\phi^\sharp u|$$

for any small $\tau > 0$ and any $u \in C_0^\infty(\tilde{W} \cap E)$.

Proof. It suffices to prove the assertion for L_ϕ. Let $u \in C_0^\infty(\tilde{W} \cap E)$. Then we have

$$\begin{aligned} |u|_{C^0(\bar{E}^\tau)} &= |u|_{C^0(\{x_1\le\tau\})} = |u_\phi|_{C^0(K(\tau))} \\ &\le C|Lu_\phi|_{C^r(K(\tau))} = C\sum_{k+l\le r} |\partial_t^k\partial_x^l Lu_\phi|_{C^0(K(\tau))} \\ &\le C'\sum_{k+l+m\le r} |x_2^{-k(1-\sigma)-m}\partial_{x_1}^k\partial_{x_2}^l (L_\phi u)|_{C^0(\{x_1\le\tau\})} \end{aligned}$$

because $|\phi^{(k)}(x_2)| \le C_k x_2^{\sigma-k}$. This proves the assertion. □

Let

$$y_1 = \lambda^{\delta p} x_1, \quad y_2 = \lambda^{\delta q} x_2, \quad \delta, p, q \in \mathbb{Q}_+$$

be a dilation of coordinates such that $p \ge \gamma q$. Let $L_{\phi,\lambda}$ be the representation of L_ϕ in the coordinates (y_1, y_2)

$$L_{\phi,\lambda}(y, D) = L_\phi(\lambda^{-\delta p} y_1, \lambda^{-\delta q} y_2, \lambda^{\delta p} D_{y_1}, \lambda^{\delta q} D_{y_2}).$$

Then we have

Proposition 3.11. *Let $B > 0, c > 0$ be given and let $p \ge \gamma q$, $\phi \in \mathscr{C}^+(h|a^\sharp|^2)$ and $1 + p > q$. Assume that the Cauchy problem for L is C^∞ well posed near the origin. Then there are $C > 0$, $r \in \mathbb{N}$, $\lambda_0 = \lambda_0(B, \sigma, \phi)$ such that*

$$\sup_{0\le y_1\le\bar{y}_1} |u| \le C\lambda^{\delta k r+\delta q r} \sup_{0\le y_1\le\bar{y}_1, |\beta|\le r} |D_y^\beta(L_{\phi,\lambda}^\sharp u)|$$

for any $u \in C_0^\infty(\{0 < y_1, c < y_2 < B\})$, $k = \max(p, q)$, $\delta = (1 + p - q)^{-1}$, $\lambda \ge \lambda_0$.

Proof. Let $u \in C_0^\infty(\{0 < y_1, c < y_2 < B\})$ and $u_\lambda(y) = u(\lambda^{\delta p} y_1, \lambda^{\delta q} y_2)$. Then there are λ_0 and M_0 so that $u_\lambda \in C_0^\infty(\tilde{W} \cap E)$ if $\lambda \geq \lambda_0$, $M \geq M_0$ and $u \in C_0^\infty(\{0 < y_1, c < y_2 < B\})$. Applying Proposition 3.10 we get

$$\sup_{0 \leq y_1 \leq \tau} |u_\lambda| \leq C\lambda^{\delta q r} \sup_{0 \leq y_1 \leq \tau, |\alpha| \leq r} |D^\alpha(L_\phi^\sharp u_\lambda)|.$$

Taking $\tau = \lambda^{-\delta p} \bar{y}_1$ we get the desired inequality. □

Proof of Theorem 3.4. From Proposition 3.9 we can construct a family of asymptotic solutions U_λ. Take $\chi(y) \in C_0^\infty(W)$ so that $\chi(y) = 1$ on a neighborhood of $\bar{y}$. Set $u_\lambda = \chi(y) U_\lambda(y)$ then we have

$$\sup_{0 \leq y_1 \leq \bar{y}_1, |\alpha| \leq r} |D^\alpha(L_{\phi,\lambda}^\sharp u_\lambda)| \leq C\lambda^{2\sigma_1 + 2\delta p + r - (\nu + N + 1)/\tau}.$$

On the other hand since $u_\lambda(\bar{y}) \geq c\lambda^\kappa$ with some $c > 0$, taking N large these two inequalities are not compatible which proves the assertion.

□

3.14 Equivalence of Conditions

The aim of this section is to prove

Proposition 3.12. *The condition* $(C^\pm; Y)$ *is equivalent to*

$$\Gamma(tY_\phi) \subset \frac{1}{2}\Gamma([h|a^\sharp|^2]_\phi), \quad \forall \phi \in \mathcal{G}^\pm(\gamma).$$

Similarly the condition $(C^\pm; Z)$ *is equivalent to*

$$\Gamma(tZ_\phi) \subset \frac{1}{2}\Gamma([h|a^\sharp|^2]_\phi), \quad \forall \phi \in \mathcal{G}^\pm(\gamma).$$

Corollary 3.3. *Assume that the Cauchy problem* (CP) *is* C^∞ *well posed near the origin. Then we have*

$$\Gamma(tY_\phi) \subset \frac{1}{2}\Gamma([h|a^\sharp|^2]_\phi), \quad \Gamma(tZ_\phi) \subset \frac{1}{2}\Gamma([h|a^\sharp|^2]_\phi)$$

for any pseudo-characteristic curve $t = \phi(x)$ *of* $\partial_t - A(t, x)\partial_x$.

Proof. It is clear since $\mathcal{C}^\pm(h|a^\sharp|^2) \subset \mathcal{G}^\pm(\gamma)$. □

Proof of Proposition 3.12. Since the proof is similar for all cases we prove that $(C^+; Z)$ is equivalent to

$$\Gamma(tZ_\phi) \subset \frac{1}{2}\Gamma([h|a^\sharp|^2]_\phi), \quad \forall \phi \in \mathcal{G}^+(\gamma). \tag{3.26}$$

Let $p, q \in \mathbb{Q}_+$, $\phi \in \mathcal{G}^+(\gamma)$, $p \geq \sigma(\phi)q$, $\mu(h_\phi; p, q) > 2q(1 - \sigma(\phi))$. Note that (3.26) implies that

$$\begin{aligned} p + \mu([Z]_\phi; p, q) &\geq \frac{1}{2}\mu([h|a^\sharp|^2]_\phi; p, q) \\ &= \frac{1}{2}\mu(h_\phi; p, q) + \frac{1}{2}\mu([|a^\sharp|^2]_\phi; p, q) \\ &= \frac{1}{2}\mu(h_\phi; p, q) + \mu(a^\sharp_{12,\phi}; p, q). \end{aligned}$$

By definition, this shows that

$$p + \mu\left(\left[\frac{Z}{a^\sharp}\right]_\phi; p, q\right) \geq \frac{1}{2}\mu(h_\phi; p, q)$$

which is $(C^+; Z)$.

Conversely we show that $(C^+; Z)$ implies (3.26). Note that

$$\begin{aligned} (tZ_\phi)(s^p x_1, s^q x_2) &= (a^\sharp)_\phi(s^p x_1, s^q x_2)\{t(\frac{Z}{a^\sharp})_\phi\}(s^p x_1, s^q x_2) \\ &= s^\nu(c^0(x) + o(1)) \end{aligned}$$

with $\nu = \mu([a^\sharp]_\phi; p, q) + \mu([Z/a^\sharp]_\phi; p, q) + p$. Let

$$[h|a^\sharp|^2]_\phi(s^p x_1, s^q x_2) = s^\kappa(d^0(x) + o(1))$$

with $\kappa = 2\mu([a^\sharp]_\phi; p, q) + \mu(h_\phi; p, q)$ and hence

$$2\nu - \kappa = 2p + 2\mu([\frac{Z}{a^\sharp}]_\phi; p, q) - \mu(h_\phi; p, q).$$

Thus $(C^+; Z)$ implies that $2\nu \geq \kappa$, that is

$$2\mu(tZ_\phi; p, q) \geq \mu([h|a^\sharp|^2]_\phi; p, q) \tag{3.27}$$

for any $p, q \in \mathbb{Q}_+$ and for any $\phi \in \mathcal{G}^+(\gamma)$ verifying the conditions $p \geq \sigma(\phi)q$ and $\mu(h_\phi; p, q) > 2q(1 - \sigma(\phi))$ (if $\phi = 0$ then $q\sigma(\phi)$ should read as p). Take $\phi \in \mathcal{G}^+(\gamma)$ and denote by

$$\{(j, \beta_j(\phi))\}_{j=0}^r, \quad \{(j, \gamma_j(\phi))\}_{j=0}^{\tilde{r}}$$

the set of points which consists in the boundary of $\frac{1}{2}\Gamma([h|a^\sharp|^2]_\phi)$ and $\Gamma(Z_\phi)$ respectively where $\beta_r(\phi) = n$, $\gamma_{\tilde{r}}(\phi) = \tilde{n}$, $n = n_1 + n_2$ and $r = m_1 + m_2$. Set

$$\epsilon_j(\phi) = \beta_{j-1}(\phi) - \beta_j(\phi), \quad 1 \le j \le r,$$
$$\delta_j(\phi) = \gamma_{j-1}(\phi) - \gamma_j(\phi), \quad 1 \le j \le \tilde{r}.$$

Note that the boundary points of $\Gamma(tZ_\phi)$ consists of $\{(j+1, \gamma_j(\phi))\}_{j=0}^{\tilde{r}}$. Then to prove the assertion it is enough to show that

$$\gamma_j(\phi) \ge \beta_{j+1}(\phi), \quad \forall j \ge 0.$$

Assume that $\sigma(\phi) > \epsilon_r(\phi)$. Let

$$\epsilon_1(\phi) \ge \cdots \ge \epsilon_\ell(\phi) \ge \sigma(\phi) > \epsilon_{\ell+1}(\phi) \ge \cdots \ge \epsilon_r(\phi)$$

and $\alpha p_j + \beta q_j = 1$ be the line passing $(j-1, \beta_{j-1}(\phi))$ and $(j, \beta_j(\phi))$, that is

$$\frac{p_j}{q_j} = \epsilon_j(\phi)$$

which is tangent to $\frac{1}{2}\Gamma((h|a^\sharp|^2)_\phi)$. Hence we have

$$\frac{p_j}{q_j} = \epsilon_j(\phi) \ge \sigma(\phi) \quad \text{for} \quad 1 \le j \le \ell$$

that is $p_j \ge \epsilon_j(\phi) q_j$ for $1 \le j \le \ell$. □

Here we note

Lemma 3.38. *We have*

$$\frac{1}{2}\Gamma((h|a^\sharp|^2)_\phi) \subset \text{convex hull of } ((r,n) + \mathbb{R}^2_+) \cup ((0, n+1) + \mathbb{R}^2_+).$$

Proof. Let us write

$$h|a^\sharp|^2 = x^{2(n_1+n_2)} \prod^{2(m_1+m_2)} (t - t_\nu(x)) \hat{e}(t,x)$$

where $n = n_1 + n_2$ and $r = m_1 + m_2$. It is clear that

$$\Gamma((h|a^\sharp|^2)_\phi) = \Gamma(x^{2n} \prod^{2r} (t + \phi(x) - t_\nu(x))).$$

Recall that there is ν_0 such that $t_{\nu_0}(x) \sim t^*(x)$ and this implies that

$$C\,|t_{\nu_0}(x)| \geq |\phi(x) - t_\nu(x)| \quad \text{for any} \quad 1 \leq \nu \leq 2r.$$

Hence we have

$$\Gamma(x^{2n}\prod_{\nu=1}^{2r}(t + \phi(x) - t_\nu(x))) \subset \Gamma(x^{2n}\prod_{\nu=1}^{2r}(t - t_{\nu_0}(x))).$$

On the other hand, from the proof of Lemma 3.9 we see that

$$t_{\nu_0}(x)^{2r} = O(|x|^2).$$

Since

$$\frac{1}{2}\Gamma(x^{2n}\prod_{\nu=1}^{2r}(t - t_{\nu_0}(x))) \subset \text{ convex hull of } ((r,n) + \mathbb{R}^2_+) \cup ((0, n+1)) + \mathbb{R}^2_+)$$

this proves the assertion. □

Lemma 3.38 shows that

$$\frac{1}{q_j} \geq n + 1$$

and hence $q_j \leq 1$. Since $\sigma(\phi) > 0$ we get $1 > q_j(1 - \sigma(\phi))$. Then the condition $(C^+; Z)$ is verified for $p = p_j, q = q_j$. Thus we get from (3.27) that

$$\Gamma(Z_\phi) \text{ lies right side of the line } (\alpha + 1)p_j + \beta q_j = 1, \quad 1 \leq j \leq \ell.$$

This proves that

$$\gamma_j(\phi) \geq \beta_{j+1}(\phi), \quad 0 \leq j \leq \ell - 1. \tag{3.28}$$

We now show that $\tilde{n} \geq n$. If $n = 0$ nothing to be proved. If $n \geq 1$ then with $\phi = 0$, $q = s/n$, $p = (1-s)/r$ one can apply (3.27) because

$$1 + p = \frac{1 - s + r}{r} > \frac{s}{n}.$$

Thus one gets

$$\tilde{n} \geq \frac{n}{s}.$$

Letting $s \uparrow 1$ we conclude that $\tilde{n} \geq n$. Then we have

$$\gamma_j(\phi) \geq \tilde{n} \geq n = \beta_{j+1}(\phi) \quad \text{for} \quad r - 1 \leq j.$$

Then it remains to prove

$$\gamma_j(\phi) \geq \beta_{j+1}(\phi) \quad \text{for} \quad \ell \leq j \leq r-2. \tag{3.29}$$

To prove (3.29) we prepare a lemma.

Lemma 3.39. *Let $f(t,x) = x^n \prod^m (t - t_\nu(x))$ and $\{(j, \beta_j(\phi))\}$ be on the boundary of $\Gamma(f_\phi)$. Assume that $\sigma(\psi - \phi) = \sigma(\psi)$. Let $\epsilon_j(\phi) = \beta_{j-1}(\phi) - \beta_j(\phi)$.*

(1) *Assume $\sigma(\psi) > \epsilon_{k+1}(\phi)$ then we have*

$$\epsilon_j(\psi) = \epsilon_j(\phi) \quad \textit{for} \quad j \geq k+1.$$

(2) *Assume $\sigma(\psi) \geq \epsilon_{k+1}(\phi)$ then we have*

$$\epsilon_j(\psi) \geq \epsilon_j(\phi) \quad \textit{for} \quad j \geq k+1.$$

Proof. (1) Take $\ell \leq k$ so that $\epsilon_\ell(\phi) > \epsilon_{\ell+1}(\phi) = \cdots = \epsilon_{k+1}(\phi)$. From the definition of $\epsilon_j(\phi)$ it is clear that $(\ell, \beta_\ell(\phi))$ is a vertex of $\Gamma(f_\phi)$. Recall that

$$f_\phi(t,x) = x^n \prod^m (t + \phi(x) - t_\nu(x)) = \sum_{j=1}^m C_j^\phi(x) t^j.$$

By definition we get

$$C_j^\phi(x) = O(|x|^{\beta_m(\phi) + \sum_{i=m}^{j+1} \epsilon_i(\phi)}).$$

When $j = \ell$ since $(\ell, \beta_\ell(\phi))$ is a vertex of $\Gamma(f_\phi)$ we see

$$|C_\ell^\phi(x)| = |x|^{\beta_m(\phi) + \sum_{i=m}^{\ell+1} \epsilon_i(\phi)} (c + o(1)) \tag{3.30}$$

with $c \neq 0$. We observe that

$$\frac{1}{\ell!} \left(\frac{\partial}{\partial t} \right)^\ell f_\phi(t,x)|_{t=\psi-\phi} = \frac{1}{\ell!} \left(\frac{\partial}{\partial t} \right)^\ell f_\psi(t,x)|_{t=0} = C_\ell^\psi(x).$$

Then we see that

$$C_\ell^\psi(x) = \sum_{j \geq \ell} \frac{j!}{(j-\ell)!} C_j^\phi(x)(\psi - \phi)^{j-\ell}. \tag{3.31}$$

Note that

$$C_j^\phi(x)(\psi - \phi)^{j-\ell} = O(|x|^{\beta_m(\phi) + \sum_{i=m}^{j+1} \epsilon_i(\phi) + (j-\ell)\sigma(\psi)}).$$

Since $\sigma(\psi) > \epsilon_{\ell+1}(\phi) \geq \epsilon_{\ell+2}(\phi) \geq \cdots$ it follows from (3.31) that

$$C_\ell^\psi(x) = |x|^{\beta_m(\phi)+\sum_{i=m}^{\ell+1}\epsilon_i(\phi)}(c + o(1))$$

with $c \neq 0$. Similarly one can show that if $(i, \beta_i(\phi))$ is a vertex of $\Gamma(f_\phi)$ and $i > \ell$, then $(i, \beta_i(\phi))$ also belongs to $\Gamma(f_\psi)$.

On the other hand, we have

$$C_j^\psi(x) = O(|x|^{\beta_m(\phi)+\sum_{i=m}^{j+1}\epsilon_i(\phi)}) \quad \text{for} \quad j \geq \ell + 1$$

in general, and hence this shows that

$$\Gamma(f_\phi) \cap \{x \leq \beta_\ell(\phi)\} = \Gamma(f_\psi) \cap \{x \leq \beta_\ell(\phi)\} \tag{3.32}$$

and hence the assertion.

(2) If there is $\ell \geq k+1$ so that $\epsilon_{k+1}(\phi) = \cdots = \epsilon_\ell(\phi) > \epsilon_{\ell+1}(\phi)$ then since $\sigma(\psi) \geq \epsilon_{k+1}(\phi) > \epsilon_{\ell+1}(\phi)$ the same arguments proving (1) shows

$$\epsilon_j(\psi) = \epsilon_j(\phi) \quad \text{for} \quad j \geq \ell + 1.$$

We turn to $\epsilon_j(\phi)$, $\epsilon_j(\psi)$ for $j < \ell + 1$. Since

$$C_j^\psi(x) = O(|x|^{\beta_m(\phi)+\sum_{i=m}^{j+1}\epsilon_i(\phi)}) \quad \text{for} \quad j \geq \ell$$

and $\Gamma(f_\phi)$ is convex this proves that

$$\epsilon_j(\psi) \geq \epsilon_j(\phi) \quad \text{for} \quad j = k+1, \ldots, \ell + 1$$

and hence the assertion. □

We prove (3.29) by contradiction. Suppose that there were j with $\ell \leq j \leq r-2$ such that

$$\gamma_j(\phi) < \beta_{j+1}(\phi).$$

Let us define $j^* = \max\{j \mid \gamma_j(\phi) < \beta_{j+1}(\phi)\}$. By definition we have

$$\gamma_{j^*+1}(\phi) \geq \beta_{j^*+2}(\phi) \quad \text{and} \quad \gamma_{j^*}(\phi) < \beta_{j^*+1}(\phi).$$

This implies that

$$\delta_{j*+1}(\phi) = \gamma_{j^*}(\phi) - \gamma_{j^*+1}(\phi) < \beta_{j^*+1}(\phi) - \beta_{j^*+2}(\phi) = \epsilon_{j^*+2}(\phi) < \sigma(\phi).$$

Take $\psi \in \mathscr{G}^+(\gamma)$ so that

$$\sigma(\psi) = \epsilon_{j^*+2}(\phi).$$

Since $\sigma(\psi - \phi) = \sigma(\phi)$, $\delta_{j^*+1}(\phi) < \sigma(\psi)$ it follows from Lemma 3.39 that

$$\delta_{j+1}(\psi) = \delta_{j+1}(\phi) \quad \text{for} \quad j \geq j^*.$$

Hence one has

$$\gamma_{j^*}(\psi) = \sum_{j=j^*+1}^{r} \delta_j(\psi) + \tilde{n} = \sum_{j=j^*+1}^{r} \delta_j(\phi) + \tilde{n} = \gamma_{j^*}(\phi).$$

Noting that $\sigma(\psi) = \epsilon_{j^*+2}(\phi) \geq \cdots \geq \epsilon_r(\phi)$ we apply Lemma 3.39 to get

$$\epsilon_j(\psi) \geq \epsilon_j(\phi) \quad \text{for} \quad j^* + 2 \leq j \leq r$$

and then

$$\beta_{j^*+1}(\psi) = \sum_{j=j^*+2}^{r} \epsilon_j(\psi) + n \geq \sum_{j=j^*+2}^{r} \epsilon_j(\phi) + n = \beta_{j^*+1}(\phi).$$

Since $\epsilon_i(\psi) \geq \epsilon_{j^*+2}(\psi) \geq \epsilon_{j^*+2}(\phi) = \sigma(\psi)$ for $0 \leq i \leq j^* + 2$ the same arguments as before give that

$$\gamma_j(\psi) \geq \beta_{j+1}(\psi) \quad \text{for} \quad 0 \leq j \leq j^* + 1.$$

This clearly gives a contradiction because

$$\gamma_{j^*}(\phi) = \gamma_{j^*}(\psi) \geq \beta_{j^*+1}(\psi) \geq \beta_{j^*+1}(\phi)$$

where the last inequality follows from

$$\beta_{j^*+1}(\psi) = \sum_{j=j^*+2}^{r} \epsilon_j(\psi) + n \geq \sum_{j=j^*+2}^{r} \epsilon_j(\phi) + n = \beta_{j^*+1}(\phi).$$

Thus we get (3.29).

When $\epsilon_r(\phi) \geq \sigma(\phi)$ taking the line given by

$$tp_j + xq_j = 1 \quad \text{with} \quad \frac{p_j}{q_j} = \sigma_j(\phi) \geq \sigma(\phi) \quad (0 \leq j \leq r - 2)$$

one can conclude that

$$\gamma_j(\phi) \geq \beta_{j+1}(\phi) \quad \text{for} \quad 0 \leq j \leq r - 2$$

and hence the result. □

3.15 Concluding Remarks

The main results Theorems 3.1 and 3.2 were proved in [55]. Extensions of these results to $m \times m$ ($m \geq 3$) systems are open while some partial results are found in [32, 49]. For the case that $h(t,x) \equiv 0$ a necessary and sufficient condition for the C^∞ well-posedness was obtained in [36,37] from somewhat different point of view.

There is a class of systems introduced in [8] and called pseudosymmetric systems which includes symmetric systems and triangular systems. The C^∞ well-posedness of the Cauchy problem for pseudosymmetric systems depending only on the time variable analytically is discussed in [8].

For scalar second order hyperbolic operators with two independent variables with real analytic coefficients, a necessary and sufficient condition in order that the Cauchy problem is C^∞ well posed was given in [46, 47]. Extension to scalar hyperbolic operators of order $m \geq 3$ with two independent variables with real analytic coefficients is also still open.

For scalar second order hyperbolic operators with coefficients depending only on the time variable analytically, a necessary and sufficient condition for C^∞ well-posedness is obtained in [70].

Problem. Look for a C^∞ version of Theorem 3.1. As for this question we can find interesting results in [7, 42] where they have studied 2×2 systems with C^∞ coefficients depending only on t and $D^\#$ or its analog plays an important role there.

Problem. Generalize Theorems 3.1 and 3.2 to the case $m \times m$ ($m \geq 3$) systems with real analytic coefficients.

Problem. We have assumed that $A(t,x)$ is real valued which is not necessary for $A(t,x)$ has only real eigenvalues. Generalize Theorems 3.1 and 3.2 to the case when $A(t,x)$ is not necessarily real valued.

Problem. In the case of first order $m \times m$ systems with constant coefficients L is strongly hyperbolic if and only if $L(\xi)$ is uniformly diagonalizable remarked in Sect. 2.10. This fact motivates the study of uniformly diagonalizable systems with variable coefficients. In Corollary 3.2 we have proved that if $A(t,x)$ is uniformly diagonalizable then 2×2 system $L = \partial_t - A(t,x)\partial_x$ is strongly hyperbolic near the origin. For $m \times m$ system $L = \partial_t - A(t)\partial_x$ with two independent variables (t,x) if $A(t)$ is real analytic and uniformly diagonalizable then L is strongly hyperbolic (see [9]). On the other hand there exists $A(t)$ which belongs to Gevrey s class for any $s > 2$ such that $A(t)$ is uniformly diagonalizable with real eigenvalues while $L = \partial_t - A(t)\partial_x$ is not strongly hyperbolic [6, 65]. If $A(t,x)$ belongs to the Gevrey class $1 \leq s \leq 2$ and uniformly diagonalizable what one can say for strong hyperbolicity of $L = \partial_t - A(t,x)\partial_x$?

Chapter 4
Systems with Nondegenerate Characteristics

Abstract In this chapter we introduce the notion of nondegenerate multiple characteristics. Simple characteristics are nondegenerate characteristics of order 1. A double characteristic ρ of L is nondegenerate if and only if the rank of the Hessian at ρ of the determinant of $L(x, \xi)$ is maximal. We prove that every hyperbolic system which is close to a hyperbolic system with nondegenerate multiple characteristic has a nondegenerate characteristic of the same order nearby. This implies that hyperbolic systems with a nondegenerate multiple characteristic can not be approximated by strictly hyperbolic systems which contrasts with the case of scalar hyperbolic operators. We also prove that if every multiple characteristic of the system L is nondegenerate then there exists a smooth symmetrizer and hence the Cauchy problem for L is C^∞ well posed for any lower order term. Finally we discuss about the stability of symmetric systems in the space of hyperbolic systems.

4.1 Nondegenerate Characteristics

Let $P(x)$ be an $m \times m$ matrix valued smooth function defined near $\bar{x} \in \mathbb{R}^n$. We assume that $P(x)$ is a polynomial in x_1 so that

$$P(x) = \sum_{j=0}^{q} A_j(x')x_1^{q-j} \tag{4.1}$$

where $x' = (x_2, \ldots, x_n)$. We adapt the definitions of hyperbolicity and characteristics in Chap. 1 to $P(x)$.

Definition 4.1. We say $P(x)$ is hyperbolic near $\bar{x}$ with respect to $\theta = (1, 0, \ldots, 0) \in \mathbb{R}^n$ if $\det A_0(x') \neq 0$ near $x' = \bar{x}'$ and

$$\det P(x + \lambda\theta) = 0 \Longrightarrow \lambda \text{ is real} \tag{4.2}$$

T. Nishitani, *Hyperbolic Systems with Analytic Coefficients*, Lecture Notes in Mathematics 2097, DOI 10.1007/978-3-319-02273-4_4,

for any x near $\bar{x}$. We say that $\bar{x}$ is a characteristic of order r of $P(x)$ if

$$\partial_x^\alpha(\det P)(\bar{x}) = 0, \ \ \forall|\alpha| < r, \ \ \partial_x^\alpha(\det P)(\bar{x}) \neq 0, \ \ \exists|\alpha| = r. \tag{4.3}$$

We now define nondegenerate characteristics. To do so we first define the localization of $P(x)$ at a characteristic.

Definition 4.2. Let $\bar{x}$ be a characteristic of $P(x)$ verifying

$$\mathrm{Ker}P(\bar{x}) \cap \mathrm{Im}\, P(\bar{x}) = \{0\}. \tag{4.4}$$

Set $\dim \mathrm{Ker}P(\bar{x}) = r$. Let $\{v_1, \ldots, v_r\}$ be a basis for $\mathrm{Ker}P(\bar{x})$ and let $\{\ell_1, \ldots, \ell_r\}$ be the dual basis vanishing on $\mathrm{Im}\, P(\bar{x})$, that is

$$\ell_i(\mathrm{Im}\, P(\bar{x})) = 0, \quad \ell_i(v_j) = \delta_{ij}$$

where δ_{ij} is the Kronecker's delta. Then we define the localization of P at $\bar{x}$, a linear transformation on $\mathrm{Ker}\, P(\bar{x})$, defined by a $r \times r$ matrix $P_{\bar{x}}(x)$ with respect to the basis $\{v_1, \ldots, v_r\}$

$$\big(\ell_i(P(\bar{x} + \mu x)v_j)\big)_{1 \le i,j \le r} = \mu[P_{\bar{x}}(x) + O(\mu)]. \tag{4.5}$$

Remark. Let $\{\tilde{v}_j\}$ be another basis for $\mathrm{Ker}P(\bar{x})$ where $\tilde{v}_j = \sum t_{kj} v_k$ with a non singular $r \times r$ matrix $T = (t_{ij})$ and let $\{\tilde{\ell}_i\}$ be the dual basis vanishing on $\mathrm{Im}\, P(\bar{x})$. Define $\tilde{P}_{\bar{x}}(x)$ by (4.5) with $\{\tilde{v}_j\}$ and $\{\tilde{\ell}_i\}$ then it is clear that $\tilde{P}_{\bar{x}}(x) = T^{-1} P_{\bar{x}}(x) T$ and hence $P_{\bar{x}}(x)$ is a well defined linear map on $\mathrm{Ker}P(\bar{x})$.

Let us denote

$$P_{\bar{x}} = \{P_{\bar{x}}(x) \mid x \in \mathbb{R}^n\} \subset M_r(\mathbb{C}) \tag{4.6}$$

which is a linear subspace of $M_r(\mathbb{C})$.

Definition 4.3. We call $\dim_{\mathbb{R}} P_{\bar{x}}$, the dimension of the linear subspace $\{P_{\bar{x}}(x) \mid x \in \mathbb{R}^n\}$ over $\mathbb{R}$, the real reduced dimension of $P_{\bar{x}}(x)$.

We first show

Lemma 4.1. *Let $T(x)$ be a smooth non singular $m \times m$ matrix near $\bar{x}$ and let $\tilde{P}(x) = T^{-1}(x)P(x)T(x)$. Then if $\bar{x}$ is a characteristic of order r of $P(x)$ verifying (4.4) then $\bar{x}$ is also a characteristic of order r of $\tilde{P}(x)$ verifying (4.4) and there is a non singular $r \times r$ matrix such that*

$$\tilde{P}_{\bar{x}} = T^{-1} P_{\bar{x}} T.$$

Proof. Since $\mathrm{Ker}\, \tilde{P}(\bar{x}) = T^{-1}(\mathrm{Ker}\, P(\bar{x}))$ and $\mathrm{Im}\, \tilde{P}(\bar{x}) = T^{-1}(\mathrm{Im}\, P(\bar{x}))$ with $T = T(\bar{x})$ it is easy to see

$$\tilde{\ell}_i\big(\tilde{P}(\bar{x}+\mu x)\tilde{v}_j\big) = \ell_i\big(P(\bar{x}+\mu x)v_j\big) + O(\mu^2)$$

where $\tilde{\ell}_i(\cdot) = \ell_i(T\cdot)$ and $\tilde{v}_j = T^{-1}v_j$. This proves the assertion. □

Lemma 4.2. *Assume that $P(x)$ is hyperbolic near $\bar{x}$. Let $\bar{x}$ be a characteristic verifying (4.4) with* $\dim \mathrm{Ker} P(\bar{x}) = r$. *Then we have*

$$\det P(\bar{x}+\mu x) = \mu^r\big(c \det P_{\bar{x}}(x) + O(\mu)\big) \tag{4.7}$$

with $c \neq 0$. Assume further that $\det P_{\bar{x}}(x) \not\equiv 0$ *then*

$$\det P_{\bar{x}}(\theta) \neq 0, \tag{4.8}$$
$$\det P_{\bar{x}}(x+\lambda\theta) = \det\big(P_{\bar{x}}(x) + \lambda P_{\bar{x}}(\theta)\big) = 0 \Longrightarrow \lambda \in \mathbb{R}, \forall x \in \mathbb{R}^n. \tag{4.9}$$

Proof. In view of (4.4) we can choose a non singular constant matrix T so that

$$T^{-1}P(\bar{x})T = \begin{bmatrix} 0 & 0 \\ 0 & G \end{bmatrix}$$

where G is a non singular $(m-r)\times(m-r)$ matrix. With $\tilde{P}(x) = T^{-1}P(x)T$ we write

$$\tilde{P}(\bar{x}+\mu x) = \tilde{P}(\bar{x}) + \mu\hat{P}(x) + O(\mu^2).$$

Denoting

$$\hat{P}(x) = \begin{bmatrix} \hat{P}_{11}(x) & \hat{P}_{12}(x) \\ \hat{P}_{21}(x) & \hat{P}_{22}(x) \end{bmatrix}$$

it is clear $\tilde{P}_{\bar{x}}(x) = \hat{P}_{11}(x)$ which follows from the definition. Since $\det \tilde{P}_{\bar{x}} = \det P_{\bar{x}}$ by Lemma 4.1 we have

$$\det P_{\bar{x}}(x) = \det \hat{P}_{11}(x). \tag{4.10}$$

Note that

$$\det P(\bar{x}+\mu x) = \det \tilde{P}(\bar{x}+\mu x) = \mu^r\big(\det G \det \hat{P}_{11}(x) + O(\mu)\big) \tag{4.11}$$

which shows the first assertion. To prove the second assertion suppose $\det P_{\bar{x}}(\theta) = 0$ so that $\det P(\bar{x}+\mu\theta) = o(\mu^r)$ by (4.7). This implies that $(\partial/\partial x_1)^j \det P(\bar{x}) = 0$ for $j = 0,\dots,r$. Since $\det P(x)$ is hyperbolic in the sense (4.2) it follows from Lemma 1.9 that

$$\big(\partial_x^\alpha \det P\big)(\bar{x}) = 0, \quad \forall|\alpha| \le r.$$

This implies $\det P_{\bar{x}}(x) \equiv 0$ which is a contradiction. We turn to the third assertion. Since

$$\det P(\bar{x} + \mu(x + \lambda\theta)) = \mu^r\big(c \det P_{\bar{x}}(x + \lambda\theta) + O(\mu)\big)$$

if $\det P_{\bar{x}}(x + \lambda\theta) = 0$ has a non real root λ, then taking $\mu \neq 0$ sufficiently small the equation

$$c \det P_{\bar{x}}(x + \lambda\theta) + O(\mu) = 0$$

admits a non real root. This contradicts (4.2). □

Definition 4.4. Denote by $M_r^h(\mathbb{C})$ the set of all $r \times r$ Hermitian matrices and by $M_r^s(\mathbb{R})$ the set of all real $r \times r$ symmetric matrices. Then r^2 and $r(r+1)/2$ is the dimension of $M_r^h(\mathbb{C})$ and $M_r^s(\mathbb{R})$ over $\mathbb{R}$ respectively.

Definition 4.5. We say that $\bar{x}$ is a nondegenerate characteristic of order r of $P(x)$ if the following conditions are verified;

$$\operatorname{Ker} P(\bar{x}) \cap \operatorname{Im} P(\bar{x}) = \{0\}, \tag{4.12}$$

$$\dim_{\mathbb{R}} P_{\bar{x}} = r^2 = \dim_{\mathbb{R}} M_r^h(\mathbb{C}), \quad (r = \dim \operatorname{Ker} P(\bar{x})), \tag{4.13}$$

$$\det P_{\bar{x}}(\theta) \neq 0, \quad P_{\bar{x}}(\theta)^{-1} P_{\bar{x}}(x) \text{ is diagonalizable } \forall x \in \mathbb{R}^n. \tag{4.14}$$

When $P(x)$ is real valued then we say that $\bar{x}$ is a nondegenerate characteristic of order r if

$$\operatorname{Ker} P(\bar{x}) \cap \operatorname{Im} P(\bar{x}) = \{0\}, \tag{4.15}$$

$$\dim_{\mathbb{R}} P_{\bar{x}} = r(r+1)/2 = \dim_{\mathbb{R}} M_r^s(\mathbb{R}), \quad (r = \dim \operatorname{Ker} P(\bar{x})), \tag{4.16}$$

$$\det P_{\bar{x}}(\theta) \neq 0, \quad P_{\bar{x}}(\theta)^{-1} P_{\bar{x}}(x) \text{ is diagonalizable } \forall x \in \mathbb{R}^n. \tag{4.17}$$

Example 4.1. Simple characteristics verify (4.12)–(4.14) with $r = 1$ and hence a simple characteristic is a nondegenerate characteristic of order 1.

Example 4.2. Let $q = 1$ and $m = 2$ so that $P(x) = x_1 + A_1(x')$ where $A_1(x')$ is a real valued 2×2 matrix with $A_1(0) = O$. As we will see in the next section that if the rank of the Hessian of $\det P(x)$ at $x = 0$ is 3 then $x = 0$ is a nondegenerate characteristic of order 2.

Example 4.3. Let us consider

$$P(x) = \xi_1 I + \sum_{j=2}^{d} F_j \xi_j$$

where $\{I, F_2, \ldots, F_d\}$ span $M_m^s(\mathbb{R})$ and $d = m(m+1)/2$. Then every characteristic of P is nondegenerate. We check this. Let $\bar{\xi}$ be a characteristic of order r of $P(\xi)$ so that 0 is an eigenvalue of $P(\bar{\xi})$ of multiplicity r. Take an orthogonal matrix T such that

$$T^{-1}P(\bar{\xi})T = \begin{bmatrix} O & O \\ O & G \end{bmatrix}$$

where G is a $(m-r) \times (m-r)$ non singular matrix. Denoting

$$\tilde{P}(\xi) = T^{-1}P(\xi)T = (\varphi_{ij}(\xi))_{1 \le i,j \le m}$$

we note that $\varphi_{ij}(\xi) = \varphi_{ji}(\xi)$ and $\varphi_{ij}(\xi)$, $i \le j$ are linearly independent. Writing

$$\tilde{P}(\xi) = \begin{bmatrix} \tilde{P}_{11} & \tilde{P}_{12} \\ \tilde{P}_{21} & \tilde{P}_{22} \end{bmatrix}$$

it is clear that $\tilde{P}_{\bar{\xi}}(\xi) = \tilde{P}_{11}(\xi)$ and $\dim \tilde{P}_{\bar{\xi}} = r(r+1)/2$ because φ_{ij}, $i \le j$ are linearly independent. Since $\tilde{P}_{11}(\xi)$ is symmetric for every ξ then (4.14) is also obvious. Thus by Lemma 4.1 we conclude that $\bar{\xi}$ is a nondegenerate characteristic of P.

To study $P(x)$ we consider the following $mq \times mq$ matrix valued function

$$\mathscr{P}(x) = x_1 I + \begin{bmatrix} 0 & -I_m & & & \\ 0 & 0 & -I_m & & \\ & & & \ddots & \\ & & & & -I_m \\ A_q(x') & \cdots & & \cdots & A_1(x') \end{bmatrix} = x_1 I + \mathscr{A}(x')$$

where I and I_m are the $mq \times mq$ and $m \times m$ identity matrix respectively. It is clear that

$$\det \mathscr{P}(x) = \det P(x). \tag{4.18}$$

Then the condition (4.2) implies that all eigenvalues of $\mathscr{A}(x')$ are real, equivalently

$$\text{all eigenvalues of } \mathscr{P}(x) \text{ are real.} \tag{4.19}$$

In the rest of this section we prove

Proposition 4.1.1 *Let $\bar{x}$ be a nondegenerate characteristic of order r of $P(x)$. Then $\bar{x}$ is also a nondegenerate characteristic of order r of $\mathscr{P}(x)$ and vice versa.*

Proof. Assume that $\bar{x}$ is a nondegenerate characteristic of order r of $P(x)$ and show that $\bar{x}$ is also a nondegenerate characteristic of order r of $\mathscr{P}(x)$. We first check

$$\left[\frac{\partial P}{\partial x_1}(\bar{x})\mathrm{Ker}P(\bar{x})\right] \oplus \mathrm{Im}\ P(\bar{x}) = \mathbb{C}^m. \tag{4.20}$$

Let $\{v_1, \ldots, v_r\}$ be a basis for $\mathrm{Ker}P(\bar{x})$ and take $\{\ell_i\}$ so that $\ell_i(\mathrm{Im}\ P(\bar{x})) = 0$ and $\ell_i(v_j) = \delta_{ij}$. Then by definition we have

$$P_{\bar{x}}(x) = \left(\ell_i\big((\sum_{k=1}^{n} \frac{\partial P}{\partial x_k}(\bar{x})x_k)v_j\big)\right) = \sum_{k=1}^{n}\left(\ell_i\big((\frac{\partial P}{\partial x_k}(\bar{x}))v_j\big)\right)x_k$$

and hence

$$P_{\bar{x}}(\theta) = \left(\ell_i\big((\frac{\partial P}{\partial x_1}(\bar{x}))v_j\big)\right).$$

Then $\det P_{\bar{x}}(\theta) \neq 0$ implies that

$$\left[\frac{\partial P}{\partial x_1}(\bar{x})\mathrm{Ker}P(\bar{x})\right] \cap \mathrm{Im}\ P(\bar{x}) = \{0\}$$

and hence (4.20).

We note that

$$\mathrm{Ker}\mathscr{P}(x) = \{{}^t(u, x_1u, \ldots, x_1^{q-1}u) \mid u \in \mathrm{Ker}\ P(x)\}$$

and $\dim \mathrm{Ker}\mathscr{P}(\bar{x}) = r$. We next describe $\mathrm{Im}\ \mathscr{P}(x)$. Write

$$\phi_k(x) = \sum_{j=0}^{q-k} A_j(x')x_1^{q-j-k}$$

then it is easy to see that

$$\mathrm{Im}\ \mathscr{P}(x) = \{{}^t(w^{(1)}, \ldots, w^{(q-1)}, P(x)v - \sum_{k=1}^{q-1}\phi_k(x)w^{(k)}) \mid w^{(1)}, \ldots, w^{(q-1)}, v \in \mathbb{C}^m\}.$$

We now show that

$$\mathrm{Ker}\mathscr{P}(\bar{x}) \cap \mathrm{Im}\ \mathscr{P}(\bar{x}) = \{0\}. \tag{4.21}$$

Let ℓ be a linear form on $\mathbb{C}^{mq}$. Writing $v = {}^t(v^{(1)}, \ldots, v^{(q)}) \in \mathbb{C}^{mq}$ with $v^{(j)} \in \mathbb{C}^m$ one can write

$$\ell(v) = \sum_{j=1}^{q} \ell^{(j)}(v^{(j)})$$

where $\ell^{(j)}$ are linear forms on $\mathbb{C}^m$. Assume $\ell(\mathrm{Im}\,\mathscr{P}(\bar{x})) = 0$. This implies that

$$\ell^{(j)}(\cdot) = \ell^{(q)}(\phi_j(\bar{x})\cdot), \ \ 1 \le j \le q-1, \ \ \ell^{(q)}(\mathrm{Im}\,P(\bar{x})) = 0 \tag{4.22}$$

and then we have

$$\begin{aligned}\sum_{j=1}^{q} \ell^{(j)}(\bar{x}_1^{j-1}u) &= \sum_{j=1}^{q-1} \ell^{(q)}(\bar{x}_1^{j-1}\phi_j(\bar{x})u) + \ell^{(q)}(\bar{x}_1^{q-1}u) \\ &= \ell^{(q)}(\sum_{j=1}^{q-1} \bar{x}_1^{j-1}\phi_j(\bar{x})u + \bar{x}_1^{q-1}u) = 0.\end{aligned} \tag{4.23}$$

From this, noting the identity

$$\sum_{j=1}^{q-1} x_1^{j-1}\phi_j(x) + x_1^{q-1} = \frac{\partial P}{\partial x_1}(x)$$

one gets

$$\ell^{(q)}(\frac{\partial P}{\partial x_1}(\bar{x})u) = 0, \quad \forall u \in \mathrm{Ker}\,P(\bar{x}). \tag{4.24}$$

From (4.20) and (4.22) it follows that $\ell^{(q)} = 0$ and hence $\ell = 0$. This proves that

$$\mathrm{Ker}\,\mathscr{P}(\bar{x}) + \mathrm{Im}\,\mathscr{P}(\bar{x}) = \mathbb{C}^{mq} \tag{4.25}$$

and hence (4.21).

We next examine (4.13), (4.14) for $\mathscr{P}(x)$. Let $U = {}^t(u, \bar{x}_1 u, \ldots, \bar{x}_1^{q-1}u) \in \mathrm{Ker}\,\mathscr{P}(\bar{x})$ where $u \in \mathrm{Ker}\,P(\bar{x})$. Consider $\mathscr{P}(x)U$

$$\begin{aligned}\mathscr{P}(x)U &= {}^t((x_1 - \bar{x}_1)u, (x_1 - \bar{x}_1)\bar{x}_1 u, \ldots, (x_1 - \bar{x}_1)\bar{x}_1^{q-2}u, v) \\ &= {}^t(w^{(1)}, w^{(2)}, \ldots, w^{(q-1)}, v)\end{aligned}$$

where the last component v is

$$\begin{aligned}v &= P(\bar{x}_1, x')u + (x_1\bar{x}_1^{q-1} - \bar{x}_1^q)u \\ &= P(x)u + [P(\bar{x}_1, x') - P(x_1, x')]u + \bar{x}_1^{q-1}(x_1 - \bar{x}_1)u.\end{aligned}$$

Now it is easy to see that this is equal to

$$P(x)u - \sum_{k=1}^{q-1} \phi_k(\bar{x})w^{(k)} + O((x_1 - \bar{x}_1)^2). \tag{4.26}$$

Let ℓ be a linear form on $\mathbb{C}^{mq}$ with $\ell(\mathrm{Im}\,\mathscr{P}(\bar{x})) = 0$. From (4.26) it follows that

$$\begin{aligned}\ell(\mathscr{P}(x)U) &= \sum_{j=1}^{q-1} \ell^{(j)}(w^{(j)}) + \ell^{(q)}\Big(P(x)u \\ &\quad - \sum_{k=1}^{q-1} \phi_k(\bar{x})w^{(k)}\Big) + O((x_1 - \bar{x}_1)^2) \\ &= \ell^{(q)}(P(x)u) + O((x_1 - \bar{x}_1)^2)\end{aligned} \tag{4.27}$$

by (4.22). Let us take $U_j = {}^t(u_j, \bar{x}_1 u_j, \ldots, \bar{x}_1^{q-1} u_j) \in \mathrm{Ker}\,\mathscr{P}(\bar{x})$ where $\{u_j\}$ is a basis for $\mathrm{Ker}\,P(\bar{x})$. Then one can write

$$\frac{\partial P}{\partial x_1}(\bar{x})u_j - \sum_{k=1}^{r} a_{jk}u_k \in \mathrm{Im}\,P(\bar{x}),$$

thanks to (4.20) with a non singular $A = (a_{jk})$. Take $\tilde{\ell}_i$ so that

$$\tilde{\ell}_i(\mathrm{Im}\,P(\bar{x})) = 0, \quad \tilde{\ell}_i(u_j) = \delta_{ij}.$$

Let us take $\ell_i^{(q)}$

$$\ell_i^{(q)} = \sum_{k=1}^{r} b_{ik}\tilde{\ell}_k, \quad B = (b_{ik}) = {}^tA^{-1}$$

so that

$$\ell_i^{(q)}\Big(\frac{\partial P}{\partial x_1}(\bar{x})u_j\Big) = \sum_{k=1}^{r} b_{ik} \sum_{p=1}^{r} a_{jp}\tilde{\ell}_k(u_p) = \delta_{ij}.$$

We now define linear forms ℓ_i on $\mathbb{C}^{mq}$ by

$$\ell_i(w^{(1)}, \ldots, w^{(q)}) = \sum_{t=1}^{q-1} \ell_i^{(q)}(\phi_t(\bar{x})w^{(t)}) + \ell_i^{(q)}(w^{(q)})$$

then we have

$$\ell_i(\mathrm{Im}\,\mathscr{P}(\bar{x})) = 0, \quad \ell_i(U_j) = \delta_{ij} \tag{4.28}$$

as observed above. From (4.27) it follows that

$$\begin{aligned}
\ell_i(\mathscr{P}(\bar{x}+\mu x)U_j) &= \ell_i^{(q)}(P(\bar{x}+\mu x)u_j) + O(\mu^2)\\
&= \sum_{k=1}^{r} b_{ik}\tilde{\ell}_k(P(\bar{x}+\mu x)u_j) + O(\mu^2)\\
&= \mu\big(BP_{\bar{x}}(x) + O(\mu)\big).
\end{aligned}$$

Since $B = ({}^tA)^{-1} = P_{\bar{x}}(\theta)^{-1}$ we conclude that

$$\mathscr{P}_{\bar{x}}(x) = P_{\bar{x}}(\theta)^{-1}P_{\bar{x}}(x). \tag{4.29}$$

Since $\mathscr{P}_{\bar{x}}(\theta) = I$ then (4.13) and (4.14) for $\mathscr{P}_{\bar{x}}(x)$ follow immediately.

Conversely assume (4.25). Let $\ell^{(q)}$ be a linear form on $\mathbb{C}^m$ with $\ell^{(q)}(\mathrm{Im}\,P(\bar{x})) = 0$, $\ell^{(q)}(\mathrm{Ker}\,P(\bar{x})) = 0$ and define $\ell^{(j)}$, $1 \le j \le q-1$ by (4.22). Then we have $\ell(\mathrm{Im}\,\mathscr{P}(\bar{x})) = 0$ and moreover (4.22) shows $\ell(\mathrm{Ker}\,\mathscr{P}(\bar{x})) = 0$ and hence $\ell = 0$ by (4.25). Thus we have $\ell^{(q)} = 0$ which proves $\mathrm{Ker}\,P(\bar{x}) \oplus \mathrm{Im}\,P(\bar{x}) = \mathbb{C}^m$ and hence (4.12). To check (4.13), (4.14) for $P(x)$ we note that $\mathrm{Ker}\,\mathscr{P}(\bar{x}) \cap \mathrm{Im}\,\mathscr{P}(\bar{x}) = \{0\}$ implies that

$$u \in \mathrm{Ker}\,P(\bar{x}),\ \frac{\partial P}{\partial x_1}(\bar{x})u \in \mathrm{Im}\,P(\bar{x}) \Longrightarrow u = 0.$$

Hence we have (4.20) again and thus (4.29). Then the rest of the proof is clear. □

Remark. Assume that $q = 1$ and $A_1(x')$ is symmetric in (4.1). Then (4.12) and (4.14) are always verified.

Remark. By definition, the order of nondegenerate characteristics never exceed m, the size of the matrix whatever q is.

4.2 Nondegenerate Double Characteristics

Nondegenerate double characteristics have a special feature.

Lemma 4.3. *Let $\bar{x}$ be a double characteristic. Then $\bar{x}$ is nondegenerate if and only if* $\dim\mathrm{Ker}\,P(\bar{x}) = 2$ *and the rank of the Hessian of* $\det P(x)$ *at $\bar{x}$ is* 4. *When $P(x)$ is real valued then $\bar{x}$ is nondegenerate if and only if* $\dim\mathrm{Ker}\,P(\bar{x}) = 2$ *and the rank of the Hessian of* $\det P(x)$ *at $\bar{x}$ is* 3.

To prove the lemma we first note

Lemma 4.4. *Let A_j be 2×2 constant matrices with* $\operatorname{Tr} A_j = 0$, $1 \le j \le m$. *Assume that the quadratic form*

$$Q(x) = \det\Big(\sum_{j=1}^{m} A_j x_j\Big)$$

is real nonpositive definite in $\mathbb{R}^m$. Then the rank of $Q(x)$ is at most 3 and if $\operatorname{rank} Q = 3$ *then there is a constant matrix N such that*

$$N^{-1} A_j N$$

is an Hermitian matrix for all j. If all A_j are real then $\operatorname{rank} Q \le 2$ *and if* $\operatorname{rank} Q = 2$ *then there is a real constant matrix N such that all*

$$N^{-1} A_j N$$

are real symmetric.

Proof. With a non singular real matrix $T = (t_{ij})$ one can assume

$$Q(Tx) = \det\Big(\sum_{j=1}^{m} H_j x_j\Big) = -\sum_{j=1}^{k} x_j^2, \quad \operatorname{Tr} H_j = 0 \tag{4.30}$$

where $H_j = \sum_{i=1}^{m} t_{ji} A_i$ and $\operatorname{rank} Q = k$. If $k \le 2$ then nothing to be proved. Thus we assume $k \ge 3$. Since $\det H_1 = -1$, $\operatorname{Tr} H_1 = 0$, one can diagonalize H_1

$$H_1' = N_1^{-1} H_1 N_1 = \begin{bmatrix} 1 & 0 \\ 0 & -1 \end{bmatrix}.$$

Denoting $H_2' = N_1^{-1} H_2 N_1 = (h_{ij})$ and taking $x_j = 0$, $j \ge 3$ it follows from (4.30) that $h_{11} = h_{22} = 0$, $h_{12} h_{21} = 1$. Setting

$$N_2^{-1} = \begin{bmatrix} 1 & 0 \\ 0 & h_{12} \end{bmatrix}$$

it follows that

$$N_2^{-1} H_1' N_2 = H_1', \quad N_2^{-1} H_2' N_2 = \begin{bmatrix} 0 & 1 \\ 1 & 0 \end{bmatrix}.$$

Let us put $N = N_1N_2$ and $N^{-1}H_jN = H_j' = (h_{pq}^{(j)})$, $j \geq 3$. Take $x_j = 0$ unless $j = 1, 3$ then we get $h_{11}^{(3)} = h_{22}^{(3)} = 0$, $h_{12}^{(3)}h_{21}^{(3)} = 1$ and taking $x_j = 0$ unless $j = 2$, 3 we get $h_{12}^{(3)} + h_{21}^{(3)} = 0$. Thus we conclude $h_{12}^{(3)} = \pm i$. The same procedure gives

$$H_j' = \epsilon_j \begin{bmatrix} 0 & i \\ -i & 0 \end{bmatrix}, \quad (\epsilon_j = 1 \text{ or } -1), \ \ 3 \leq j \leq k.$$

Repeating similar arguments we obtain $H_j = O$ for $j > k$. We summarize

$$N^{-1}(\sum_{j=1}^{k} H_j x_j)N = \begin{bmatrix} 1 & 0 \\ 0 & -1 \end{bmatrix} x_1 + \begin{bmatrix} 0 & 1 \\ 1 & 0 \end{bmatrix} x_2$$

$$+ \begin{bmatrix} 0 & i \\ -i & 0 \end{bmatrix} (\sum_{j=3}^{k} \epsilon_j x_j), \quad H_j = O, \ j > k \tag{4.31}$$

and from (4.30)

$$-\det(\sum_{j=1}^{k} H_j x_j) = x_1^2 + x_2^2 + (\sum_{j=3}^{k} \epsilon_j x_j)^2 = \sum_{j=1}^{k} x_j^2. \tag{4.32}$$

The identity (4.32) holds only if $k = 3$ and all $N^{-1}H_jN$ are Hermitian. Since T is real then $N^{-1}A_jN$ are also Hermitian. This proves the assertion. If all A_j are real, we can take N real and the proof is similar. □

Proof of Lemma 4.3. Take T so that

$$T^{-1}P(\bar{x})T = \begin{bmatrix} A & O \\ O & G \end{bmatrix} \tag{4.33}$$

where G is a non singular matrix of order $m - 2$ and all eigenvalues of A are zero. Assume that dimKer $P(\bar{x}) = 2$. Then it follows that $A = O$ and hence Ker $P(\bar{x}) \cap$ Im $P(\bar{x}) = \{0\}$.

Assume that rank $\text{Hess}_{\bar{x}} \det P = 4$ and hence $\det P_{\bar{x}}(x) \not\equiv 0$ by Lemma 4.2. From Lemma 4.2 again we have $\det P_{\bar{x}}(\theta) \neq 0$ and $P_{\bar{x}}(\theta)^{-1}P_{\bar{x}}(x)$ has only zero eigenvalues for every x. Then writing

$$P_{\bar{x}}(\theta)^{-1}P_{\bar{x}}(x) = x_1 I_2 + \sum_{j=2}^{n} A_j x_j$$

$$= \Big(x_1 - \frac{1}{2}\text{Tr}\,(\sum_{j=2}^{n} A_j x_j)\Big) I_2 + \sum_{j=2}^{n} \tilde{A}_j x_j$$

it follows that $\det(\sum_{j=2}^{n} \tilde{A}_j x_j)$ is a real nonpositive quadratic form on $\mathbb{R}^{n-1}$ of which rank is 3 since the rank of the real quadratic form $\det(P_{\bar{x}}(\theta)^{-1} P_{\bar{x}}(x))$ is 4. Note $\mathrm{Tr}\, \tilde{A}_j = 0$. From Lemma 4.4 there exists a constant 2×2 matrix T such that $T^{-1} \tilde{A}_j T$ is Hermitian for every j so that one can write

$$T^{-1}(P_{\bar{x}}(\theta)^{-1} P_{\bar{x}}(x))T \tag{4.34}$$
$$= \phi_1(x) \begin{bmatrix} 1 & 0 \\ 0 & 1 \end{bmatrix} + \phi_2(x) \begin{bmatrix} 1 & 0 \\ 0 & -1 \end{bmatrix} + \phi_3(x) \begin{bmatrix} 0 & 1 \\ 1 & 0 \end{bmatrix} + \phi_4(x) \begin{bmatrix} 0 & i \\ -i & 0 \end{bmatrix}$$

with real linear forms $\phi_i(x)$ and obviously $P_{\bar{x}}(\theta)^{-1} P_{\bar{x}}(x)$ is diagonalizable for every x. Since $\phi_i(x)$, $i = 1, 2, 3, 4$ are linearly independent it is clear that $\dim_{\mathbb{R}} P_{\bar{x}} = 4$.

Conversely we assume that a double characteristic $\bar{x}$ is nondegenerate. Take T so that (4.33) holds. From $\mathrm{Ker} P(\bar{x}) \cap \mathrm{Im} P(\bar{x}) = \{0\}$ it follows that $A = O$ and hence $\dim \mathrm{Ker}\, P(\bar{x}) = 2$. Assume $\dim P_{\bar{x}} = 4$ and $\det P_{\bar{x}}(\theta) \neq 0$. Let us write

$$P_{\bar{x}}(\theta)^{-1} P_{\bar{x}}(x) = (x_1 - \psi(x)) I_2 + \sum_{j=2}^{4} A_j x_j$$

where $\mathrm{Tr}\, A_j = 0$ and $\{I_2, A_2, A_3, A_4\}$ are linearly independent by assumption. Since $P_{\bar{x}}(\theta)^{-1} P_{\bar{x}}(x)$ has only real eigenvalues for every x then $\det(\sum_{j=2}^{4} A_j x_j)$ is nonpositive definite so that one can write

$$\det(\sum_{j=2}^{4} A_j x_j) = -\sum_{i=1}^{k} \ell_j(x)^2$$

with linearly independent $\ell_j(x)$ where $k \leq 3$ by Lemma 4.4. Assume that $\ell_i(x) = 0$, $i = 1, \ldots, k$ then $\sum_{j=2}^{4} A_j x_j$ has only zero eigenvalues because $\mathrm{Tr}\, A_j = 0$. Since $\sum_{j=2}^{4} A_j x_j$ is diagonalizable by assumption then we conclude that $\sum_{j=2}^{4} A_j x_j = O$ so that

$$\sum_{j=2}^{4} A_j x_j = \sum_{i=1}^{k} H_j \ell_j(x)$$

which proves $k = 3$. Thus $\det(P_{\bar{x}}(\theta)^{-1} P_{\bar{x}}(x))$ has rank 4 and from Lemma 4.2 it follows that $\mathrm{rank}\, \mathrm{Hess}_{\bar{x}} \det P = 4$. This proves the assertion.

The case that $P(x)$ is real valued, the proof is just a repetition with obvious modifications. □

Proposition 4.1. *Let $m = 2$ and $q = 1$. Assume that $P(\bar{x}) = O$ and the rank of* $\mathrm{Hess} \det P$ *is 4 at $\bar{x}$ (3 if $P(x)$ is real valued). Then $\Sigma = \{x \mid \partial_x^{\alpha}(\det P)(x) = 0,$*

$|\alpha| \le 1\}$ *is a* C^∞ *manifold near* $\bar{x}$ *with* $\operatorname{codim}\Sigma = \operatorname{rank}\operatorname{Hess}_{\bar{x}}\det P$ *on which* $P(x) = O$.

In fact, in Sect. 4.5, we prove this proposition in much more generality (Proposition 4.3). The smoothness of the characteristic set is closely related to the existence of smooth symmetrizers (see [48]). Indeed we have

Proposition 4.2 ([17, 48]). *Let* $m = 2$ *and* $q = 1$*. Assume that* $P(\bar{x}) = O$ *and the rank of* Hess det P *is 4 at* $\bar{x}$ *(3 if* $P(x)$ *is real valued). Then* $P(x)$ *has a smooth symmetrizer near* $\bar{x}$*, that is there is a smooth* 2×2 *matrix valued* $S(x')$ *defined near* $\bar{x}'$ *such that*

$$S^*(x') = S(x') \text{ and } S(x') \text{ is positive definite},$$
$$S(x')P(x) = P^*(x)S(x')$$

where $P^*(x)$ *denotes the adjoint matrix of* $P(x)$.

Example 4.4. Let us consider second order differential operator $P(D) = (p_{ik}(D))$ with 3×3 constant matrix coefficients

$$p_{ik}(\tau,\xi) = (\tau^2 - \sigma_i|\xi|^2)\delta_{ik} - (1-\sigma_i)\xi_i\xi_k$$

which is called the modified elasticity operator in [25] where $\xi = (\xi_1, \xi_2, \xi_3)$ and

$$0 < \sigma_1 < \sigma_2 < \sigma_3 < 1.$$

Note that the excluded case where $\sigma_1 = \sigma_2 = \sigma_3$ yields the elasticity equations. We follow the arguments in John [25]. Let $Q(\tau,\xi) = \det P(\tau,\xi)$ then $Q(\tau,\xi)$ can be written

$$Q(\tau,\xi) = (\tau^2 - q_0(\xi))(\tau^4 - 2q_1(\xi)\tau^2 + q_0(\xi)q_2(\xi))$$

where q_0, q_1, q_2 are the definite quadratic forms given by

$$q_0 = |\xi|^2,\ \ q_2 = \sigma_1\sigma_2\sigma_3\sum_{j=1}^{3}\frac{1}{\sigma_j}\xi_j^2,\ \ q_1 = \frac{1}{2}(\sigma_1+\sigma_2+\sigma_3)|\xi|^2 - \frac{1}{2}\sum_{j=1}^{3}\sigma_j\xi_j^2.$$

Taking the homogeneity into account we consider multiple characteristics (τ,ξ) with $|\xi| = 1$. It is shown in [25] that (τ,ξ), $|\xi| = 1$ is a multiple characteristic if and only if

$$D(\xi) = 4(q_1^2 - q_0q_2) = 0,\quad |\xi| = 1$$

which gives 4 points

$$\pm(\beta_3/\beta_2, 0, \beta_1/\beta_2),\ \ \pm(-\beta_3/\beta_2, 0, \beta_1/\beta_2) \tag{4.35}$$

where $\beta_1 = (\sigma_3 - \sigma_2)^{1/2}$, $\beta_2 = (\sigma_3 - \sigma_1)^{1/2}$, $\beta_3 = (\sigma_2 - \sigma_1)^{1/2}$. If we set

$$D^*(\xi) = D(\xi) + \beta_1^2\beta_3^2(q_0(\xi) - 1)^2$$

then we have at a double characteristic (4.35) which we denote $\hat{\xi}$

$$\frac{\partial^2 D^*(\hat{\xi})}{\partial \xi_i \xi_k} = 8\beta_1^2\beta_3^2(\delta_{ik} - \frac{1}{2}(\delta_{i1}\delta_{k3} + \delta_{k1}\delta_{i3})\hat{\xi}_i\hat{\xi}_k)$$

and hence Hessian of D^* is positive definite. This shows that the Hessian of $D(\xi)$ has at least rank 2 and then the Hessian of $(\tau^2 - q_1)^2 + D(\xi)$ has rank 3 which proves that the double characteristic (4.35) are nondegenerate.

We find similar second order differential operators $P(D) = (p_{ik}(D))$ with 3×3 constant matrix coefficients in [64] in the studies of relativistic elastodynamics.

Example 4.5. We have

Theorem 4.1 ([22]). *In the set $\mathscr{P}$ of all positive definite real symmetric 3×3 matrix valued quadratic forms*

$$A(\xi) = \sum_{j,k=1}^{3} A_{jk}\xi_j\xi_k$$

the subset for which the characteristics of $\det(\tau^2 I - A(\xi))$ *are at most double and the double characteristics are nondegenerate is an open and dense subset.*

We have also

Theorem 4.2 ([22]). *One can choose a positive definite real symmetric 3×3 matrix valued quadratic form A such that the characteristics of* $\det(\tau^2 I - A(\xi))$ *are at most double, the double characteristics are nondegenerate, and there are at least* 12 *of them.*

4.3 Symmetrizability (Special Case)

We first note that, considering $-A_0(x')^{-1}P(x)$, we may assume that $P(1, 0, \ldots, 0) = -I_2$ so that

$$P(x) = -x_1 I_2 + A'(x'), \quad A'(x') \in C^\infty(\Omega, M_2(\mathbb{C}))$$

which is also written

$$P(x) = -(x_1 - \frac{1}{2}\mathrm{Tr}\, A'(x'))I_2 + A(x'), \ \mathrm{Tr}\, A(x') = 0.$$

Note that

$$g(x') = \det A(x') \leqq 0$$

and $\operatorname{Tr} A'(x')$ is real which follows from the hyperbolicity of $\det P(x)$. Let us denote

$$A(x') = \begin{bmatrix} a(x') & b(x') \\ c(x') & -a(x') \end{bmatrix}.$$

We denote by $da(x')$ the differential of a at $\bar{x}'$ so that $a(x'+\bar{x}') = da(x') + O(|x'-\bar{x}'|^2)$ and by $\operatorname{Re} a$ and $\operatorname{Im} a$ the real part and the imaginary part of a respectively.

We first assume that $P(x)$ is real valued and $\operatorname{rank} \operatorname{Hess}_{\bar{x}} \det P = 3$. The assumption is reduced to $\operatorname{rank} \operatorname{Hess}_{\bar{x}} g = 2$. From Proposition 4.1 it follows that $\Sigma' = \{x' \mid g(x') = 0\}$ is a smooth manifold of codimension 2. Then there are $\ell_i(x'), i = 1, 2$ such that $\Sigma' = \{\ell_1(x') = 0, \ell_2(x') = 0\}$ and

$$A(x') = H_1(x')\ell_1 + H_2(x')\ell_2, \quad g(x') = -\ell_1^2 - \ell_2^2$$

where $d\ell_i(x')$ are linearly independent. Let K_1 be the restriction of H_1 to $\ell_2 = 0$ then it is clear that $\det K_1 = -1$ and $\operatorname{Tr} K_1 = 0$. Hence there is a real 2×2 matrix $N(x')$ such that

$$N^{-1} K_1 N = \begin{bmatrix} 1 & 0 \\ 0 & -1 \end{bmatrix}$$

and then we have

$$N^{-1}AN = \begin{bmatrix} 1 & 0 \\ 0 & -1 \end{bmatrix} (\ell_1 + \alpha\ell_2) + \begin{bmatrix} 0 & \beta \\ \gamma & 0 \end{bmatrix} \ell_2.$$

From the Taylor expansion of $\det A(x')$ around $\bar{x}'$ it is easy to see that $\alpha(\bar{x}') = 0$, $\beta(\bar{x}')\gamma(\bar{x}') = 1$ and consequently the matrix

$$M = \begin{bmatrix} 1 & 0 \\ 0 & 1/\gamma(x') \end{bmatrix}$$

is well defined near $\bar{x}'$. Putting $T(x') = N(x')M(x')$ and writing $\beta(x')\gamma(x') = 1 + \psi$ we have

$$T^{-1}AT = \begin{bmatrix} 1 & 0 \\ 0 & -1 \end{bmatrix} (\ell_1 + \alpha\ell_2) + \begin{bmatrix} 0 & 1+\psi \\ 1 & 0 \end{bmatrix} \ell_2.$$

We now define S by

$$S = \begin{bmatrix} 1 & 0 \\ 0 & 1+\psi \end{bmatrix}.$$

Since $\psi(\bar{x}') = 0$ it is easy to see that S is a desired symmetrizer of $T^{-1}AT$. Since the symmetrizability is invariant under similar transformations we get the desired assertion.

We next prove the proposition assuming that $\operatorname{rank} \operatorname{Hess}_{\bar{x}} \det P = 4$. Since the hypothesis $\operatorname{rank} \operatorname{Hess}_{\bar{x}} \det P = 4$ reduces to $\operatorname{rank} \operatorname{Hess}_{\bar{x}} g = 3$ we may assume that

$$Q = (d\operatorname{Re} a)^2 - (d\operatorname{Im} a)^2 + (d\operatorname{Re} b)(d\operatorname{Re} c) - (d\operatorname{Im} b)(d\operatorname{Im} c)$$

is nonnegative definite and has rank 3. Here we note that a real quadratic form Q which is nonnegative definite can not vanish on a linear subspace V unless $\operatorname{codim} V \geqq \operatorname{rank} Q$. We first remark that $d\operatorname{Re} a \neq 0$. If it were not true we would have

$$\begin{aligned} 0 \ll Q &= -(d\operatorname{Im} a)^2 + (d\operatorname{Re} b)(d\operatorname{Re} c) - (d\operatorname{Im} b)(d\operatorname{Im} c) \\ &\ll (d\operatorname{Re} b)(d\operatorname{Re} c) - (d\operatorname{Im} b)(d\operatorname{Im} c). \end{aligned}$$

It is clear that there is a linear subspace V $(\subset \mathbb{R}^{n-1})$ with $\operatorname{codim} V \leqq 2$ on which Q vanishes and hence $\operatorname{rank} Q \leqq 2$. This contradicts the assumption.

Set $\varphi = \operatorname{Re} a$ and denote by $b|_{\varphi=0}$ the restriction of b to the surface $\{\varphi = 0\}$.

Lemma 4.5. *Let $b = \beta\varphi + \tilde{b}$, $c = \gamma\varphi + \tilde{c}$ with $\tilde{b} = b|_{\varphi=0} = \tilde{b}_1 + i\tilde{b}_2$, $\tilde{c} = c|_{\varphi=0} = \tilde{c}_1 + i\tilde{c}_2$ where $\tilde{b}_i, \tilde{c}_i$ are real. Then we have*

$$d\tilde{b}_i \neq 0, \;\; d\tilde{c}_i \neq 0 \text{ at } \bar{x}', \; i = 1, 2.$$

Proof. Denoting $\operatorname{Im} a = \alpha\varphi + \tilde{\alpha}$ with $\tilde{\alpha} = a|_{\varphi=0}$ one can write

$$A(x') = \varphi \begin{bmatrix} (1+i\alpha) & \beta \\ \gamma & -(1+i\alpha) \end{bmatrix} + \begin{bmatrix} i\tilde{\alpha} & \tilde{b}_1 + i\tilde{b}_2 \\ \tilde{c}_1 + i\tilde{c}_2 & -i\tilde{\alpha} \end{bmatrix}.$$

From the non-positivity of g on $\{\varphi = 0\}$ it follows that

$$\tilde{b}_1\tilde{c}_1 - \tilde{b}_2\tilde{c}_2 - \tilde{\alpha}^2 \geqq 0, \tag{4.36}$$

$$\tilde{b}_1\tilde{c}_2 + \tilde{b}_2\tilde{c}_1 = 0 \tag{4.37}$$

near $\bar{x}'$. Suppose, for instance, that $d\tilde{b}_1(\bar{x}') = 0$ and hence $d\tilde{b}_2 = 0$ or $d\tilde{c}_1 = 0$ (at $\bar{x}'$) by (4.37). If $d\tilde{b}_2 = 0$ then $d\tilde{\alpha} = 0$ by (4.36) and then Q vanishes on $\{x' \mid d\varphi(x') = 0\}$ because $da = (1+i\alpha)d\varphi$ at $\bar{x}'$. This is a contradiction. The other cases will be proved similarly. □

Lemma 4.6. *$d\tilde{b}_1$ is not proportional to $d\tilde{b}_2$ at $\bar{x}'$. There is a positive function $m(x')$ defined near $\bar{x}'$ such that*

$$\tilde{c}_1(x') = m(x')\tilde{b}_1(x'), \quad \tilde{c}_2(x') = -m(x')\tilde{b}_2(x').$$

Proof. Suppose that $d\tilde{b}_2 = kd\tilde{b}_1$ at $\bar{x}'$ with some $k \in \mathbb{R}$ and hence $d\tilde{c}_2 = -kd\tilde{c}_1$ by (4.37) at $\bar{x}'$. Since from (4.36) we see

$$d\tilde{b}_1 d\tilde{c}_1 - d\tilde{b}_2 d\tilde{c}_2 - d\tilde{\alpha} d\tilde{\alpha} = (1+k^2)d\tilde{b}_1 d\tilde{c}_1 - d\tilde{\alpha} d\tilde{\alpha} \gg 0,$$

and hence $d\tilde{b}_1$ and $d\tilde{c}_1$ must be proportional to $d\tilde{\alpha}$ at $\bar{x}'$ if $d\tilde{\alpha} \neq 0$. Then it is clear that Q vanishes on $\{x' \mid d\tilde{\alpha}(x') = d\varphi(x') = 0\}$ which is a contradiction. If $d\tilde{\alpha} = 0$ (at $\bar{x}'$) then Q vanishes on $\{x' \mid d\varphi(x') = d\tilde{c}_1(x') = 0\}$ which also gives a contradiction. This proves the first assertion. The second assertion easily follows from the first one and (4.36), (4.37). □

We can put A in a special form.

Lemma 4.7. *Let $\beta = \beta_1 + i\beta_2$, $\gamma = \gamma_1 + i\gamma_2$, β_i, γ_i real. Set $\psi_i = \tilde{b}_i + \beta_i\varphi$ $(i = 1, 2)$, $B = \gamma_2 + m\beta_2$, $C = \gamma_1 - m\beta_1$. Then we have*

$$A = \varphi\begin{bmatrix} 1 & 0 \\ C+iB & -1 \end{bmatrix} + \psi_1\begin{bmatrix} -iB/2 & 1 \\ m & iB/2 \end{bmatrix} + \psi_2\begin{bmatrix} -iC/2 & i \\ -im & iC/2 \end{bmatrix}.$$

Moreover $d\varphi, d\psi_i$ are linearly independent at $\bar{x}'$ and the set $\{x' \mid A(x') = O\}$ is given by

$$S = \{x' \mid \varphi(x') = \psi_1(x') = \psi_2(x') = 0\}.$$

Proof. Recall that

$$A = \varphi\begin{bmatrix} 1+i\alpha & \beta \\ \gamma & -(1+i\alpha) \end{bmatrix} + \begin{bmatrix} i\tilde{\alpha} & \tilde{b}_1 + i\tilde{b}_2 \\ m(\tilde{b}_1 - i\tilde{b}_2) & -i\tilde{\alpha} \end{bmatrix}.$$

We observe the imaginary part of g

$$\mathrm{Im}\, g = 2\alpha\varphi^2 + 2\tilde{\alpha}\varphi + \mathrm{Im}\,(\beta\gamma)\varphi^2 + \mathrm{Im}\,(\gamma + \beta m)\varphi\tilde{b}_1 + \mathrm{Re}\,(\gamma - \beta m)\varphi\tilde{b}_2.$$

Since $\mathrm{Im}\, g = 0$ near $\bar{x}'$ and $d\varphi \neq 0$ at $\bar{x}'$ it follows that

$$2\alpha\varphi + 2\tilde{\alpha} + \mathrm{Im}\,(\beta\gamma)\varphi + \mathrm{Im}\,(\gamma + \beta m)\tilde{b}_1 + \mathrm{Re}\,(\gamma - \beta m)\tilde{b}_2 = 0 \tag{4.38}$$

near $\bar{x}'$. Now we set

$$D = \mathrm{Im}\,(\beta\gamma), \quad B = \mathrm{Im}\,(\gamma + \beta m) = \gamma_2 + \beta_2 m,$$
$$C = \mathrm{Re}\,(\gamma - \beta m) = \gamma_1 - \beta_1 m.$$

Noticing $D = \beta_1 B + \beta_2 C$ it follows from (4.38) that

$$(\alpha\varphi + \tilde{\alpha}) = -\frac{1}{2}(\psi_1 B + \psi_2 C) \tag{4.39}$$

which shows that $a = (1 + i\alpha)\varphi + i\tilde{\alpha} = \varphi - i(\psi_1 B + \psi_2 C)/2$. On the other hand it is easy to see

$$m(\tilde{b}_1 - i\tilde{b}_2) + \gamma\varphi = (C + iB)\varphi + m(\psi_1 - i\psi_2), \quad \tilde{b}_1 + i\tilde{b}_2 + \beta\varphi = \psi_1 + i\psi_2$$

because $\gamma_1 = C + m\beta_1$ and $\gamma_2 = B - m\beta_2$. These prove the first part. The rest of the assertion is obvious. □

Lemma 4.8. *We have*

$$4m - (B^2 + C^2) > 0 \;\; at \;\; \bar{x}'.$$

Proof. Let us set $\tilde{B} = B|_{\varphi=0}$, $\tilde{C} = C|_{\varphi=0}$. From (4.39) it follows that

$$\tilde{\alpha} = -(\tilde{B}\tilde{b}_1 + \tilde{C}\tilde{b}_2)/2.$$

On the other hand (4.36) and Lemma 4.6 give that

$$m(\tilde{b}_1^2 + \tilde{b}_2^2) - \tilde{\alpha}^2 \geqq 0 \;\; \text{near} \;\; \bar{x}'.$$

Since the quadratic form $m((d\tilde{b}_1)^2 + (d\tilde{b}_2)^2) - (\tilde{B}d\tilde{b}_1 + \tilde{C}d\tilde{b}_2)^2/4$ is the restriction of Q to $\{x' \mid d\varphi(x') = 0\}$ this must have rank 2 and then positive definite. This shows that $4m - (\tilde{B}^2 + \tilde{C}^2) > 0$ at $\bar{x}'$ and hence the result. □

To finish the proof of Proposition 4.2 we give a required smooth symmetrizer $S(x')$ for P by

$$S(x') = \begin{bmatrix} 2m(x') & -C(x') + iB(x') \\ -C(x') - iB(x') & 2 \end{bmatrix}$$

which satisfies $S(x') = S^*(x')$ clearly. Using Lemma 4.7 it is easy to check that $S(x')A(x') = A^*(x')S(x')$ and hence

$$S(x')P(x) = P^*(x)S(x').$$

The positivity of S follows from Lemma 4.8.

4.4 Stability and Smoothness of Nondegenerate Characteristics

In this section we discuss the stability of nondegenerate characteristics and the smoothness of nondegenerate characteristic set.

Theorem 4.3. *Assume that $P(x)$ is an $m \times m$ (resp. real) matrix valued smooth function of the form (4.1) verifying (4.2) in a neighborhood U of $\bar{x}$ and let $\bar{x}$ be a nondegenerate characteristic of order r of P. Let $\tilde{P}(x)$ be another $m \times m$ (resp. real) matrix valued smooth function of the form (4.1) verifying (4.2) which is sufficiently close to $P(x)$ in C^{q+2}, then $\tilde{P}(x)$ has a nondegenerate characteristic of the same order close to $\bar{x}$. Moreover, near $\bar{x}$, the characteristics of order r are nondegenerate and they form a smooth manifold of codimension r^2 (resp. $r(r+1)/2$). In particular, near $\bar{x}$ the set of characteristics of order r of $P(x)$ itself consists of nondegenerate ones which form a smooth manifold of codimension r^2(resp. $r(r+1)/2$).*

To prove Theorem 4.3, taking Proposition 4.1.1 into account, we study $P(x)$ of the form

$$P(x) = x_1 I + P^{\#}(x') \tag{4.40}$$

where we assume that

$$\det P(x) = 0 \Longrightarrow x_1 \text{ is real near } x' = \bar{x}'. \tag{4.41}$$

This is equivalent to say that all eigenvalues of $P^{\#}(x')$ are real. Now to prove Theorem 4.3 it suffices to prove

Proposition 4.3. *Assume that $P(x)$ is an $m \times m$ (resp. real) matrix valued smooth function of the form (4.40) verifying (4.41) and $\bar{x}$ is a nondegenerate characteristic of order r of $P(x)$. Let $\tilde{P}(x)$ be another $m \times m$ (resp. real) matrix valued smooth function of the form (4.40) verifying (4.41) which is sufficiently close to $P(x)$ in C^2 near $\bar{x}$. Then $\tilde{P}(x)$ has a nondegenerate characteristic of the same order close to $\bar{x}$. Moreover, near $\bar{x}$, the characteristics of order r of $\tilde{P}(x)$ are nondegenerate and form a smooth manifold of codimension r^2 (resp. $r(r+1)/2$). In particular, the characteristics of order r of $P(x)$ itself consists of nondegenerate ones which form a smooth manifold of codimension r^2 (resp. $r(r+1)/2$).*

The rest of this section is devoted to the proof of Proposition 4.3. We first show that the proof is reduced to the case that P and $\tilde{P}$ are $r \times r$ matrix valued function. Without restrictions we may assume that $\bar{x} = 0$. As in the previous section, we take T so that one has

$$T^{-1} P(0) T = \begin{bmatrix} 0 & 0 \\ 0 & G \end{bmatrix}$$

where G is non singular. Denote $T^{-1}P(x)T$ and $T^{-1}\tilde{P}(x)T$ by $P(x)$ and $\tilde{P}(x)$ again. Writing

$$P(x) = \begin{bmatrix} P_{11}(x) & P_{12}(x) \\ P_{21}(x) & P_{22}(x) \end{bmatrix}$$

we have

$$P_{11}(x) = x_1 I + \sum_{j=2}^{n} A_j x_j + O(|x|^2) = P_0(x) + O(|x|^2). \tag{4.42}$$

From the assumption $P_0(x)$ is diagonalizable for every x and $\{I, A_2, \ldots, A_n\}$ span a r^2 (resp. $r(r+1)/2$) dimensional subspace over $\mathbb{R}$ in $M_r(\mathbb{C})$ (resp. $M_r(\mathbb{R})$). By Lemma 4.2 all eigenvalues of $P_0(x)$ are real then one can apply

Lemma 4.9 ([52, 66, 68, 69]). *Let us consider*

$$L(x) = \sum_{j=1}^{n} A_j x_j, \quad A_1 = I$$

where A_j are $r \times r$ constant matrices. Assume that the real reduced dimension of $L(x)$, that is the dimension of the space spanned by $\{A_j\}$ over $\mathbb{R}$, is at least $r^2 - 2$ $((r(r+1)/2) - 1$ if all A_j are real) and $L(x)$ is diagonalizable with real eigenvalues for every x. Then there is a constant matrix T such that

$$T^{-1}L(x)T$$

is Hermitian (symmetric) for every $x \in \mathbb{R}^n$.

Thus we conclude that there is a constant matrix S such that

$$S^{-1}(x_1 + \sum_{j=2}^{n} A_j x_j)S = x_1 + \sum_{j=2}^{n} \tilde{A}_j x_j$$

where $\tilde{A}_j$ are Hermitian (resp. symmetric) and $\{I, \tilde{A}_2, \ldots, \tilde{A}_n\}$ span $M_r^h(\mathbb{C})$ (resp. $M_r^s(\mathbb{R})$). We still denote

$$\begin{bmatrix} S^{-1} & 0 \\ 0 & I \end{bmatrix} P(x) \begin{bmatrix} S & 0 \\ 0 & I \end{bmatrix}, \quad \begin{bmatrix} S^{-1} & 0 \\ 0 & I \end{bmatrix} \tilde{P}(x) \begin{bmatrix} S & 0 \\ 0 & I \end{bmatrix}$$

by $P(x)$ and $\tilde{P}(x)$ again so that writing

$$P(x) = \begin{bmatrix} P_{11}(x) & P_{12}(x) \\ P_{21}(x) & P_{22}(x) \end{bmatrix}$$

we may assume that

$$P_{11}(x) = x_1 I + \sum_{j=2}^{n} A_j x_j + O(|x|^2) \tag{4.43}$$

where

$$\{I, A_2, \ldots, A_n\} \text{ span } M_r^h(\mathbb{C}) \quad (\text{resp. } M_r^s(\mathbb{R})). \tag{4.44}$$

Let $\{F_1, F_2, \ldots, F_k\}$, $F_1 = I$ be a basis for $M_r^h(\mathbb{C})$ (resp. $M_r^s(\mathbb{R})$) where $k = r^2$ (resp. $k = r(r+1)/2$). Writing

$$x_1 I + \sum_{j=2}^{n} A_j x_j = \sum_{j=1}^{k} F_j \ell_j(x)$$

we make a linear change of coordinates $\tilde{x}_j = \ell_j(x)$, $j = 1, \ldots, n$ so that denoting $x_j = \tilde{x}_j$, $1 \le j \le k$ again and $(\tilde{x}_{k+1}, \ldots, \tilde{x}_n) = (y_1, \ldots, y_l)$ we have

$$P_{11}(x, y) = \sum_{j=1}^{k} F_j x_j + O((|x| + |y|)^2). \tag{4.45}$$

Note that the coefficient of x_1 in $\tilde{P}_{11}(x, y)$ is the identity matrix I. We now prepare the next lemma.

Lemma 4.10. *Let $P(x)$ be an $m \times m$ matrix valued C^∞ function defined near $x = 0$. With a blocking*

$$P(0) = \begin{bmatrix} A_{11} & A_{12} \\ A_{21} & A_{22} \end{bmatrix}$$

assume that A_{11} and A_{22} has no common eigenvalue. Then there is $\epsilon = \epsilon(A_{11}, A_{22}) > 0$ such that if $\|A_{21}\| + \|A_{12}\| < \epsilon$ then one can find a smooth matrix $T(x)$ defined in $|x| < \epsilon$ such that

$$T(x)^{-1} P(x) T(x) = \begin{bmatrix} \hat{P}_{11}(x) & 0 \\ 0 & \hat{P}_{22}(x) \end{bmatrix}$$

where $T(x) = I + T_1(x)$ and $\|T_1(0)\| \to 0$ as $\|A_{21}\| + \|A_{12}\| \to 0$.

Proof. We first show that there are G_{12}, G_{21} such that

$$\begin{bmatrix} A_{11} & A_{12} \\ A_{21} & A_{22} \end{bmatrix}\begin{bmatrix} I & G_{12} \\ G_{21} & I \end{bmatrix} = \begin{bmatrix} I & G_{12} \\ G_{21} & I \end{bmatrix}\begin{bmatrix} A_{11}+X_{11} & 0 \\ 0 & A_{22}+X_{22} \end{bmatrix} \tag{4.46}$$

provided $\|A_{12}\| + \|A_{21}\|$ is small. Equation (4.46) is written as

$$\begin{bmatrix} A_{11}+A_{12}G_{21} & A_{11}G_{12}+A_{12} \\ A_{21}+A_{22}G_{21} & A_{21}G_{12}+A_{22} \end{bmatrix} = \begin{bmatrix} A_{11}+X_{11} & G_{12}A_{22}+G_{12}X_{22} \\ G_{21}A_{11}+G_{21}X_{11} & A_{22}+X_{22} \end{bmatrix}.$$

This gives $A_{12}G_{21} = X_{11}$, $A_{21}G_{12} = X_{22}$. Plugging these relations into the remaining two equations we have

$$A_{12} + A_{11}G_{12} = G_{12}A_{22} + G_{12}A_{21}G_{12},$$
$$A_{21} + A_{22}G_{21} = G_{21}A_{11} + G_{21}A_{12}G_{21}.$$

Let us set

$$F_1(G_{12}, G_{21}, A_{12}, A_{21}) = G_{12}A_{22} - A_{11}G_{12} + G_{12}A_{21}G_{12} - A_{12},$$
$$F_2(G_{12}, G_{21}, A_{12}, A_{21}) = G_{21}A_{11} - A_{22}G_{12} + G_{21}A_{12}G_{21} - A_{21}$$

then the equations become

$$\begin{cases} F_1(G_{12}, G_{21}, A_{12}, A_{21}) = 0, \\ F_2(G_{12}, G_{21}, A_{12}, A_{21}) = 0. \end{cases} \tag{4.47}$$

It is well known that (see [71] for example)

$$\frac{\partial(F_1, F_2)}{\partial(G_{12}, G_{21})}(0,0,0,0)$$

is non singular if A_{11} and A_{22} have no common eigenvalue. Then by the implicit function theorem there exist smooth $G_{12}(A_{12}, A_{21})$ and $G_{21}(A_{12}, A_{21})$ defined for small $\|A_{12}\| + \|A_{21}\|$ with $G_{12}(0,0) = 0$, $G_{21}(0,0) = 0$ verifying (4.47). This proves the assertion.

We next look for $T(x)$ in the form

$$T(x) = T_0 + T_1(x), \quad T_0(x) = \begin{bmatrix} I & G_{12} \\ G_{21} & I \end{bmatrix}, \quad T_1(0) = 0.$$

The equation which is verified by $T(x)$ is

$$(P_0 + P_1(x))(T_0 + T_1(x)) = (T_0 + T_1(x))(\tilde{P}_0 + \tilde{P}_1(x)) \tag{4.48}$$

where $P_0 = P(0)$, $P_0 T_0 = T_0 \tilde{P}_0$ and

$$\tilde{P}_1(x) = \begin{bmatrix} \tilde{P}_{11}(x) & 0 \\ 0 & \tilde{P}_{22}(x) \end{bmatrix}.$$

Recall that

$$\tilde{P}_0 = \begin{bmatrix} A_{11} + A_{12}G_{21} & 0 \\ 0 & A_{22} + A_{21}G_{12} \end{bmatrix}, \quad P_1(x) = \begin{bmatrix} P_{11}(x) & P_{12}(x) \\ P_{21}(x) & P_{22}(x) \end{bmatrix}.$$

Look for $T_1(x)$ in the form

$$T_1(x) = \begin{bmatrix} 0 & T_{12}(x) \\ T_{21}(x) & 0 \end{bmatrix}.$$

Equating the off diagonal entries of both sides of (4.48) we get

$$\begin{cases} A_{11}T_{12} + P_{12}(x) + P_{11}(x)G_{12} + P_{11}(x)T_{12} \\ \qquad = (G_{12} + T_{12})\tilde{P}_{22}(x) + T_{12}(A_{22} + A_{21}G_{12}), \\ A_{22}T_{21} + P_{21}(x) + P_{22}(x)G_{21} + P_{22}(x)T_{21} \\ \qquad = (G_{21} + T_{21})\tilde{P}_{11}(x) + T_{21}(A_{11} + A_{12}G_{21}). \end{cases} \tag{4.49}$$

On the other hand, equating the diagonal entries of both sides we have

$$\begin{cases} \tilde{P}_{11}(x) = A_{12}T_{21} + P_{11}(x) + P_{12}(x)(G_{21} + T_{21}), \\ \tilde{P}_{22}(x) = A_{21}T_{12} + P_{22}(x) + P_{21}(x)(G_{12} + T_{12}). \end{cases} \tag{4.50}$$

Plugging (4.50) into (4.49) we obtain

$$\begin{aligned} f_1(T_{12}, x) = {} & A_{11}T_{12} - T_{12}(A_{22} + A_{21}G_{12}) \\ & + P_{11}(x)G_{12} + P_{12}(x) + P_{11}(x)T_{12} \\ & - (G_{12} + T_{12})(A_{21}T_{12} + P_{21}(x)(G_{12} + T_{12}) + P_{22}(x)) = 0 \end{aligned}$$

and

$$\begin{aligned} f_2(T_{21}, x) = {} & A_{22}T_{21} - T_{21}(A_{11} + A_{12}G_{21}) \\ & + P_{22}(x)G_{21} + P_{21}(x) + P_{22}(x)T_{12} \\ & - (G_{21} + T_{21})(A_{12}T_{21} + P_{12}(x)(G_{21} + T_{21}) + P_{11}(x)) = 0. \end{aligned}$$

Since

$$f_1(T_{12}, 0) = A_{11}T_{12} - T_{12}A_{22}, \quad f_2(T_{21}, 0) = A_{22}T_{21} - T_{21}A_{11}$$

when $A_{21} = 0$, $A_{12} = 0$, $x = 0$, it is clear that

$$\frac{\partial f_1}{\partial T_{12}}(0,0), \quad \frac{\partial f_2}{\partial T_{21}}(0,0)$$

are non singular if $\|A_{12}\| + \|A_{21}\|$ is small. Then by the implicit function theorem there exist smooth $T_{12}(x)$ and $T_{21}(x)$ with $T_{12}(0) = 0$, $T_{21}(0) = 0$ such that

$$f_1(T_{12}(x), x) = 0, \quad f_2(T_{21}(x), x) = 0.$$

This proves the assertion. □

We return to the proof of Proposition 4.3. Since $\tilde{P}(x, y)$ is sufficiently close to $P(x, y)$ and

$$P(0,0) = \begin{bmatrix} 0 & 0 \\ 0 & G \end{bmatrix}, \quad \det G \neq 0$$

one can apply Lemma 4.10 to $\tilde{P}(x, y)$ and find a $G(x, y)$ such that

$$G(x,y)^{-1}\tilde{P}(x,y)G(x,y) = \begin{bmatrix} \tilde{P}_{11}(x,y) & 0 \\ 0 & \tilde{P}_{22}(x,y) \end{bmatrix}. \tag{4.51}$$

Denote $G(x, y)^{-1}P(x, y)G(x, y)$ and $G(x, y)^{-1}\tilde{P}(x, y)G(x, y)$ by $P(x, y)$ and $\tilde{P}(x, y)$ again. We summarize our arguments in

Proposition 4.4. *Assume that P_{orig} and $\tilde{P}_{orig}$ verify the assumption in Proposition 4.3. Then we may assume that P_{orig} and $\tilde{P}_{orig}$ have the form*

$$\tilde{P}(x,y) = \begin{bmatrix} \tilde{P}_{11}(x,y) & 0 \\ 0 & \tilde{P}_{22}(x,y) \end{bmatrix}, \quad P(x,y) = \begin{bmatrix} P_{11}(x,y) & P_{12}(x,y) \\ P_{21}(x,y) & P_{22}(x,y) \end{bmatrix}$$

with

$$P_{11}(x,y) = \sum_{j=1}^{k} A_j x_j + \sum_{j=1}^{l} B_j y_j + R(x,y), \quad R(x,y) = O(|(x,y)|^2)$$

where the following properties are verified; for any neighborhood U of the origin there is a neighborhood $W \subset U$ of the origin such that for any $\epsilon > 0$ one can find $\tilde{\epsilon} > 0$ so that if $|\tilde{P}_{orig} - P_{orig}|_{C^2(U)} < \tilde{\epsilon}$ then we have

$$|\tilde{P}_{11}(x,y) - P_{11}(x,y)|_{C^2(W)} < \epsilon, \tag{4.52}$$

$$|\sum_{j=1}^{k} A_j x_j + \sum_{j=1}^{l} B_j y_j - \sum_{j=1}^{k} F_j x_j| < C\epsilon(|x| + |y|). \tag{4.53}$$

Moreover one has

$$\det(\lambda + \tilde{P}_{11}(x, y)) = 0 \Longrightarrow \lambda \textit{ is real.}$$

Proof. Since $P(x, y)$ and $\tilde{P}(x, y)$ are obtained from P_{orig} and $\tilde{P}_{orig}$ by a smooth change of basis and a linear change of coordinates then (4.52) is clear. Let us recall

$$G(x, y) = \begin{bmatrix} I & G_{12}(x, y) \\ G_{21}(x, y) & I \end{bmatrix}$$

which verifies (4.51) where $\|G_{12}(0, 0)\| + \|G_{21}(0, 0)\|$ becomes as small as we please if $\tilde{\epsilon}$ is small. Hence $G(x, y)$ is enough close to the identity and then (4.53) follows from (4.45). Note that

$$\det(\lambda + \tilde{P}_{orig}) = \det(\lambda + \tilde{P}_{11}(x, y))\det(\lambda + \tilde{P}_{22}(x, y)).$$

Then the last assertion follows immediately. □

We proceed to the next step. Write

$$\tilde{P}_{11}(x, y) = \tilde{P}_{11}(0, y) + (\tilde{\phi}^i_j(x, y))_{1 \le i, j \le r} \tag{4.54}$$

so that $\tilde{\phi}^i_j(0, y) = 0$. Let us define $t^i_j(x, y)$ by

$$\tilde{\phi}^i_j(x, y) = \phi^i_j(x) + t^i_j(x, y)$$

where

$$F(x) = \sum_{j=1}^{k} F_j x_j = (\phi^i_j(x))_{1 \le i, j \le r}.$$

Lemma 4.11. *Assume that* $|\tilde{P}_{11}(x, y) - P_{11}(x, y)|_{C^2(W)} < \epsilon$ *and* $\{(x, y) \mid |x|, |y| < \epsilon\} \subset W$. *Then for* $|x|$, $|y| < \epsilon$ *we have*

$$|t^i_j(x, y)| \le C|x|, \quad |\partial_{x_\mu} t^i_j(x, y)| \le C\epsilon, \quad \mu = 1, \dots, k.$$

Proof. Write

$$\tilde{P}_{11}(x, y) = \tilde{P}_{11}(0, y) + \sum_{j=1}^{k} \tilde{A}_j(y) x_j + \tilde{R}(x, y), \quad \tilde{R}(x, y) = O(|x|^2)$$

so that

$$T = (t^i_j(x, y)) = \sum_{j=1}^{k} \tilde{A}_j(y)x_j - \sum_{j=1}^{k} F_j x_j + \tilde{R}(x, y). \tag{4.55}$$

Noting $\partial_{x_j} \tilde{P}_{11}(0, y) = \tilde{A}_j(y)$, $\partial_{x_j} P_{11}(0, y) = A_j + \partial_{x_j} R(0, y)$ and

$$|\partial_{x_j} R(0, y)| \le C|y| \le C\epsilon \quad \text{if } |y| < \epsilon$$

with C independent of $\tilde{P}$, one gets

$$|\tilde{A}_j(y) - A_j| \le C\epsilon \quad \text{if } |y| < \epsilon. \tag{4.56}$$

Now it is clear that

$$|\tilde{A}_j(y) - F_j| \le C'\epsilon \quad \text{if } |y| < \epsilon \tag{4.57}$$

because of (4.53) and (4.56). On the other hand from

$$P_{11}(0, y) = \sum_{j=1}^{l} B_j y_j + R(0, y)$$

and (4.53) it follows that

$$|P_{11}(0, y)| \le C\epsilon|y| + C|y|^2 \le C\epsilon|y| \quad \text{if } |y| < \epsilon.$$

Moreover $|\tilde{P}_{11}(0, y) - P_{11}(0, y)|_{C^2(W)} < \epsilon$ shows

$$|\tilde{P}_{11}(0, y)| < \epsilon + C\epsilon|y| < C'\epsilon \quad \text{if } |y| < \epsilon. \tag{4.58}$$

We now estimate $T(x, y) = (t^i_j(x, y))$ and $\partial_{x_j} T(x, y)$. Note that $|\partial_{x_j} \tilde{R}(x, y)| \le C|x|$ since $\partial_{x_j} \tilde{R}(0, y) = 0$ and $|\partial^\alpha_x \tilde{R}(x, y)| \le C$ for $|\alpha| = 2$ with C independent of $\tilde{P}$. Then by (4.55) and (4.57) one sees

$$\begin{cases} |T(x, y)| \le C\epsilon|x| + C|x|^2 \le C'\epsilon|x| & \text{if } |x| < \epsilon, \\ |\partial_{x_j} T(x, y)| \le C\epsilon + C|x| \le C'\epsilon & \text{if } |x|, |y| < \epsilon \end{cases} \tag{4.59}$$

which proves the assertion. □

Recall

$$F(x) = \sum_{j=1}^{k} F_j x_j = (\phi^i_j(x))_{1 \le i, j \le r}$$

where $F_1 = I$ and $\{F_1, \ldots, F_k\}$ be a basis for $M_r^h(\mathbb{C})$ over $\mathbb{R}$ (resp. $M_r^s(\mathbb{R})$) and hence $k = r^2$ (resp. $k = r(r+1)/2$).

Proposition 4.5. *Assume that $P(x)$ is a $r \times r$ matrix valued smooth function defined in a neighborhood of the origin of $\mathbb{R}^n$. Assume that all eigenvalues of $P(x)$ are real and*

$$\sum_{j=1}^{n} \frac{\partial P}{\partial x_j}(0) x_j \tag{4.60}$$

is sufficiently close to $F(x)$ in C^1. Then there is a $\delta > 0$ such that $P(x)$ is diagonalizable for every x with $|x| < \delta$.

Proof. Let us write

$$P(\omega + x) = P(\omega) + Q(x, \omega)$$

so that $Q(0, \omega) = 0$. For $T \in U(r)$, a unitary matrix of order r we consider

$$\begin{aligned} T^* P(\omega + x) T &= T^* P(\omega) T + T^* Q(x, \omega) T \\ &= P^T(\omega) + Q^T(x, \omega) = P^T(\omega) + (\phi_j^i(x, \omega; T))_{1 \le i, j \le m}. \end{aligned}$$

We show that there exist a $\delta > 0$ and a neighborhood W of the origin of $\mathbb{R}^k$ such that with $x = (x_a, x_b)$, $x_a = (x_1, \ldots, x_k)$, $x_b = (x_{k+1}, \ldots, x_n)$ the map

$$W \ni x_a \mapsto \big((\mathrm{Re}\, \phi_j^i(x, \omega; T))_{i \ge j}, (\mathrm{Im}\, \phi_j^i(x, \omega; T))_{i > j} \big) \in \mathbb{R}^k$$

is a diffeomorphism from W into $\{y \in \mathbb{R}^k \mid |y| < \delta\}$ for every $T \in U(r)$ and every x_b, ω with $|x_b|$, $|\omega| < \delta$. To see this we write

$$\begin{aligned} Q(x, \omega) = P(x + \omega) - P(\omega) &= \sum_{j=1}^{n} \frac{\partial P}{\partial x_j}(\omega) x_j + \tilde{R}(x, \omega) \\ &= \sum_{j=1}^{k} F_j x_j + \sum_{j=1}^{k} \left(\frac{\partial P}{\partial x_j}(\omega) - F_j \right) x_j + \sum_{j=k+1}^{n} \frac{\partial P}{\partial x_j}(\omega) x_j + \tilde{R}(x, \omega) \\ &= \sum_{j=1}^{k} F_j x_j + R(x, \omega) \end{aligned}$$

then it is clear that for any $\epsilon > 0$ one can find $\delta' > 0$ such that

$$\| R(x, \omega) \| \le \epsilon |x| \tag{4.61}$$

if $|x|$, $|\omega| < \delta'$ and if (4.60) is sufficiently close to $F(x)$. Let us study

$$Q^T(x,\omega) = \sum_{j=1}^{k} F_j^T x_j + R^T(x,\omega) = \sum_{j=1}^{k} \ell_j(x_a;T)F_j + R^T(x,\omega)$$

where $\ell_j(x_a;T)$ are linear in x_a. Since $U(r) \subset \mathbb{R}^{r^2}$ is compact it is clear that we have

$$\left|\frac{\partial(\ell_1,\ldots,\ell_k)}{\partial(x_1,\ldots,x_k)}(x_a;T)\right| \geq c > 0$$

with some $c > 0$ for every $T \in U(r)$. In view of (4.61), taking $\epsilon > 0$ so small we conclude that

$$\left|\frac{\partial\big((\mathrm{Re}\,\phi_j^i)_{i\geq j}, (\mathrm{Im}\,\phi_j^i)_{i>j}\big)}{\partial(x_a)}(0,0,0;T)\right| \geq c' > 0$$

with some $c' > 0$ for every $T \in U(r)$. By the implicit function theorem and the compactness of $U(r)$ there exists a smooth $x_a(y_a, x_b, \omega; T)$ defined in $|y_a|$, $|x_b|$, $|\omega| < \delta''$ and $T \in U(r)$ such that

$$\begin{cases} \mathrm{Re}\,\phi_j^i(x_a(y_a,x_b,\omega;T),x_b,\omega;T) = y_j^i & \text{for } \ i \geq j, \\ \mathrm{Im}\,\phi_j^i(x_a(y_a,x_b,\omega;T),x_b,\omega;T) = \tilde{y}_j^i & \text{for } \ i > j \end{cases}$$

where we have set $y_a = \big((y_j^i)_{i\geq j}, (\tilde{y}_j^i)_{i>j}\big) \in \mathbb{R}^k$. This proves the assertion.

We now show that $P(\omega)$ is diagonalizable for every $\omega \in \mathbb{R}^n$ with $|\omega| < \delta = \min\{\delta', \delta''\}$. Take $T \in U(r)$ so that

$$P^T(\omega) = (\bigoplus_{i=1}^{s} \lambda_i I_{r_i}) + (A_{ij})_{1\leq i,j\leq s} \tag{4.62}$$

where $\{\lambda_i\}$ are different from each other and A_{ij} are $r_i \times r_j$ matrices such that $A_{ij} = 0$ if $i > j$ and A_{ii} are upper triangular with zero diagonal entries. Let us set

$$J = \bigcup_{p=1}^{s-1} \{(i,j) \mid r_p < i \leq m, r_{p-1} < j \leq r_p\}$$

where $r_0 = 0$. As observed above one can take $((y_j^i)_{i\geq j}, (\tilde{y}_j^i)_{i>j}, x_b)$ as a new system of local coordinates around the origin of $\mathbb{R}^n$. Denote

$$y_{II} = \Big((y_j^i)_{(i,j)\in J}, (\tilde{y}_j^i)_{((i,j)\in J, i>j)}\Big), \quad y_a = (y_I, y_{II})$$

and, putting $y_{II} = 0$, $x_b = 0$, consider

$$\det(\lambda + P(\omega + x)) = \det(\lambda + P^T(\omega + x)) = \prod_{i=1}^{s} \det(\lambda + K_i(y_I, \omega; T))$$

where

$$K_i(y_I, \omega; T) = \lambda_i I_{r_i} + A_{ii} + (\phi_q^p(y_I, \omega; T))_{s_{i-1} < p,q \le s_i}$$

with $s_i = r_1 + \cdots + r_i$, $s_0 = 0$. Note that we have

$$\begin{cases} \phi_q^p(y_I, \omega; T) = y_q^p + i\tilde{y}_q^p \quad \text{if} \quad p > q, \\ \phi_p^p(y_I, \omega; T) = y_p^p + \operatorname{Im} \phi_p^p(y_I, \omega; T). \end{cases}$$

We will conclude $A_{ii} = 0$ repeating the same arguments proving the next lemma.

Lemma 4.12. *Let A be a constant matrix of order r such that $A = \alpha I_r + \tilde{A}$ where α is a real constant and $\tilde{A}$ is upper triangular with zero diagonal entries. Let $P(x) = A + (\phi_j^i(x))$ where $\phi_j^i(x)$ are linear in x and $\operatorname{Re} \phi_j^i(x)$, $i \ge j$, $\operatorname{Im} \phi_j^i(x)$, $i > j$ are linearly independent over $\mathbb{R}$. Suppose that all eigenvalues of $P(x)$ are real. Then A is necessarily diagonal matrix.*

Proof. Let us set $y_a = (y_j^i) = (\operatorname{Re} \phi_j^i)_{i \ge j}$, $y_b = (\operatorname{Im} \phi_j^i)_{i > j}$ and let (y_a, y_b, y_c) is a new system of local coordinates of $\mathbb{R}^n$ which is related to x by a non singular linear transformation. Let $A = (a_{pq})$ and we first show that $a_{p,p+1} = 0$ for $p = 1, \ldots, r-1$. Take $y_j^i = 0$ for $i \ge j$ unless $(i, j) = (p+1, p)$ and $y_b = 0$, $y_c = 0$. Then it is clear that

$$\begin{aligned} &\det(\lambda + A + (\phi_j^i)) \\ &= \prod_{j \ne p,p+1} (\lambda + \alpha + \operatorname{Im} \phi_j^j) \\ &\quad \times \Big((\lambda + \alpha + \operatorname{Im} \phi_p^p)(\lambda + \alpha + \operatorname{Im} \phi_{p+1}^{p+1}) - y_p^{p+1}(a_{p,p+1} + \phi_{p+1}^p) \Big). \end{aligned}$$

Since $\operatorname{Im} \phi_i^i(x)$ and $\phi_{p+1}^p(x)$ are constant times y_p^{p+1} then we see

$$\begin{aligned} &(\lambda + \alpha + \operatorname{Im} \phi_p^p)(\lambda + \alpha + \operatorname{Im} \phi_{p+1}^{p+1}) - y_p^{p+1}(a_{p,p+1} + \phi_{p+1}^p) \\ &= (\lambda + \alpha)^2 + O(|y_p^{p+1}|)(\lambda + \alpha) - y_p^{p+1} a_{p,p+1} + O(|y_p^{p+1}|^2) = 0 \end{aligned}$$

would have a non-real root for small y_p^{p+1} unless $a_{p,p+1} = 0$.

We now proceed by induction on $q - p$. Suppose that

$$a_{pq} = 0 \quad \text{for} \quad p + 1 \le q \le p + r - 1.$$

Let $q = p + r$ and take $y^i_j = 0$ for $i \geq j$ unless $(i, j) = (q, p)$ and $y_b = 0$, $y_c = 0$. We note that

$$\det(\lambda + A + (\phi^i_j)) = \prod_{j \neq p, p+1, \ldots, q} (\lambda + \alpha + \phi^j_j) Q(\lambda)$$

where $Q(\lambda)$ has the form

$$(\lambda + \alpha)^{r+1} - (\lambda + \alpha)^{r-1} y^q_p a_{pq} + \sum_{j=0}^{r-1} O(|y^q_p|^{r-1-j})(\lambda + \alpha)^j.$$

If we have set $\lambda + \alpha = \sqrt{|y^q_p|} z$ then $Q(\lambda) = 0$ is reduced to

$$z^{r+1} - z^{r-1} a_{pq} + O(|y^q_p|) R(z, y^q_p) = 0 \tag{4.63}$$

where R is a polynomial in z of degree $r - 1$. Thus if $a_{pq} \neq 0$ then (4.63) has a non real root for small y^q_p and hence $Q(\lambda) = 0$ would have a non real root. This proves that $a_{pq} = 0$. By induction we get the desired assertion.

For the real matrix case the proof is similar. □

Since ω, $|\omega| < \delta$ is arbitrary to prove Proposition 4.5 it suffices to show that $P^T(\omega)$ is diagonalizable which follows from the next lemma.

Lemma 4.13. *Let $A = (A_{ij})_{1 \leq i, j \leq s}$ be a constant matrix of order m where A_{ij} are $r_i \times r_j$ matrices. Suppose that $A_{ii} = \lambda_i I_{r_i}$ where $\lambda_i \neq \lambda_j$ if $i \neq j$ and $A_{ij} = O$ if $i > j$. Then A is diagonalizable.*

Proof. It suffices to construct S so that $S^{-1} A S = D = \lambda_1 I_{r_1} \oplus \cdots \oplus \lambda_s I_{r_s}$. Let us set $S = (S_{ij})$ where the blocking corresponds to that of A and $S_{ij} = O$ if $i > j$ and $S_{ii} = I_{r_i}$. From $AS = SD$ it follows that

$$(\lambda_i - \lambda_j) S_{ij} = - \sum_{k \geq i+1} A_{ik} S_{kj} \quad (i < j).$$

In particular $S_{r-1,r} = -(\lambda_{r-1} - \lambda_r)^{-1} A_{r-1,r}$ is determined by the above equation. Inductively S_{ir} are determined for $1 \leq i \leq r - 1$. Then we proceed to

$$(\lambda_i - \lambda_{r-1}) S_{i,r-1} = - \sum_{k \geq i+1} A_{ik} S_{k,r-1}.$$

Repeating this argument we obtain $S_{i,r-1}$ and hence the desired assertion. □

We now prove that near $(0, 0)$ the set of characteristics of order r of $\tilde{P}(x, y)$ is a smooth manifold. We first show that near $y = 0$ there is a unique smooth $g(y)$ such that

$$\tilde{P}_{11}(g(y), y) = O.$$

To see this let us study the map

$$\Phi : B_a \ni x \mapsto \big((\mathrm{Re}\,\tilde{\phi}^i_j(x,y))_{i\geq j}, (\mathrm{Im}\,\tilde{\phi}^i_j(x,y))_{i>j}\big) \in \mathbb{R}^k$$

where $B_a = \{x \in \mathbb{R}^k \mid |x| \leq a\}$. Let

$$A : \mathbb{R}^k \ni x \mapsto \big((\mathrm{Re}\,\phi^i_j(x))_{i\geq j}, (\mathrm{Im}\,\phi^i_j(x))_{i>j}\big) \in \mathbb{R}^k$$

which is a linear transformation on $\mathbb{R}^k$. Since $(\mathrm{Re}\,\phi^i_j(x))_{i\geq j}, (\mathrm{Im}\,\phi^i_j(x))_{i>j}$ are linearly independent then A is non singular. From Lemma 4.11 one can choose $\epsilon > 0$ so that

$$|A^{-1}\Phi'_x(x,y) - I| < 1/2 \quad \text{if } |x|, |y| < \epsilon.$$

Let us write $\tilde{P}_{11}(0,y) = (b^i_j(y))$ and note that $|b^i_j(y)| \leq C'\epsilon$ for $|y| < \epsilon$ by (4.58). Then choosing $\epsilon > 0$ sufficiently small we can apply the implicit function theorem to conclude that there exists a unique smooth $g(y,\theta,\kappa)$ defined in $|(\theta,\kappa)| \leq \epsilon$, $|y| \leq \epsilon$ such that

$$\begin{cases} \mathrm{Re}\,\tilde{\phi}^i_j(g(y,\theta,\kappa),y) = \theta^i_j - \mathrm{Re}\,b^i_j(y), \; i \geq j, \\ \mathrm{Im}\,\tilde{\phi}^i_j(g(y,\theta,\kappa),y) = \kappa^i_j - \mathrm{Im}\,b^i_j(y), \; i > j \end{cases} \tag{4.64}$$

and in the real case

$$\tilde{\phi}^i_j(g(y,\theta),y) = \theta^i_j - b^i_j(y), \quad i \geq j \tag{4.65}$$

such that

$$|g(y,\theta,\kappa)| < C\epsilon. \tag{4.66}$$

Set

$$(\psi^i_j(y,\theta,\kappa)) = \tilde{P}_{11}(g(y,\theta,\kappa),y) \tag{4.67}$$

then from (4.54) and (4.64) it follows that

$$\begin{cases} \mathrm{Re}\,\psi^i_j(y,\theta,\kappa) = \theta^i_j, \quad i \geq j, \\ \mathrm{Im}\,\psi^i_j(y,\theta,\kappa) = \kappa^i_j, \quad i > j. \end{cases} \tag{4.68}$$

Let us write

$$\psi^i_j(y,\theta,\kappa) = c^i_j(y) + \chi^i_j(y,\theta,\kappa)$$

where $c^i_j(y) = \psi^i_j(y,0,0)$ and $\chi^i_j(y,\theta,\kappa) = O(|(\theta,\kappa)|)$. Let us put $h(\lambda) = \det(\lambda + \tilde{P}_{11}(g(y,\theta,\kappa),y))$. From Proposition 4.4 it follows that $h(\lambda) = 0$ implies

λ is real. Repeating the same arguments as in the proof of Lemma 4.12 we conclude that $c_q^p(y) = 0$ for $p < q$ and $\operatorname{Im} c_p^p(y) = 0$. This, together with (4.68), implies

$$\tilde{P}_{11}(g(y,0,0),y) = O. \tag{4.69}$$

The proof for the real case is similar.

We now prove that near $(0,0)$ the set of characteristics of order r of $\tilde{P}(x,y)$ is a smooth manifold given by $x = g(y,0,0)$. Let $(\bar{x},\bar{y})$ be a characteristic of order r of $\tilde{P}(x,y)$ close to $(0,0)$. Then it is clear that $(\bar{x},\bar{y})$ is a characteristic of the same order for $\tilde{P}_{11}(x,y)$ because $\det \tilde{P}_{22}(x,y) \neq 0$ near $(0,0)$. Recalling that $\tilde{P}_{11}(x,y)$ has the form

$$\tilde{P}_{11}(x,y) = x_1 + \tilde{P}_{11}^{\#}(x',y), \quad x' = (x_2,\ldots,x_k)$$

we see that $\det \tilde{P}_{11}(x_1,\bar{x}',\bar{y}) = (x_1 - \bar{x}_1)^r$ and hence

$$\det(\lambda + \tilde{P}_{11}(\bar{x},\bar{y})) = \lambda^r.$$

Thus the zero is an eigenvalue of multiplicity r of $\tilde{P}_{11}(\bar{x},\bar{y})$. On the other hand Proposition 4.4 gives

$$\left|\frac{\partial \tilde{P}_{11}}{\partial x_j}(0) - F_j\right|, \quad \left|\frac{\partial \tilde{P}_{11}}{\partial y_j}(0)\right| < C\epsilon. \tag{4.70}$$

Then one can apply Proposition 4.5 to conclude that $\tilde{P}(\bar{x},\bar{y})$ is diagonalizable. This shows that

$$\tilde{P}_{11}(\bar{x},\bar{y}) = O$$

and hence one gets $\bar{x} = g(\bar{y},0,0)$.

Finally we show that the characteristics $(g(y,0,0),y)$ are nondegenerate. From (4.69) we have

$$\tilde{P}(g(y,0,0),y) = \begin{bmatrix} 0 & 0 \\ 0 & \tilde{P}_{22}(g(y,0,0),y) \end{bmatrix}$$

and hence

$$\operatorname{Ker} \tilde{P}(g(y,0,0),y) \cap \operatorname{Im} \tilde{P}(g(y,0,0),y) = \{0\}. \tag{4.71}$$

It is also clear that $\tilde{P}_{(g(y,0,0),y)}(x,y)$ is given by

$$\sum_{j=1}^{k} \frac{\partial \tilde{P}_{11}}{\partial x_j}(g(y,0,0),y)x_j + \sum_{j=1}^{l} \frac{\partial \tilde{P}_{11}}{\partial y_j}(g(y,0,0),y)y_j.$$

On the other hand since $|\tilde{P}_{11} - P_{11}|_{C^2(W)} < \epsilon$ it follows from Proposition 4.5 and (4.66) that

$$\left|\frac{\partial \tilde{P}_{11}}{\partial x_j}(g(y,0,0),y) - F_j\right|, \quad \left|\frac{\partial \tilde{P}_{11}}{\partial y_j}(g(y,0,0),y)\right| < C\epsilon \tag{4.72}$$

if $|y| < \epsilon$. This clearly shows that

$$\dim \tilde{P}_{(g(y,0,0),y)} = r^2. \tag{4.73}$$

To finish the proof, taking $\tilde{P}_{(g(y,0,0),y)}(\theta) = I$ into account, it is enough to show that $\tilde{P}_{(g(y,0,0),y)}(x,y)$ is diagonalizable for every (x,y). Note that from Lemma 4.2 all eigenvalues of $\tilde{P}_{(g(y,0,0),y)}(x,y)$ are real. Then from Proposition 4.5 and (4.72) it follows that $\tilde{P}_{(g(y,0,0),y)}(x,y)$ is diagonalizable for every (x,y) near $(0,0)$ and hence for all (x,y).

The proof for the real case is similar. Thus the proof of Proposition 4.3 is completed. □

Example 4.6. Consider a second order differential operator $P(x,D) = (p_{ik}(x,D))$ with 3×3 matrix coefficients

$$p_{ik}(x,\tau,\xi) = (\tau^2 - \sigma_i(x)|\xi|^2)\delta_{ik} - (1 - \sigma_i(x))\xi_i\xi_k$$

where $\sigma_i(x)$ are real smooth and close to σ_i in Example 4.4. We assume that $Q(x,\tau,\xi) = \det P(x,\tau,\xi) = 0$ has only real roots for any x and ξ. Then from Theorem 4.3 it follows that every characteristic of $P(x,\xi)$ are at most double and the double characteristics are nondegenerate.

Example 4.7. Let $A(\xi) = \sum_{j,k=1}^{3} A_{jk}\xi_j\xi_k$ be one of them discussed in Example 4.5, that is the characteristics of $\det(\tau^2 I - A(\xi))$ are at most double and the double characteristics are nondegenerate. Let $A_{jk}(x)$ be real smooth 3×3 matrices which are close to A_{jk} and set

$$A(x,\xi) = \sum_{j,k=1}^{3} A_{jk}(x)\xi_j\xi_k$$

and assume that $\det(\tau^2 I - A(x,\xi)) = 0$ has only real roots for any x and ξ. Then from Theorem 4.3 we see that every characteristic of $\tau^2 I - A(x,\xi)$ are at most double and the double characteristics are nondegenerate.

Example 4.8. As in Example 4.3 take $P(\xi) = \xi_1 I + \sum_{j=2}^{d} F_j\xi_j$ where $\{I, F_2, \ldots, F_d\}$ is a basis for $M_m^s(\mathbb{R})$. Consider

$$P(x,\xi) = \xi_1 I + \sum_{j=2}^{d} A_j(x)\xi_j$$

where $A_j(x)$ are real smooth $m \times m$ matrices which are enough close to F_j in C^2 and we assume that $P(x, \xi)$ has only real eigenvalues for any x and any ξ. Then from Theorem 4.3 it follows that every multiple characteristic of $P(x, \xi)$ is nondegenerate.

4.5 Symmetrizability (General Case)

In this section to simplify notations let us write $\mathscr{P}(x)$, $x = (x_0, x_1, \dots, x_n)$ which is a real analytic $m \times m$ matrix valued function defined near the origin of $\mathbb{R}^{n+1}$. We assume that all eigenvalues of $\mathscr{P}(x)$ are real near $x = 0$. We also denote by d_m the dimension of $M_m^h(\mathbb{C})$ (resp. $M_m^s(\mathbb{R})$) over $\mathbb{R}$, that is

$$d_m = m^2 \quad (resp.\ d_m = m(m+1)/2).$$

Our main concern in this section is to prove

Theorem 4.4. *Assume that all eigenvalues of $\mathscr{P}(x)$ are real near a nondegenerate characteristic $x = 0$ of order m and $\mathscr{P}_0(\Theta) = I_m$ with some $\Theta \in \mathbb{R}^{n+1}$. Then there is a real analytic symmetrizer near $x = 0$, that is there is a real analytic positive definite $H(x)$, $H^*(x) = H(x)$, defined near $x = 0$ such that*

$$\mathscr{P}(x)H(x) = H(x)\mathscr{P}^*(x).$$

Corollary 4.1. *Assume that $\mathscr{P}(x)$ has the form $x_0 I + P(x')$ with $x' = (x_1, \dots, x_n)$ and all eigenvalues of $P(x')$ are real near $x' = 0$. Suppose that $x = 0$ is a nondegenerate characteristic of order m of $\mathscr{P}(x)$. Then there is a real analytic positive definite $H(x')$, $H^*(x') = H(x')$, defined near $x' = 0$ such that*

$$P(x')H(x') = H(x')P^*(x').$$

We first give another proof, based on Theorem 4.4, for that the set of nondegenerate characteristics is a smooth manifold of codimension d_m.

Proposition 4.6. *Assume the same assumptions as in Theorem 4.4. Then we can choose a new system of local coordinates X and a real analytic $T(X)$ defined near $X = 0$ so that*

$$T(X)^{-1}\mathscr{P}(x(X))T(X) = \sum_{j=0}^{k-1} F_j X_j$$

with $k = d_m$ where $F_0 = I$ and $\{F_j\}$ span $M_m^h(\mathbb{C})$ over $\mathbb{R}$ (resp. $M_m^s(\mathbb{R})$).

Proof. From Theorem 4.4 there is a positive definite $H(x)$ such that $\mathscr{P}(x)H(x) = H(x)\mathscr{P}^*(x)$. This shows that

$$S(x) = H(x)^{-1/2}\mathscr{P}(x)H(x)^{1/2}$$

is Hermitian. Let us write $S(x) = (\phi^i_j(x))$ and hence $\phi^i_j(0) = 0$. With $\phi^i_j(x) = d\phi^i_j(x) + O(|x|^2)$ we note that $\{(\mathrm{Re}\, d\phi^i_j(x))_{i\geq j}, (\mathrm{Im}\, d\phi^i_j(x))_{i>j}\}$ are linearly independent over $\mathbb{R}$. Then taking a new system of local coordinates X so that $X_0 = \mathrm{Re}\,\phi^1_1(x)$, $X_i = \mathrm{Re}\,\phi^i_i(x) - \mathrm{Re}\,\phi^1_1(x)$, $2 \leq i \leq m$, $(X_{m+1}, \dots, X_{p-1}) = (\mathrm{Re}\,\phi^i_j(x))_{i>j}$, $(X_p, \dots, X_{k-1}) = (\mathrm{Im}\,\phi^i_j(x))_{i>j}$ we get the assertion with $T(X) = H(x(X))^{1/2}$. □

From Proposition 4.6 it is clear that, near $x = 0$, the set $\mathscr{P}(x) = O$ is given by $\Sigma = \{X_j = 0 \mid j = 0, \dots, d_m - 1\}$ which is a smooth manifold of codimension d_m. It is also clear that for $x \in \Sigma$ the properties (4.13) and (4.14) hold, that is Σ consists of nondegenerate characteristics. On the other hand let $\bar{x}$ be a characteristic of order m for $\mathscr{P}(x)$ so that 0 is the eigenvalue of $\mathscr{P}(\bar{x})$ of multiplicity m. Then 0 is the eigenvalue of $\sum_{j=0}^{d_m-1} F_j \bar{X}_j$ of multiplicity m where $\bar{x} = x(\bar{X})$. Since $\sum_{j=0}^{d_m-1} F_j \bar{X}_j$ is Hermitian we see $\sum_{j=0}^{d_m-1} F_j \bar{X}_j = O$ and hence $\bar{X}_j = 0$ for $j = 0, \dots, d_m - 1$. Thus we conclude $\bar{X} \in \Sigma$.

We start to prove Theorem 4.4. Choosing a system of local coordinates so that $\Theta = (1, 0, \dots, 0)$ we can assume that $\mathscr{P}_0(x)$ verifies the assumption of Lemma 4.9. Then one can assume that $T^{-1}\mathscr{P}_0(x)T$ is Hermitian (resp. symmetric) for every x with some constant matrix T. By a linear change of coordinates x one may assume that

$$T^{-1}\mathscr{P}_0(x)T = x_0 I + \sum_{j=1}^{k} F^j x_j$$

with $k = d_m - 1$ where $\{I, F^j\}$ span the space $M^h_m(\mathbb{C})$ (resp. $M^s_m(\mathbb{R})$) over $\mathbb{R}$. Since $\mathscr{P}(x) = \mathscr{P}_0(x) + R(x)$, $R(x) = O(|x|^2)$ as $x \to 0$, to prove Theorem 4.4, writing $\mathscr{P}(x) = x_0 I + P(x)$, it is enough to show the following theorem.

Theorem 4.5. *Let $P(x) = \sum_{j=1}^{k} F^j x_j + R(x)$ where $x = (x_0, \dots, x_n)$, and $R(x)$ is real analytic near the origin so that $R(x) = O(|x|^2)$ as $x \to 0$. Assume that $\{F^j\}$ are Hermitian (resp. symmetric) $l \times l$ constant matrices such that $\{I, F^j\}$ span the space $M^h_l(\mathbb{C})$ (resp. $M^s_l(\mathbb{R})$) over $\mathbb{R}$ and $k = d_l - 1$. Suppose that all eigenvalues of $P(x)$ are real near the origin. Then there is a positive definite real analytic $G(x)$ with $G(0) = I$ defined near the origin verifying*

$$P(x)G(x) = G(x)P^*(x), \quad G^*(x) = G(x). \tag{4.74}$$

Remark. Assume, for instance, that a positive definite $G(x)$ verifying (4.74) exists. Expanding both sides of (4.74) in the Taylor expansions around the origin and equating the first order terms we see that

$$\sum_{j=1}^{k} F^j G(0) x_j = \sum_{j=1}^{k} G(0) F^j x_j$$

so that $G(0)$ commutes with all Hermitian (resp. symmetric) matrices of order m and hence $G(0) = \alpha I$ with $\alpha \neq 0$. Since $G(0)$ is positive definite and hence $\alpha > 0$ we may suppose that $G(0) = I$ considering $\alpha^{-1} G(x)$ which also verifies (4.74).

To prove Theorem 4.5, we proceed by induction on the size of matrices. When $l = 2$, since $\mathscr{P}(0) = O$ and $x = 0$ is a nondegenerate double characteristic thanks to Proposition 4.2 there is a real analytic symmetrizer $G(x)$ verifying (4.74). Let the assumption of Theorem 4.5 be verified for $l < m$. Since $\{I, F^j\}$ span $M_l^h(\mathbb{C})$ (resp. $M_l^s(\mathbb{R})$), choosing a new system of local coordinates x we may suppose that the Hermitian (resp. symmetric) part of $R(x)$ can be removed so that

$$\mathscr{P}(x) = x_0 I + \sum_{j=1}^{k} F^j x_j + R(x)$$

with $k = d_l - 1$ where $R(x)$ is anti-Hermitian (resp. anti-symmetric). Since the all eigenvalues of $\mathscr{P}(x)$ are real it follows that

$$R(x_0, 0, \ldots, 0, x_{k+1}, \ldots, x_n) = O.$$

Changing notations slightly we write $x = (x_1, x_2, \ldots, x_k)$, $y = (x_0, x_{k+1}, \ldots, x_n)$ with $k = d_l - 1$ and

$$P(x, y) = \sum_{j=1}^{k} F^j x_j + R(x, y)$$

so that $P(0, y) = O$. We divide the proof of the assertion for $l = m$ into two steps. In the first step, introducing the polar coordinates $x = r\omega$, we blow up $P(x, y)$ at $x = 0$ so that

$$Q(r, \omega, y) = r^{-1} P(r\omega, y)$$

will be studied. We prove

Proposition 4.7. *Suppose that the assertion of Theorem 4.5 holds for $l < m$. Let $P(x, y) = \sum_{j=1}^{k} F^j x_j + R(x, y)$, $k = d_m - 1$ be a real analytic $m \times m$ matrix valued function with real eigenvalues near the origin such that $R(x, y) = O(|(x, y)|^2)$ as $(x, y) \to 0$ and $R(0, y) = O$. Assume that $\{I, F^j\}$ span $M_m^h(\mathbb{C})$ (resp. $M_m^s(\mathbb{R})$). Then for every $\omega \neq 0$ there is a positive definite $H(r, \phi, y)$ with diagonal entries 1 which is real analytic near $(0, \omega, 0)$ such that*

$$P(r\phi, y) H(r, \phi, y) = H(r, \phi, y) P^*(r\phi, y), \quad H^*(r, \phi, y) = H(r, \phi, y). \quad (4.75)$$

Thus we can construct a symmetrizer with diagonal entries 1 of the blown up $P(r\phi, y)$ in a neighborhood of every $(0, \omega, 0)$ with $\omega \neq 0$. In the second step we first observe that such symmetrizers can be continued analytically to a neighborhood of $\{0\} \times S^{k-1} \times \{0\}$.

Lemma 4.14. *Suppose that at every $(0, \omega, 0)$ with $\omega \neq 0$ there is a positive definite real analytic symmetrizer $H(r, \phi, y)$ with diagonal entries 1 verifying (4.75). Then there is $H(r, \phi, y)$ with diagonal entries 1 which is real analytic in $I \times S^{k-1} \times J$ such that*

$$P(r\phi, y)H(r, \phi, y) = H(r, \phi, y)P^*(r\phi, y), \quad H^*(r, \phi, y) = H(r, \phi, y) \tag{4.76}$$

holds for $(r, \phi, y) \in I \times S^{k-1} \times J$ where I, J are open intervals containing the origin.

We next show that the symmetrizer obtained in Lemma 4.14 is the blown up of a real analytic $G(x, y)$ defined near the origin $(x, y) = (0, 0)$.

Proposition 4.8. *Assume that $H(r, \phi, y)$ verifies (4.76) where $H(r, \phi, y)$ is real analytic in $I \times S^{k-1} \times J$ with diagonal entries 1. Then $H(r, \phi, y)$ is a blown up of a real analytic $G(x, y)$, that is*

$$H(r, \phi, y) = G(r\phi, y).$$

In particular we have

$$P(x, y)G(x, y) = G(x, y)P^*(x, y), \quad G^*(x, y) = G(x, y).$$

Combining Propositions 4.7 and 4.8, Theorem 4.5 follows immediately by induction on l.

First step: We prove Proposition 4.7. Assume that the assertion of Theorem 4.5 holds for $l < m$. We study the case $l = m$. Let us recall

$$P(x, y) = L(x) + R(x, y), \quad L(x) = \sum_{j=1}^{k} F^j x_j$$

where $k = d_m - 1$ and $\{I, F^j\}$ span the space $M_m^h(\mathbb{C})$ (resp. $M_m^s(\mathbb{R})$) over $\mathbb{R}$. Let $S(a) = \epsilon I + \mathrm{diag}(a_1, \dots, a_m)$, $|a_i| < \epsilon$ where $a = (a_1, \dots, a_m) \in \mathbb{R}^m$ and set

$$P_1(x, y, a) = S(a)^{-1}P(x, y)S(a).$$

Introducing the polar coordinates $x = r\omega$ we study

$$\tilde{P}(r, \omega, y, a) = r^{-1}P_1(r\omega, y, a)$$

near $(r, \omega, y, a) = (0, \omega, 0, 0)$.

Lemma 4.15. *All eigenvalues of $\tilde{P}(r,\omega,y,a)$ are real near $(0,\omega,0,0)$ with $\omega \neq 0$. The multiplicity of eigenvalues of $\tilde{P}(0,\omega,0,0)$ are less than m if $\omega \neq 0$.*

Proof. The first assertion is clear. Recall that

$$\tilde{P}(r,\omega,y,a) = S(a)^{-1}\left(L(\omega) + P^2(\omega; r, y) + O(|(r,y)|^2)\right) S(a)$$

with some $m \times m$ matrix $P^2(\omega; r, y)$ which is linear in (r, y) so that

$$\tilde{P}(0,\omega,0,0) = L(\omega). \tag{4.77}$$

If $L(\omega)$, $\omega \neq 0$ has an eigenvalue $\lambda \in \mathbb{R}$ of multiplicity m then it follows that $L(\omega) - \lambda I = O$ because $L(\omega)$ is Hermitian (resp. symmetric). This contradicts the fact that $\{I, F^j\}$ are linearly independent. Hence the assertion. □

We fix $\omega \neq 0$ and choose a unitary (resp. an orthogonal) T_0 so that

$$T_0^{-1} L(\omega) T_0 = \oplus_{j=1}^{p} \lambda_j I_{s_j}$$

where λ_j are different from each other and $p \geq 2$ as was seen above. Taking into account

$$S(a) = \epsilon I + O(|a|), \quad S^{-1}(a) = \epsilon^{-1} I + O(|a|),$$
$$P^2(\omega + \theta; r, y) = P^2(\omega; \theta, y) + O(|(r,\theta,y)|^2),$$
$$L(\omega + \theta) = L(\omega) + L(\theta)$$

we set

$$\begin{aligned} Q(r,\theta,y,a) &= r^{-1} T_0^{-1} P_1(r(\omega+\theta), y, a) T_0 = T_0^{-1} \tilde{P}(r, \omega+\theta, y, a) T_0 \\ &= \tilde{L}(\omega) + \tilde{L}(\theta) + \tilde{P}^2(\omega; r, y, a) + O(|(r,\theta,y,a)|^2) \end{aligned}$$

where $\tilde{L}(\omega) = \oplus \lambda_j I_{s_j}$, $\tilde{L}(\theta) = T_0^{-1} L(\theta) T_0$ and $\tilde{P}^2(\omega; r, y, a)$ is linear in (r, y, a). It is also clear that with $\tilde{L}(\theta) = \sum_{j=1}^{k} \tilde{F}^j \theta_j$, the matrices $\{I, \tilde{F}^j\}$ span $M_m^h(\mathbb{C})$ (resp. $M_m^s(\mathbb{R})$).

Note that the coefficients of a_j in $\tilde{P}^2(\omega; r, y, a)$ are anti-Hermitian (resp. anti-symmetric) although the fact is not used in the sequel.

Set $Q(r,\theta,y,a) = (Q_{ij}(r,\theta,y,a))$ then it is well known that there is a real analytic $T(r,\theta,y,a)$ defined near the origin with $T(0) = I$ such that

$$QT = T(\oplus_{j=1}^{p} \tilde{Q}_j) \tag{4.78}$$

(see for example [72]). We need a little bit more information on $\tilde{Q}_j$. Let $T = (T_{ij})$ with $T_{ii} = I_{s_i}$ then (4.78) yields

$$\sum Q_{ik}T_{ki} = \tilde{Q}_i, \quad \sum Q_{ik}T_{kj} = T_{ij}\tilde{Q}_j, \ i \neq j. \tag{4.79}$$

Plugging the first term of (4.79) into the second, we get

$$Q_{ii}T_{ij} - T_{ij}Q_{jj} = \sum_{k\neq j} T_{ij}Q_{jk}T_{kj} - \sum_{k\neq i} Q_{ik}T_{kj}$$

and hence for $i \neq j$

$$(\lambda_i - \lambda_j)T_{ij} = -Q_{ij} + O(|(r,\theta,y,a)|\sum_{k\neq l}|T_{kl}|). \tag{4.80}$$

By the implicit function theorem one can solve (4.80) so that $T_{ij} = T_{ij}(r,\theta,y,a)$, $T_{ij}(0) = 0$. Plugging T_{ij} into (4.79) we get $\tilde{Q}_i$ to be

$$\tilde{Q}_i(r,\theta,y,a) = Q_{ii}(r,\theta,y,a) + O(|(r,\theta,y,a)|^2).$$

We summarize what we have proved; there is a real analytic $T(r,\theta,y,a)$ defined near the origin with $T(0) = I$ such that

$$Q(r,\theta,y,a)T(r,\theta,y,a) = T(r,\theta,y,a)(\oplus_{j=1}^{p}\tilde{Q}_j(r,\theta,y,a))$$

where $\tilde{Q}_j(r,\theta,y,a)$ verifies

$$\tilde{Q}_j(r,\theta,y,a) = \lambda_j I_{s_j} + \tilde{L}_{jj}(\theta) + \tilde{P}^2_{jj}(\omega;r,y,a) + O(|(r,\theta,y,a)|)^2.$$

Here we have written $\tilde{L}(\theta) = (\tilde{L}_{ij}(\theta))$, $\tilde{P}^2(\omega;r,y,a) = (\tilde{P}^2_{ij}(\omega;r,y,a))$ and the blocking corresponds to that of $\oplus\lambda_j I_{s_j}$.

Lemma 4.16. *All eigenvalues of $\tilde{Q}_j(r,\theta,y,a)$ are real near $(r,\theta,y,a) = (0,0,0,0)$. In a new system of local coordinates (r,ψ,y,a), where ψ is linear in (r,θ,y,a), $\tilde{Q}_j$ takes the form*

$$\tilde{Q}_j(r,\psi,y,a) = (\lambda_j + b_j(r,\psi,y,a))I_{s_j} + \sum_{i=1}^{r_j}\tilde{F}^i_{jj}\psi_i + O(|(r,\psi,y,a)|^2)$$

with $r_j = d_{s_j} - 1$ where $b_j(r,\psi,y,a)$ is linear in (r,ψ,y,a) and $\{I_{s_j},\tilde{F}^i_{jj}\}$ span $M^h_{s_j}(\mathbb{C})$ (resp. $M^s_{s_j}(\mathbb{R})$).

Proof. It is clear that all eigenvalues of $\tilde{Q}_j(r,\theta,y,a)$ are real near the origin because so are those of $Q(r,\theta,y,a)$ by Lemma 4.15. We next show that $\tilde{P}^2_{jj}(\omega;r,y,a)$ is Hermitian (resp. symmetric). Recall that

$$\tilde{L}_{jj}(\theta)=\sum_i \tilde{F}^i_{jj}\theta_i$$

where $\{\tilde{F}^i_{jj}\}$ are Hermitian (resp. symmetric) matrices and, together with I_{s_j}, span the space $M^h_{s_j}(\mathbb{C})$ (resp. $M^s_{s_j}(\mathbb{R})$) over $\mathbb{R}$ since $\tilde{F}^i = T_0^{-1}F^iT_0$ and I span the $M^h_m(\mathbb{C})$ (resp. $M^s_m(\mathbb{R})$). Take $\bar{\theta}$, $\tau\in\mathbb{R}$ so that

$$\tilde{L}_{jj}(\bar{\theta})+\tilde{P}^2_{jj}(\omega;r,y,a)=\tilde{P}^{2(ah)}_{jj}(\omega;r,y,a)+\tau I_{s_j}$$

where $\tilde{P}^{2(ah)}_{jj}(\omega;r,y,a)$ denotes the anti-Hermitian (resp. anti-symmetric) part of $\tilde{P}^2_{jj}(\omega;r,y,a)$. Then we have

$$\tilde{Q}_j(\mu r,\mu\bar{\theta},\mu y,\mu a)=(\lambda_j+\mu\tau)I_{s_j}+\mu\tilde{P}^{2(ah)}_{jj}(\omega;r,y,a)+O(\mu^2).$$

If $\tilde{P}^{2(ah)}_{jj}(\omega;r,y,a)\neq O$ then $\tilde{Q}_j(\mu r,\mu\bar{\theta},\mu y,\mu a)$ has non-real eigenvalues, taking μ small enough, and hence a contradiction. Thus we can write

$$\tilde{P}^2_{jj}(\omega;r,y,a)=\sum_i c_i(r,y,a)\tilde{F}^i_{jj}+c_0(r,y,a)I_{s_j}$$

where $c_i(r,y,a)$ are linear functions of (r,y,a) so that

$$\tilde{Q}_j(r,\theta,y,a)=(\lambda_j+c_0(r,y,a))I_{s_j}+\sum_i \tilde{F}^i_{jj}(\theta_i+c_i(r,y,a))+O(|(r,\theta,y,a)|^2).$$

Renumbering $\{\tilde{F}^i_{jj}\}$, if necessary, we may suppose that $\{I,\tilde{F}^1_{jj},\ldots,\tilde{F}^{r_j}_{jj}\}$ are linearly independent so that

$$\sum_i \tilde{F}^i_{jj}(\theta_i+c_i(r,y,a))=\sum_{i=1}^{r_j}\tilde{F}^i_{jj}\psi_i(r,\theta,a).$$

This proves the assertion. □

By Lemma 4.16, each $\tilde{Q}_j(r,\theta,y,a)-(\lambda_j+b_j)I_{s_j}$ verifies the hypothesis of Theorem 4.5 with $l=r_j<m$ and hence there are positive definite $K_j(r,\theta,y,a)$ which are real analytic near the origin such that

$$\tilde{Q}_j(r,\theta,y,a)K_j(r,\theta,y,a)=K_j(r,\theta,y,a)\tilde{Q}^*_j(r,\theta,y,a),$$
$$K^*_j(r,\theta,y,a)=K_j(r,\theta,y,a)$$

with $K_j(0) = I_{s_j}$. Let us define $K(r, \theta, y, a)$ as

$$K(r, \theta, y, a) = \oplus_{j=1}^{p} K_j(r, \theta, y, a)^{1/2}$$

so that

$$K(r, \theta, y, a)^{-1}(\oplus_{j=1}^{p} \tilde{Q}_j(r, \theta, y, a))K(r, \theta, y, a) = \text{Hermitian (resp. symmetric).}$$

With $V = T(r, \theta, y, a)K(r, \theta, y, a)$, this shows that

$$V^{-1}Q(r, \theta, y, a)V = \text{Hermitian (resp. symmetric).}$$

Setting $U = ST_0TK$ we conclude that $U^{-1}P(r(\omega + \theta), y)U$ becomes Hermitian (resp. symmetric) and hence

$$P(r(\omega + \theta), y)UU^* = UU^*P^*(r(\omega + \theta), y).$$

Since $UU^* = ST_0T(KK^*)T^*T_0^*S$, noting that

$$KK^* = \oplus K_j = I + O(|(r, \theta, y, a)|), \quad T_0TT^*T_0^* = I + O(|(r, \theta, y, a)|)$$

we see that

$$UU^* = S(a)(I + K')S(a)$$

where $K' = O(|(r, \theta, y, a)|)$. Hence every diagonal entry of UU^* takes the form

$$\epsilon^2 + 2\epsilon a_i + a_i^2 + O(\epsilon^2|(r, \theta, y, a)|) + O(|(r, \theta, y, a)|^2).$$

Now taking $\epsilon > 0$ small enough, by the implicit function theorem one can solve $a(r, \theta, y) = (a_1(r, \theta, y), \ldots, a_m(r, \theta, y))$ so that $a_i(0) = 0$ and

$$\text{every diagonal entry of } UU^* = \epsilon^2$$

where $a(r, \theta, y)$ is real analytic near the origin. With

$$H(r, \phi, y) = \epsilon^{-2}U(r, \phi - \omega, y, a(r, \phi - \omega, y))U^*(r, \phi - \omega, a(r, \phi - \omega, y))$$

which is real analytic near $(0, \omega, 0)$ we conclude that

$$P(r\phi, y)H(r, \phi, y) = H(r, \phi, y)P^*(r\phi, y)$$

where all diagonal entries of $H(r, \phi, y)$ are 1. Since $\omega \neq 0$ is arbitrary the proof of Proposition 4.7 is completed.

Second step: We prove Proposition 4.8. We begin with proving Lemma 4.14. Recall that

$$r^{-1}P(r\omega, y) = L(\omega) + \sum_{j+|\alpha|\geq 1} r^j y^\alpha R_{j\alpha}(\omega), \quad R_{j\alpha}(\omega) = \sum_{|\beta|=j+1} R_{j\alpha\beta}\omega^\beta$$

with constant $m \times m$ matrices $R_{j\alpha\beta}$ so that $R_{j\alpha}(\omega)$ is a homogeneous polynomial in ω of degree $j+1$.

Lemma 4.17. *Let $H_i(r, \omega, y)$, $i = 1, 2$ be real analytic Hermitian (resp. symmetric) $m \times m$ matrix with diagonal entries 1 defined in open neighborhoods $\mathscr{U}_i = I \times U_i \times J$ of $(0, \omega_i, 0)$ such that*

$$P(r\omega, y)H_i(r, \omega, y) = H_i(r, \omega, y)P^*(r\omega, y) \quad \textit{in } \mathscr{U}_i. \tag{4.81}$$

Then we have $H_1(r, \omega, y) = H_2(r, \omega, y)$ in $\mathscr{U}_1 \cap \mathscr{U}_2$.

Proof. We expand $H_i(r, \omega, y)$ around $(r, y) = (0, 0)$

$$H_i(r, \omega, y) = \sum_{j,\alpha} r^j y^\alpha H_{ij\alpha}(\omega), \quad H_{ij\alpha}(\omega) \in \mathscr{A}(U_i).$$

Then (4.81) yields

$$\sum_{j+k=p,\alpha+\beta=\gamma} R_{j\alpha}(\omega)H_{ik\beta}(\omega) = \sum_{j+k=p,\alpha+\beta=\gamma} H_{ik\beta}(\omega)R^*_{j\alpha}(\omega)$$

where $R_{00}(\omega) = L(\omega) = R^*_{00}(\omega)$. Hence we get

$$[L(\omega), H_{ip\gamma}(\omega)] = \sum_{j+k=p,\alpha+\beta=\gamma,j+|\alpha|\geq 1} H_{ik\beta}(\omega)R^*_{j\alpha}(\omega) - R_{j\alpha}(\omega)H_{ik\beta}(\omega). \tag{4.82}$$

Note that the right-hand side of (4.82) is anti-Hermitian (resp. anti-symmetric).

For the time being we stop to continue the proof and we make more detailed look on (4.82) than needed here, which will give a key of the proof of Proposition 4.8.

Let $L \in M_m(\mathbb{C})$. We consider the mapping from $H \in M^h_m(\mathbb{C})$ with the zero diagonal entries to the space consisting of off diagonal entries of $m \times m$ matrices defined by

$$H \mapsto \text{off diagonal entries of } [L, H].$$

This is a linear mapping from the real $m(m-1)$ dimensional linear space V consisting of H to the linear space W of real dimension $m(m-1)$ consisting of off diagonal entries of $m \times m$ matrices. These vector spaces admit complex structures and we are naturally identifying $\mathbb{C}^{m(m-1)/2}$ to $\mathbb{R}^{m(m-1)}$. We denote by S the representation matrix with respect to fixed bases of V and W.

Lemma 4.18. *Let* $L = (l_{ij})$. *Then there is a real polynomial* $f \in \mathbb{R}[\mathrm{Re}\, l_{ij}, \mathrm{Im}\, l_{ij}]$ *such that* $\det S = f^2$.

Proof. [1] In the proof we are regarding components l_{ij} and $l_{ji} = \overline{l_{ij}}$ are independent variables. We write $H = (h_{ij})$ where $h_{ii} = 0$ and $h_{ji} = \overline{h_{ij}}$. We identify $(h_{ij}) \in V$ with the complex vector $(h_{12}, h_{13}, \ldots, h_{m-1,m}, \overline{h_{12}}, \overline{h_{13}}, \ldots, \overline{h_{m-1,m}})$ and $(c_{ij}) \in W$ with $(c_{12}, c_{13}, \ldots, c_{m-1,m}, -\overline{c_{12}}, -\overline{c_{13}}, \ldots, -\overline{c_{m-1,m}})$. Sometimes we write $z = (h_{12}, \ldots, h_{m-1,m}) \in \mathbb{C}^{m(m-1)/2}$ and $V = \{(z, \bar{z}) \mid z \in \mathbb{C}^{m(m-1)/2}\}$ and also write $Z = (c_{12}, \ldots, c_{m-1,m})$ and $W = \{(Z, -\bar{Z}) \mid Z \in \mathbb{C}^{m(m-1)/2}\}$. We represent S with respect to these bases and write

$$S\begin{bmatrix} z \\ \bar{z} \end{bmatrix} = \begin{bmatrix} S_{11} & S_{12} \\ S_{21} & S_{22} \end{bmatrix}\begin{bmatrix} z \\ \bar{z} \end{bmatrix} = \begin{bmatrix} Z \\ -\bar{Z} \end{bmatrix}$$

where $S_{ij} \in M_{m(m-1)/2}(\mathbb{C})$. Since we have

$$-\overline{(S_{11}z + S_{12}\bar{z})} = S_{21}z + S_{22}\bar{z}$$

for any $z \in \mathbb{C}^{m(m-1)/2}$ we have $S_{22} = -\overline{S_{11}}$ and $S_{21} = -\overline{S_{12}}$. We now show that S is a Hermitian matrix. This is checked by direct calculation. Let $L = (a_{ij})_{1 \le i,j \le m}$. (Here we use the letter a since the letter l seems confusing.) We may write $S = (s_{(i,j)}, s_{(k,l)})$ $(1 \le i \ne j \le m,\ 1 \le k \ne l \le m)$ since components of V and W are indexed by (i, j) $(1 \le i \ne j \le m)$. We compare $s_{(i,j),(k,l)}$ and $s_{(k,l),(i,j)}$ and show that

$$\overline{s_{(k,l),(i,j)}} = s_{(i,j),(k,l)}.$$

We determine $s_{(i,j),(k,l)}$. Since

$$c_{ij} = \sum_{p=1}^{m} a_{ip}h_{pj} - \sum_{p=1}^{m} h_{ip}a_{pj}$$

then $s_{(i,j),(k,l)}$ is the coefficient of h_{kl} of c_{ij}.

(i) If $(i, j) = (k, l)$, then we have

$$s_{(i,j),(i,j)} = a_{ii} - a_{jj}$$

which is a real number.

[1] We owe the proof of this lemma to T. Ibukiyama.

(ii) If $i = k$ and $j \neq l$, then

$$s_{(i,j),(k,l)} = -a_{lj}$$

so that $s_{(k,l),(i,j)} = -a_{jl} = -\overline{a_{lj}} = \overline{s_{(i,j),(k,l)}}$.

(iii) If $j = l$ and $i \neq k$, then

$$s_{(i,j),(k,l)} = a_{ik}$$

so that $s_{(k,l),(i,j)} = a_{ki} = \overline{a_{ik}} = \overline{s_{(i,j),(k,l)}}$.

(iv) If $i \neq k$ and $j \neq l$, then we have

$$s_{(i,j),(k,l)} = s_{(k,l),(i,j)} = 0.$$

These proves $S^* = S$. We summarize what we have checked.

(1) S is Hermitian and moreover

$$S = \begin{bmatrix} S_{11} & S_{12} \\ -\overline{S_{12}} & -\overline{S_{11}} \end{bmatrix}.$$

(2) If we write $S_{11} = A_1 + iB_1$ and $S_{12} = A_2 + iB_2$ with A_j, $B_j \in M_{m(m-1)/2}(\mathbb{R})$ then A_1 is symmetric and A_2, B_1, B_2 are anti-symmetric, that is ${}^tA_1 = A_1$, ${}^tA_2 = -A_2$, ${}^tB_j = -B_j$ for $j = 1, 2$.

Indeed the relation ${}^t\overline{S_{11}} = {}^tA_1 - i\,{}^tB_1 = S_{11} = A_1 + iB_1$ shows that ${}^tA_1 = A_1$ and ${}^tB_1 = -B_1$. Since S is Hermitian and hence $-\overline{S_{12}} = {}^t S_{12}$ it follows that ${}^tA_2 = -A_2$ and ${}^tB_2 = -B_2$.

We now prove that a representation matrix of S can be taken to be an anti-symmetric matrix by a suitable change of basis. We write down matrices with respect to the real coordinates. Recall

$$\begin{bmatrix} Z \\ -\bar{Z} \end{bmatrix} = S \begin{bmatrix} z \\ \bar{z} \end{bmatrix}.$$

So writing $z = x + iy$ and $Z = X + iY$ for real vectors x, y, X, Y we have

$$\begin{bmatrix} E_m & iE_m \\ -E_m & iE_m \end{bmatrix} \begin{bmatrix} X \\ Y \end{bmatrix} = S \begin{bmatrix} E_m & iE_m \\ E_m & -iE_m \end{bmatrix} \begin{bmatrix} x \\ y \end{bmatrix}.$$

We put

$$T = \begin{bmatrix} E_m & iE_m \\ -E_m & iE_m \end{bmatrix}^{-1} S \begin{bmatrix} E_m & iE_m \\ E_m & -iE_m \end{bmatrix}.$$

Then we have

$$T = \frac{1}{2}\begin{bmatrix} S_{11} + \overline{S_{11}} + S_{12} + \overline{S_{12}} & i(S_{11} - \overline{S_{11}} + \overline{S_{12}} - S_{12}) \\ -i(S_{11} - \overline{S_{11}} + S_{12} - \overline{S_{12}}) & S_{11} + \overline{S_{11}} - S_{12} - \overline{S_{12}} \end{bmatrix}$$

$$= \begin{bmatrix} A_1 + A_2 & -B_1 + B_2 \\ B_1 + B_2 & A_1 - A_2 \end{bmatrix}.$$

Then the matrix

$$T_1 = \begin{bmatrix} O & -E_m \\ E_m & O \end{bmatrix} T = \begin{bmatrix} -B_1 - B_2 & -A_1 + A_2 \\ A_1 + A_2 & -B_1 + B_2 \end{bmatrix}$$

is an anti-symmetric matrix, in fact since B_i are anti-symmetric and ${}^t(-A_1 + A_2) = -{}^tA_1 + {}^tA_2 = -A_1 - A_2 = -(A_1 + A_2)$. So $\det T_1$ is a square of the Pfaffian, that is $\det T_1 = f^2$ where f is a real polynomial in components of T_1, that is a real polynomial in components of A_i and B_i and hence a real polynomial in $(\mathrm{Re}\, l_{ij}, \mathrm{Im}\, l_{ij})$. Thus $\det S$ is also a square of a real polynomial in $(\mathrm{Re}\, l_{ij}, \mathrm{Im}\, l_{ij})$. □

We now check

Lemma 4.19. *Let f be in Lemma 4.18. Then f is irreducible in $\mathbb{R}[\mathrm{Re}\, l_{ij}, \mathrm{Im}\, l_{ij}]$ and $\{f = 0\}$ contains a regular point.*

We postpone the proof until stating the next lemma. We now consider the real symmetric case, that is $L \in M_m^s(\mathbb{R})$ and study the mapping from $H \in M_m^s(\mathbb{R})$ with the zero diagonal entries to the space consisting of off diagonal entries of $m \times m$ real matrices defined by

$$H \mapsto \text{off diagonal entries of } [L, H].$$

This is a linear mapping from the real $m(m-1)/2$ dimensional linear space V consisting of such H to the linear space W of real dimension $m(m-1)/2$ consisting of off diagonal entries of $m \times m$ real matrices. We denote by S again the representation matrix with respect to fixed bases of V and W.

Lemma 4.20. *Let us write $L = (l_{ij})$. Then $\det S$ is irreducible in $\mathbb{R}[l_{ij}]$ and $\{\det S = 0\}$ contains a regular point.*

Proof. [2] We write $H = (h_{ij})$ where $h_{ii} = 0$ and $h_{ji} = h_{ij}$. We identify $(h_{ij}) \in V$ with $(h_{12}, h_{13}, \ldots, h_{m-1,m})$ and $(c_{ij}) \in W$ with $(c_{12}, c_{13}, \ldots, c_{m-1,m})$. We represent S with respect to these bases. In the proof of Lemma 4.18, putting $B_j = O$ and $\mathrm{Im}\, h_{ij} = 0$, we easily see that

$$S = A_1 + A_2.$$

[2] Another proof is found in [53].

Let us write $A_1 + A_2 = (X_{ij})$ then it is clear that $X_{ii} = l_{ii} - l_{jj}$ and for (i, j), $i \neq j$ we have either $X_{ij} = 0$ or $X_{ij} = l_{pq}$ with some (p, q) with $p \neq q$. Suppose that $\det S$ is reducible so that $\det S = fg$ where f, g are homogeneous polynomials in l_{ij} of degree greater than or equal to one. Assume that f contains X_{ii} and assume that the i-th row of S consists of $\{l_{pq}\}_{(p,q)\in J}$ and 0. Note that the i-th column consists of the same $\{l_{pq}\}_{(p,q)\in J}$ and 0 because ${}^tA_1 = A_1$ and ${}^tA_2 = -A_2$. Replace these l_{pq}, $(p, q) \in J$, by λl_{pq} with $\lambda > 0$. Then, in the $\det S$, the coefficient of X_{ii} is multiplied by λ. This proves that g is independent of these l_{pq}, $(p, q) \in J$. Renumbering if necessary we may assume that f contains $X_{11},\ldots,X_{rr}$ and g contains $X_{r+1,r+1},\ldots,X_{NN}$ with $N = m(m-1)/2$. From the above arguments it follows that f is a polynomial in $(X_{ij})_{1\leq i,j\leq r}$ and g is a polynomial in $(X_{ij})_{r+1\leq i,j\leq N}$ so that $\det S$ is independent of X_{ij} with $1 \leq i \leq r$, $r+1 \leq j \leq N$. This is a contradiction. Indeed it is easy to check that there is (i^*, j^*) with $1 \leq i^* \leq r, r+1 \leq j^* \leq N$ such that $X_{i^*j^*} = l_{pq}$ and this shows that $\det S$ contains the term

$$l_{pq}^2 \prod_{i\neq i^*,j^*} X_{ii}$$

up to the sign. Thus we have proved that $\det S$ is irreducible in $\mathbb{R}[l_{ij}]$. Let us set

$$S'(l_{ij}) = S(l_{ij})\big|_{l_{ij}=0, i\neq j}$$

then it is obvious that $\det S' = \prod_{i<j}(l_{ii} - l_{jj})$ which clearly shows $\{\det S' = 0\}$ contains a regular point. This proves that $\{\det S = 0\}$ contains a regular point clearly. □

Proof of Lemma 4.19. Let us put $\mathrm{Im}\, l_{ij} = 0$. Then it follows that

$$\det S = (\det(A_1 + A_2))^2$$

up to the sign. This shows that $f(\mathrm{Im}\, l_{ij} = 0) = \det(A_1 + A_2)$ up to the sign. Thus the assertion follows from Lemma 4.20. □

Completion of the Proof of Lemma 4.17. Assume that

$$[L(\omega), H(\omega)] = C(\omega).$$

Introduce a new system of coordinates $\theta = \big((\mathrm{Re}\, l_{ij}(\omega))_{i\leq j}, (\mathrm{Im}\, l_{ij}(\omega))_{i<j}\big)$. From Lemmas 4.18 and 4.19 it follows that the above equation can be written with

$$\check{H}(\theta) = (h_{12}, h_{13}, \ldots, h_{m-1,m}, \overline{h_{12}}, \overline{h_{13}}, \ldots, \overline{h_{m-1,m}}),$$
$$\check{C}(\theta) = (c_{12}, c_{13}, \ldots, c_{m-1,m}, -\overline{c_{12}}, -\overline{c_{13}}, \ldots, -\overline{c_{m-1,m}})$$

so that

$$S_L(\theta)\check{H}(\theta) = \check{C}(\theta)$$

where $\det S_L(\theta) = f(\theta)^2$ with an irreducible f. We turn to (4.82). Let $H(r,\omega,y) = H_1(r,\omega,y) - H_2(r,\omega,y)$. Then with

$$H(r,\theta,y) = \sum_{j+|\alpha|\geq 0} r^j y^\alpha H_{j\alpha}(\theta), \tag{4.83}$$

$$C_{p\gamma}(\theta) = \sum_{j+k=p,\alpha+\beta=\gamma,j+|\alpha|\geq 0} H_{k\beta}(\theta)R^*_{j\alpha}(\theta) - R_{j\alpha}(\theta)H_{k\beta}(\theta)$$

(4.82) can be written as

$$S_L(\theta)\check{H}_{00}(\theta) = O, \quad S_L(\theta)\check{H}_{p\gamma}(\theta) = \check{C}_{p\gamma}(\theta), \quad p+|\gamma|\geq 1. \tag{4.84}$$

Note that $H_{j\alpha}(\theta)$ are Hermitian and the diagonal entries of $H_{j\alpha}(\theta)$ are 0 by the assumption. Since $\det S_L(\theta) \neq 0$ on a dense subset we conclude that $\check{H}_{00}(\theta,\phi) = O$ and hence $H_{00}(\theta) = O$. Then $C_{p\gamma}(\theta) = O$ for $p+|\gamma| = 1$ from (4.83). By induction on $p+|\gamma|$ it follows that $H_{p\gamma}(\theta) = O$ for all $p+|\gamma|\geq 0$. □

Completion of the Proof of Lemma 4.14. Suppose that at every $(0,\bar{\omega},0)$, $\bar{\omega}\neq 0$ there is an positive definite real analytic symmetrizers $H(r,\omega,y)$ with all diagonal entries 1 defined near $(0,\bar{\omega},0)$. By Lemma 4.17 these symmetrizers are continued analytically and yields $H(r,\omega,y)$ which is positive definite with all diagonal entries 1 and real analytic in a neighborhood of $\{0\}\times S^{k-1}\times\{0\}$. □

Remark. Note that $H_{j\alpha}(\theta)$ are Hermitian (resp. symmetric) and the diagonal entries of $H_{00}(\theta)$ and $H_{j\alpha}(\theta)$, $j+|\alpha|\geq 1$ are 1 and 0 respectively and $\check{H}_{j\alpha}(\theta)$ verifies (4.84). Since $S_L(\theta)$ is linear in θ and $R_{j\alpha}(\theta)$ are homogeneous of degree $j+1$ in θ, then by the homogeneity, $\check{H}_{j\alpha}(\theta)$ extends uniquely to a homogeneous function in $\mathbb{R}^k\setminus\{0\}$ of degree j with respect to θ. Then $H_{j\alpha}(\theta)$ extends there as a homogeneous function of degree j in θ.

Proof of Proposition 4.8. We prove the case that $L(\omega)$ is Hermitian since the real case is similar. Let $H(r,\omega,y)$ be positive definite and satisfy (4.76). With the coordinates (r,θ,y), we again expand $H(r,\theta,y)$ around $(r,y)=(0,0)$

$$H(r,\theta,y) = \sum_{j,\alpha} r^j y^\alpha H_{j\alpha}(\theta), \quad H_{j\alpha}(\theta)\in\mathscr{A}(S^{k-1})$$

where $H_{j\alpha}(\theta)$ are Hermitian and all diagonal entries of $H_{00}(\theta)$ and $H_{j\alpha}(\theta)$, $j+|\alpha|\geq 1$ are 1 and 0 respectively. As before $H_{p\gamma}(\theta)$ verifies

$$[L(\theta), H_{p\gamma}(\theta)] = C_{p\gamma}(\theta)$$

where $C_{p\gamma}(\theta)$ is given by (4.83). The same argument as in the proof of Lemma 4.17 gives that

$$H_{00}(\theta) = I.$$

Then it follows that $C_{p\gamma}(\theta) = R^*_{p\gamma}(\theta) - R_{p\gamma}(\theta)$ for $p + |\gamma| = 1$ which is a homogeneous polynomial of θ of degree 2. Recall that there is an $m(m-1) \times m(m-1)$ matrix $S_L(\theta)$ whose entries are linear functions of θ such that $H_{p\gamma}(\theta)$ satisfies

$$S_L(\theta)\check{H}_{p\gamma}(\theta) = \check{C}_{p\gamma}(\theta).$$

Moreover $\det S_L(\theta) = f(\theta)^2$ where $f(\theta)$ is irreducible in $\mathbb{R}[\theta]$ and $\{f(\theta) = 0\}$ contains a regular point. Let us denote by ${}^{co}S_L(\theta)$ the cofactor matrix of $S_L(\theta)$. It is clear that $\check{H}_{p\gamma}(\theta)$, $p + |\gamma| = 1$ verifies

$$f(\theta)\check{H}_{p\gamma}(\theta) = {}^{co}S_L(\theta)\check{C}_{p\gamma}(\theta)/f(\theta) = (f_{ij}(\theta)/f(\theta)).$$

Recalling that $C_{p\gamma}(\theta) = R^*_{p\gamma}(\theta) - R_{p\gamma}(\theta)$ for $p + |\gamma| = 1$ we see that $f_{ij}(\theta)$ are homogeneous polynomials of degree $m(m-1)+1$ in θ. Since $\check{H}_{p\gamma}(\theta)$ is real analytic in $\mathbb{R}^k \setminus \{0\}$ as remarked after the proof of Lemma 4.17 it follows that $f_{ij}(\theta)$ vanishes on $\{f(\theta) = 0\}$. Since $f(\theta)$ is irreducible and $\{f(\theta) = 0\}$ contains a regular point from Lemma 4.19 we can apply Lemma 2.5 in [41] (for example) to conclude that $f_{ij}(\theta)/f(\theta)$ are homogeneous polynomials of degree $m(m-1)/2+1$ in θ. Thus $f(\theta)\check{H}_{p\gamma}(\theta)$ is a homogeneous polynomial in θ of degree $m(m-1)+1$. Repeating the same arguments we conclude that $\check{H}(\theta)$ is a homogeneous polynomial in θ of degree 1 for $p + |\gamma| = 1$ and so is $H_{p\gamma}(\theta)$ because $H_{p\gamma}(\theta)$ is Hermitian and whose diagonal entries are 0. By (4.83), $C_{p\gamma}(\theta)$, $p + |\gamma| = 2$ becomes a homogeneous polynomial in θ of degree 3. By induction on $j + |\alpha|$ we prove that $H_{j\alpha}(\theta)$ is a homogeneous polynomial of degree j in θ. In the coordinates ω, $H_{j\alpha}$ is a homogeneous polynomial in ω of degree j. Then one can write

$$r^j H_{j\alpha}(\omega) = G_{j\alpha}(r\omega).$$

where $G_{j\alpha}(x)$ is a homogeneous polynomial of degree j in x. Let us define

$$G(x,y) = \sum_{j,\alpha} y^\alpha G_{j\alpha}(x).$$

Since the convergence follows from that of $\sum_{j,\alpha} r^j y^\alpha H_{j\alpha}(\omega)$ then $G(x,y)$ becomes real analytic near $(0,0)$ and the proof is complete. □

Remark. The arguments proving that $H_{p\gamma}(\theta)$ is a homogeneous polynomial in θ can be applied under less restrictive hypotheses. Let $f(\theta)$, $g(\theta)$ be homogeneous polynomials in θ of degree n, m respectively where $n \geq m$. Let

$$g(\theta) = \prod_{j=1}^{s} g_j(\theta)^{r_j}$$

be the irreducible factorization of $g(\theta)$ in $\mathbb{R}[\theta]$. We assume that $f(\theta)/g(\theta)$ is C^∞ apart from the origin and $V_j = \{\theta | g_j(\theta) = 0\}$, $1 \le j \le s$ contains a regular point. Then applying Lemma 2.5 in [41] again, we conclude that $f(\theta)$ is a homogeneous polynomial in θ of degree $n - m$.

4.6 Well Posed Cauchy Problem

Let us study a differential operator of order q

$$P(x, D) = \sum_{|\alpha| \le q} A_\alpha(x) D^\alpha, \quad D_j = \frac{1}{i} \frac{\partial}{\partial x_j} \tag{4.85}$$

where $A_\alpha(x)$ are $m \times m$ matrix valued smooth functions defined in a neighborhood Ω of the origin of $\mathbb{R}^n$. We assume that $x_1 = const.$ are non characteristic and without restrictions we may assume that

$$A_{(q,0,\dots,0)}(x) = I. \tag{4.86}$$

We are concerned with the following Cauchy problem

$$\begin{cases} P(x, D)u = f, \ \operatorname{supp} f \subset \{x_1 \ge 0\}, \\ \operatorname{supp} u \subset \{x_1 \ge 0\}. \end{cases} \tag{4.87}$$

Let $P_q(x, \xi)$ be the principal symbol of $P(x, D)$

$$P_q(x, \xi) = \sum_{|\alpha| = q} A_\alpha(x) \xi^\alpha$$

and we assume that

$$\det P_q(x, \xi) = 0 \Longrightarrow \xi_1 \text{ is real } \forall x \in \Omega, \forall \xi' = (\xi_2, \dots, \xi_n) \in \mathbb{R}^{n-1}. \tag{4.88}$$

We first study the case that P_q is of constant coefficients.

Theorem 4.6 ([22, 26]). *Let $P(\xi)$ be a homogeneous polynomial of degree q in $\xi \in \mathbb{R}^n$ with real $m \times m$ matrix values such that* det $P(\xi)$ *satisfies (4.88) and every multiple characteristic of* det $P(\xi)$ *is at most double and nondegenerate. Then the Cauchy problem for $P(D) + R(x, D)$ is C^∞ well posed for every R of order $q - 1$ with C^∞ $m \times m$ matrix coefficients.*

In [22, 26] their proof is based on the estimate of $P(\xi + i\tau N)^{-1}$ such that

$$|\tau|(|\xi| + |\tau|)^{m-1}|P(\xi + i\tau N)^{-1}| \le C, \quad \text{if } 0 \ne (\tau, \xi) \in \mathbb{R}^{n+1} \tag{4.89}$$

which is derived from the assumption of the non degeneracy of double characteristics. For $f \in C_0^\infty$ with supp $f \subset \{x_1 \ge 0\}$ we look for a solution

$$(P(D) + R(x, D))u = f$$

such that supp $u \subset \{x_1 \ge 0\}$. We set $u = u_\tau e^{\tau N x_1}$ and $f = f_\tau e^{\tau N x_1}$ and obtain the equivalent equation

$$(P(D - i\tau N) + R(x, D - i\tau N))u_\tau = f_\tau. \tag{4.90}$$

Let E_τ be the inverse Fourier transform of $P(\xi - i\tau N)^{-1}$ and set $u_\tau = E_\tau * v$. Then (4.90) becomes

$$v + R(x, D - i\tau N)E_\tau * v = f_\tau. \tag{4.91}$$

On the other hand assuming (4.89) we have

$$\tau\|E_\tau * w\|_{s+m-1,\tau} \le C\|w\|_{s,\tau}, \quad \|w\|_{s,\tau}^2 = (2\pi)^{-n}\int |\hat{u}(\xi)|^2(|\xi|^2 + \tau^2)^s d\xi.$$

Thus from this estimate it follows that

$$\|R(x, D - i\tau N)E_\tau * v\|_{s,\tau} \le C'\tau^{-1}\|f_\tau\|_{s,\tau}.$$

Choosing $\tau > 2C'$ we conclude that (4.91) has a unique solution $v \in H^s$.

We prove the following result which extends Theorem 4.6. Our proof is completely different from that in [22, 26] and based on the smooth symmetrizability of corresponding first order system (Proposition 4.1) and hence can be applicable to differential operators with variable coefficients.

Theorem 4.7. *Assume that every characteristic* $(0, \xi_1, \xi')$, $|\xi'| = 1$ *of* $P_q(x, \xi)$ *is at most double and nondegenerate. Then the Cauchy problem for* $P(x, D)$ *is* C^∞ *well posed near the origin for arbitrary lower order term. Moreover if* $\tilde{P}(x, D)$ *is another system of the form (4.85) verifying (4.88) with the principal symbol* $\tilde{P}_q(x, \xi) = \sum_{|\alpha|=q} \tilde{A}_\alpha(x)\xi^\alpha$ *of which* $\tilde{A}_\alpha$ *are sufficiently close to* $A_\alpha(x)$ *in* $C^2(\Omega)$ *then the Cauchy problem for* $\tilde{P}(x, D)$ *is* C^∞ *well posed near the origin for any lower order term.*

Assuming the analyticity of the coefficients we have

Theorem 4.8. *Assume that* $A_\alpha(x)$, $|\alpha| = q$ *are real analytic in* Ω *and every characteristic* $(0, \xi_0, \xi')$, $|\xi'| = 1$ *of* $P_q(x, \xi)$ *is nondegenerate. Then the Cauchy* *problem for $P(x, D)$ is C^∞ well posed near the origin for arbitrary lower order term.*

The proof is very simple. We reduce the Cauchy problem for $P(x, D)$ to that for a first order system $\mathscr{P}(x, D)$. Taking the invariance of non degeneracy of characteristics proved in Proposition 4.1.1, to prove Theorems 4.7 and 4.8, it suffices to apply Proposition 4.2 and Theorem 4.4 respectively which asserts the existence of a smooth symmetrizer $\mathscr{S}(x, D')$ for $\mathscr{P}(x, D)$ defined near the origin.

Let us write

$$P(x, D)u = D_1^q u + \sum_{j=1}^{q} A_j(x, D')D_1^{q-j} u = f. \tag{4.92}$$

Put

$$u^{(k)} = \langle D'\rangle^{q-k} D_1^{k-1} u, \quad k = 1, \dots, q$$

where $\langle D'\rangle^2 = 1 + \sum_{j=2}^{n} D_j^2$. Then (4.92) is reduced to

$$D_1 U + \begin{bmatrix} 0 & -I & & & \\ 0 & 0 & -I & & \\ & & & \ddots & \\ 0 & & & & -I \\ A_q^\#(x, D') & & \cdots & & A_1^\#(x, D') \end{bmatrix} \langle D'\rangle U = F$$

where $U = {}^t(u^{(1)}, \dots, u^{(q)})$, $F = {}^t(0, \dots, 0, f)$ and

$$A_j^\#(x, D') = A_j(x, D')\langle D'\rangle^{-j}.$$

Let us denote by $A_j^0(x, \xi')$ the principal symbol of $A_j^\#(x, \xi')$ and set

$$\mathscr{A}(x, \xi') = \begin{bmatrix} 0 & -I & & & \\ 0 & 0 & -I & & \\ & & & \ddots & \\ 0 & & & & -I \\ A_q^0(x, \xi') & & \cdots & & A_1^0(x, \xi') \end{bmatrix}. \tag{4.93}$$

Fix $(0, \bar{\xi}')$, $|\bar{\xi}'| = 1$. Let $(0, \lambda_i, \bar{\xi}')$, $i = 1, \dots, p$, be characteristics of $\xi_1 + \mathscr{A}(x, \xi')$ where $(0, \lambda_i, \bar{\xi}')$ are nondegenerate and λ_i are different from each other. Then there exists a smooth $\mathscr{T}(x, \xi')$ defined near $(0, \bar{\xi}')$, homogeneous of degree 0, such that

$$\mathscr{T}(x, \xi')^{-1}\mathscr{A}(x, \xi')\mathscr{T}(x, \xi') = \mathscr{A}_1(x, \xi') \oplus \cdots \oplus \mathscr{A}_p(x, \xi')$$

where $(0, \lambda_i, \bar{\xi}')$ is a nondegenerate characteristic of $\mathscr{P}_i(x,\xi) = \xi_1 + \mathscr{A}_i(x,\xi')$. Then one can apply Proposition 4.2 or Theorem 4.4 to get a smooth symmetrizer $\mathscr{S}_i(x,\xi')$ of $\mathscr{A}_i(x,\xi')$ defined near $(0,\bar{\xi}')$, homogeneous of degree 0 such that

$$\mathscr{S}_i(x,\xi')^{-1}\mathscr{A}_i(x,\xi')\mathscr{S}_i(x,\xi')$$

is Hermitian. This proves that $\mathscr{A}(x,\xi')$ is smoothly symmetrizable near $(0,\bar{\xi}')$ by $\mathscr{S}_1(x,\xi') \oplus \cdots \oplus \mathscr{S}_p(x,\xi')$. By the usual argument of partition of unity one can prove that there is a smooth $\mathscr{S}(x,\xi')$ which symmetrizes $\mathscr{A}(x,\xi')$. Thus the Cauchy problem for $\mathscr{P}(x,D)$ is C^∞ well posed for arbitrary lower order term and hence so is for $P(x,D)$.

We turn to prove the second assertion of Theorem 4.7. Recall that $(0,\lambda_i,\bar{\xi}')$, $i = 1,\ldots,p$ are characteristics of $\xi_1 + \mathscr{A}(x,\xi')$ where $\{\lambda_i\}$ are different from each other. By assumption each $(0,\lambda_i,\bar{\xi}')$ is either simple characteristic or double nondegenerate characteristic. Let $\xi_1 + \tilde{\mathscr{A}}(x,\xi')$ be the symbol of first order system associated to $\tilde{P}(x,D)$. Let $(0,\lambda_i,\bar{\xi}')$ be a double nondegenerate characteristic of $\xi_1 + \mathscr{A}(x,\xi')$. Since $\tilde{\mathscr{A}}(x,\xi')$ is enough close to $\mathscr{A}(x,\xi')$, as for characteristics of $\xi_1 + \tilde{\mathscr{A}}(0,\xi')$ enough close to $(0,\lambda_i,\bar{\xi}')$, we have either two simple characteristics $(0,\tilde{\lambda}_{ik},\bar{\xi}')$ or a double characteristic $(0,\tilde{\lambda}_i,\bar{\xi}')$. From Proposition 4.3 it follows that the double characteristic $(0,\tilde{\lambda}_i,\bar{\xi}')$ is nondegenerate. Thus we conclude that every characteristic of $\xi_1 + \tilde{\mathscr{A}}(0,\bar{\xi}')$ is nondegenerate and then repeating the same arguments as above we get the assertion.

Example 4.9. Consider the second order differential operator $P(x,D) = (p_{ik}(x,D))$ with 3×3 matrix coefficients

$$p_{ik}(x,\tau,\xi) = (\tau^2 - \sigma_i(x)|\xi|^2)\delta_{ik} - (1-\sigma_i(x))\xi_i\xi_k$$

in Example 4.6. Then from Theorem 4.7 it follows that the Cauchy problem for $P(x,D) + R(x,D)$ is C^∞ well posed for every R of first order with C^∞ 3×3 matrix coefficients. Let $A(x,\xi)$ be in Example 4.7. Then the Cauchy problem for $D_0^2 - A(x,D) + R(x,D)$ is C^∞ well posed for every R of first order with C^∞ 3×3 matrix coefficients.

Example 4.10. Let $P(\xi) = \xi_1 I + \sum_{j=2}^{d} F_j\xi_j$ be the symbol in Example 4.3. Consider

$$P(x,\xi) = \xi_1 I + \sum_{j=2}^{d} A_j(x)\xi_j$$

where $A_j(x)$ are real valued real analytic $m \times m$ matrices which are enough close to F_j in C^2 and $P(x,\xi)$ has only real eigenvalues for any x and any ξ. Then Theorem 4.8 shows that the Cauchy problem for $P(x,D) + B(x)$ is C^∞ well posed for every smooth $m \times m$ matrix $B(x)$.

4.7 Nondegenerate Characteristics of Symmetric Systems

Let P be a first order system with constant coefficients

$$P(x) = x_1 + \sum_{j=2}^{n} A_j x_j \tag{4.94}$$

where A_j are real $m \times m$ constant matrices. We always assume that $P(x)$ is hyperbolic with respect to $\theta = (1, 0, \ldots, 0)$. Then from [34] (see also [13]) $P(x)$ can not be strictly hyperbolic if $n > 3$ and $m \equiv 2$ modulo 4, that is $P(x)$ has necessarily multiple characteristics $x \neq 0$. We want to check whether these multiple characteristics are nondegenerate.

For symmetric systems with constant coefficients the description of non degeneracy of characteristics becomes simple. Consider

$$\mathscr{L}(x) = \sum_{j=1}^{n} A_j x_j$$

where $A_j \in M_m^s(\mathbb{R})$. In this and the following sections we identify a symmetric system $\mathscr{L}(x)$ with the image of $\mathscr{L}(x)$ when x varies in $\mathbb{R}^n$

$$\mathscr{L} = \{\mathscr{L}(x) \mid x \in \mathbb{R}^n\}$$

which is a linear subspace in $M_m^s(\mathbb{R})$. Indeed if $\mathscr{L}$ is a linear subspace of dimension q in $M_m^s(\mathbb{R})$ which contains the identity then choosing a basis $\{I, A_2, \ldots, A_q\}$, $A_j \in M_m^s(\mathbb{R})$ for $\mathscr{L}$ we have a symmetric system

$$x_1 I + \sum_{j=2}^{q} A_j x_j$$

and vice versa.

We denote by $M_m^s(k; \mathbb{R})$ the set of all $A \in M_m^s(\mathbb{R})$ with rank $m - k$. Then we have

Lemma 4.21. *In order that $\bar{x}$ is a nondegenerate characteristic of $\mathscr{L}(x)$ of order k if and only if the image $\mathscr{L}$ intersects with $M_m^s(k; \mathbb{R})$ at $\mathscr{L}(\bar{x})$ transversally.*

Proof. Since $\mathscr{L}(\bar{x})$ and $\mathscr{L}_{\bar{x}}(x)$ are symmetric, the conditions (4.15) and (4.17) in Definition 4.5 are automatically satisfied. Without restrictions we may assume that $\bar{x} = (0, \ldots, 0, 1)$. Then A_n is of rank $m - k$. We can make an orthogonal transformation of the matrices so that with a block matrix notation we have

$$A_n = \begin{bmatrix} O & O \\ O & G \end{bmatrix}$$

where G is a $(m-k)\times(m-k)$ non singular matrix. The tangent space of $M_m^s(k;\mathbb{R})$ at A_n consists of matrices of the form

$$\begin{bmatrix} O & * \\ * & * \end{bmatrix} \tag{4.95}$$

with the corresponding block decomposition. On the other hand, with the same block decomposition of $\mathscr{L}(x)$

$$\mathscr{L}(x) = \begin{bmatrix} L_{11}(x) & L_{12}(x) \\ L_{21}(x) & L_{22}(x) \end{bmatrix}$$

it is clear that $\mathscr{L}_{\bar{x}}(x) = L_{11}(x)$. Thus the transversality of intersection means that $\dim L_{11} = d_k$ that is, $\dim \mathscr{L}_{\bar{x}} = d_k$ and hence $\bar{x}$ is nondegenerate. The converse follows in the same way. □

We start with the special case that $\dim \mathscr{L} = d_m - 1$. Since $\mathscr{L}$ has codimension one in $M_m^s(\mathbb{R})$ then $\mathscr{L}$ is defined by

$$\mathscr{L} = \{X = (x_{ij}), x_{ij} = x_{ji} \mid \mathrm{Tr}\,(AX) = 0\} \tag{4.96}$$

with some $A \in M_m^s(\mathbb{R})$. Note that $\mathrm{Tr}\,A = 0$ because $\mathscr{L}$ contains the identity. Now we have

Proposition 4.9. *Assume that $\mathscr{L}$ is given by (4.96) with $O \neq A \in M_m^s(\mathbb{R})$ and that the rank of A is greater than k. Then every characteristic of order k of $\mathscr{L}(x)$ is nondegenerate.*

Proof. Let $\bar{x}$ be a characteristic of order k of $\mathscr{L}(x)$ and hence $H = \mathscr{L}(\bar{x}) \in \mathscr{L} \cap M_m^s(k;\mathbb{R})$. Here we note that $\dim T_H(M_m^s(k;\mathbb{R})) = d_m - d_k$ which is seen by the proof of Lemma 4.21. To show $\bar{x}$ is nondegenerate it suffices to prove that

$$\dim\,(\mathscr{L} \cap T_H(M_m^s(k;\mathbb{R})) = d_m - d_k - 1 \tag{4.97}$$

by Lemma 4.21. As in the proof of Lemma 4.21, considering $T^{-1}\mathscr{L}T$ with a suitable $T \in O(m)$ we may assume that

$$H = \begin{bmatrix} O & O \\ O & G \end{bmatrix} \tag{4.98}$$

where G is a $(m-k)\times(m-k)$ non singular matrix. Set $x_{ij} = 0$ for $1 \le i \le j \le k$. Then $\mathrm{Tr}\,(AX) = 0$, $X = (x_{ij})$ implies that

$$\sum_{k+1 \le i \le j \le m} a_{ij}x_{ij} = 0$$

where $A = (a_{ij})$. Recalling that the tangent space $T_H(M_m^s(k;\mathbb{R}))$ spanned by matrices of the form (4.95) we see that $\mathscr{L} \cap T_H(M_m^s(k;\mathbb{R}))$ consists of the matrices of the form

$$X = \begin{bmatrix} O & x_{ij} \\ x_{ij} & x_{ij} \end{bmatrix}, \quad \mathrm{Tr}\,(AX) = \sum_{k+1 \le j, i \le j} (2 - \delta_{ij}) a_{ij} x_{ij} = 0$$

where δ_{ij} is the Kronecker's delta. Since A is symmetric and the rank of A is greater than k by assumption then it follows that $(a_{ij})_{k+1 \le j, i \le j} \neq O$. This proves (4.97) and hence the assertion. □

We turn to the case that $1 \le \dim \mathscr{L} \le d_m - 1$. We first give a parametrization of the Grassmannian of l dimensional subspaces of $M_m^s(\mathbb{R})$ containing the identity. Take a map

$$\sigma : \{1, \ldots, \nu\} \to \{(i, j) | 1 \le i \le j \le m, (i, j) \neq (m, m)\}$$

which is injective. Denote by U_σ the set of all ν-tuple of $m \times m$ symmetric matrices $A = (A_1, \ldots, A_\nu)$ such that $\mathrm{Tr}\, A_j = 0$ and the $\sigma(k)$-th entry of A_j is zero unless $k = j$ and the $\sigma(j)$-th entry of A_j is 1. It is clear that U_σ can be identified with $\mathbb{R}^{\nu(d_m - \nu - 1)}$. Taking all such injective σ, U_σ and the inverse of the map

$$\phi_\sigma : U_\sigma \ni A \mapsto \mathscr{L}, \quad \mathscr{L} = \{X \in M_m^s(\mathbb{R}) | \mathrm{Tr}\,(A_j X) = 0, 1 \le j \le \nu\}$$

then $\{(\phi_\sigma^{-1}, \Omega_\sigma = \phi_\sigma(U_\sigma))\}$ give charts of the Grassmannian of $l = d_m - \nu$ dimensional subspaces of $M_m^s(\mathbb{R})$ containing I, which we denote by $G_{d_m, I}^l$.

Proposition 4.10. *In the Grassmannian $G_{d_m,I}^l$ consisting of l dimensional subspaces of $M_m^s(\mathbb{R})$ containing the identity I, the subset for which every characteristic of order less than m is nondegenerate is an open and dense subset.*

Let $\mathbf{P}^N(\mathbb{R})$ be the N dimensional real projective space and let $X \subset \mathbf{P}^N(\mathbb{R})$ be a non-singular algebraic manifold of dimension r and assume that $x_0 \notin T_x X$ for all $x \in X$. Let us denote

$$\tilde{G}_{N,x_0}^s = \{W \subset \mathbf{P}^N(\mathbb{R}) | W \text{ is a linear space}, \dim W = s, x_0 \in W\}$$

and set $s' = N - s$. Then we have

Lemma 4.22. *A generic $W \in \tilde{G}_{N,x_0}^s$ intersects X transversally.*

Proof. [3] Let $Y = \{(x, W) \in X \times \tilde{G}_{N,x_0}^s | x \in W\}$ and denote by p_1, p_2 the projections onto X and $\tilde{G}_{N,x_0}^s$ respectively. Note that $\dim Y = s's - s' + r$ and

[3] The author owes this simple proof to A. Gyoja.

$\dim \tilde{G}^s_{N,x_0} = s's$. Then if $r < s'$ a generic $W \in \tilde{G}^s_{N,x_0}$ does not intersect X and hence the result. Thus it is enough to study the case $r \geq s'$. Let us set

$$Z = \{(x, W) \in Y \,|\, \dim (T_x X + W) \leq N - 1\}.$$

It is not difficult to see that

$$\dim (p_1|Z)^{-1}(x) = ss' - r - 1, \quad x \in X$$

so that $\dim Z = ss' - 1 = \dim \tilde{G}^s_{N,x_0} - 1$. Thus for every W belonging to the open dense subset $\tilde{G}^s_{N,x_0} \setminus \overline{p_2(Z)}$, W intersects X transversally. This proves the assertion. □

Proof of Proposition 4.10. Take X and $\tilde{G}^s_{N,x_0}$ as the projective spaces $M^s_m(k;\mathbb{R})^{pr}$ and $(G^{s+1}_{d_m,I})^{pr}$ based on $M^s_m(k;\mathbb{R})$ and $G^{s+1}_{d_m,I}$ respectively. Applying Lemma 4.22 with $N = d_m - 1$, $r = N - d_k$, $x_0 = I$ we get the desired result. □

4.8 Hyperbolic Perturbations of Symmetric Systems

In this section, we discuss hyperbolic perturbations, of which definition is given below, of symmetric systems with constant coefficients near multiple characteristics which are not necessarily nondegenerate. To motivate our study in this section let us consider

$$L(x, D) = \sum_{j=0}^{n} A_j(x) D_j, \quad A_0(x) = I$$

where $A_j(x)$ are real $m \times m$ real analytic matrices and let ρ be a multiple characteristic of order m with involutive $\Lambda(\rho)$. If $L(x, D)$ is strongly hyperbolic near the origin we have $\dim \mathrm{Ker} L(\rho) = m$ by Theorem 2.2 which implies $L(\rho) = O$. We can assume $\rho = (0, e_n)$ so that $A_n(0) = O$ then one can write

$$L(x, \xi) = \xi_0 I + \sum_{j=1}^{n-1} A_j(x)\xi_j + \sum_{j=0}^{n} A_{nj}(x) x_j \xi_n$$

$$= \xi_n \Big\{(\xi_0/\xi_n) I + \sum_{j=1}^{n-1} A_j(x)(\xi_j/\xi_n) + \sum_{j=0}^{n} A_{nj}(x) x_j\Big\}$$

and note that

$$L_\rho(x, \xi') = \xi_0 I + \sum_{j=1}^{n-1} A_j(0)\xi_j + \sum_{j=0}^{n} A_{nj}(0) x_j.$$

From Lemma 4.2 $L_\rho(x, \xi')$ is hyperbolic with respect to $\theta = (0, \dots, 0, 1, 0, \dots, 0)$. Assume that $L_\rho(x, \xi')$ is diagonalizable for every (x, ξ'). If $\dim_{\mathbb{R}} L_\rho = d_m$ so that ρ is nondegenerate then by Theorem 4.4 we see that $L(x, \xi)$ is symmetrizable near ρ. Moreover under the assumption

$$\dim_{\mathbb{R}} L_\rho \geq d_m - 1$$

it follows from Lemma 4.9 that there exists $T \in M_m(\mathbb{R})$ such that $T^{-1} L_\rho(x, \xi') T$ is symmetric for every (x, ξ'). Considering $T^{-1} L(x, \xi) T$ from the beginning we can assume that $L_\rho(x, \xi')$ is symmetric. Thus we can write

$$L(x, \xi) = \xi_n \{L_\rho(x, \xi'/\xi_n) + R(x, \xi'/\xi_n)\}, \quad R(x, \xi'/\xi_n) = O(|x|^2 + |\xi'/\xi_n|^2)$$

where $L_\rho(x, \xi') + R(x, \xi')$ is hyperbolic with respect to $(0, \dots, 0, 1, 0, \dots, 0)$.
Let us consider symmetric systems with constant coefficients

$$\mathcal{L}(x) = x_1 I + \sum_{j=2}^{q} F^j x_j = x_1 I + L(x') \tag{4.99}$$

where $F^j \in M_m^s(\mathbb{R})$ and $\{I, F^j\}$ are linearly independent. Note that if $q \leq d_m - 1$ then $x = 0$ is a degenerate characteristic of $\mathcal{L}(x)$.

We perturb $\mathcal{L}(x)$ near $x = 0$ by adding $R(x) = O(|x|^2)$ as $x \to 0$. We start with

Definition 4.6. We say that $M_m(\mathbb{R})$ valued real analytic $R(x) = O(|x|^2)$, $x \to 0$ is a hyperbolic perturbation to $\mathcal{L}(x)$ near $x = 0$ if the perturbed system

$$\mathcal{P}(x) = \mathcal{L}(x) + R(x)$$

remains to be hyperbolic near $x = 0$, that is

$$\text{all eigenvalues of } \mathcal{P}(x + \lambda\Theta) \text{ are real near } x = 0 \tag{4.100}$$

where $\Theta = (1, 0, \dots, 0)$ and

$$R(x) = O \quad \text{if} \quad \mathcal{L}(x) = O. \tag{4.101}$$

Example 4.11. Let $\mathcal{L}(x)$ be as in (4.99) and let $T(x)$ be real analytic $m \times m$ matrix defined near $x = 0$ with $T(0) = I$. Then it is clear that

$$T^{-1}(x)\mathcal{L}(x)T(x) = x_1 I + \sum_{j=2}^{q} T^{-1}(x) F_j T(x) x_j = \mathcal{L}(x) + R(x)$$

is a hyperbolic perturbation, while it is never trivial to find $T(x)$ starting from $\mathcal{L}(x) + R(x)$.

As before, we define $S_{\mathscr{L}}(x)$ as the representation matrix of the linear map sending $M_m^s(\mathbb{R}) \ni H$ with zero diagonal entries to the anti-symmetric matrix $[\mathscr{L}(x), H]$. Note that

$$S_{\mathscr{L}}(x) = S_{\tilde{\mathscr{L}}}(x) \tag{4.102}$$

if $\tilde{\mathscr{L}}(x) - \mathscr{L}(x)$ is a scalar matrix. Let

$$g(x) = \prod_{j=1}^{s} g_j(x)^{r_j}$$

be the irreducible factorization of $\det S_{\mathscr{L}}(x)$ in $\mathbb{R}[x]$. We assume that

$$\{x | g_j(x) = 0\} \text{ contains a regular point} \tag{4.103}$$

for $1 \le j \le s$. Then we have

Theorem 4.9. *Assume that every characteristic of $\mathscr{L}(x)$ of order less than m is nondegenerate. Suppose that* $\det S_{\mathscr{L}}(x)$ *satisfies (4.103). Then for every perturbed $\mathscr{P}(x) = \mathscr{L}(x) + R(x)$ with a hyperbolic perturbation $R(x)$ we can find real analytic $A(x)$, $B(x)$ defined near the origin with $A(0) = B(0) = I$ such that*

$$A(x)\mathscr{P}(x)B(x)$$

becomes symmetric.

Proof. By a preparation theorem proved in [11], generalizing the Weierstrass preparation theorem to matrix valued functions, one can write

$$\mathscr{P}(x + \lambda\Theta) = C(x, \lambda)(\lambda I + Q(x)) \tag{4.104}$$

where $C(x, \lambda)$ is real analytic near $(0, 0)$ with $\det C(0, 0) \neq 0$ and $Q(x)$, $Q(0) = O$ is real analytic with values in $M_m(\mathbb{R})$. Comparing the first order term in the Taylor expansion at $(x, \lambda) = (0, 0)$ of both sides we see that $C(0, 0) = I$ and $Q(x) = \mathscr{L}(x) + \tilde{R}(x)$ where $\tilde{R}(x) = O(|x|^2)$. Since $\mathscr{L}(0, \dots, 0, x_{q+1}, \dots, x_n) = O$ taking $\lambda = -x_1$, $x_j = 0$, $2 \le j \le q$ in (4.104) it follows from (4.101) that $O = C(x_1, 0, \dots, 0, x_{q+1}, \dots, x_n, -x_1)\tilde{R}(x_1, 0, \dots, 0, x_{q+1}, \dots, x_n)$ and hence

$$\tilde{R}(x_1, 0, \dots, 0, x_{q+1}, \dots, x_n) = O.$$

Since $C(x, 0)^{-1}\mathscr{P}(x) = \mathscr{L}(x) + \tilde{R}(x)$ it is enough to study a perturbation term $R(x)$ which verifies $R(x_1, 0, \dots, 0, x_{q+1}, \dots, x_n) = O$. Changing notations we set $x = (x_2, \dots, x_q)$, $y = (x_1, x_{q+1}, \dots, x_n)$ and

$$P(x,y) = L(x) + R(x,y), \quad L(x) = \sum_{j=2}^{q} F^j x_j$$

where $S_L(x)$ verifies the assumptions because of (4.102). As in Sect. 4.5 we set

$$\tilde{P}(r,\omega,y,a) = r^{-1} S(a)^{-1} P(r\omega, y) S(a).$$

Since $\tilde{P}(0,\omega,0,0) = L(\omega)$ and $\{I, F^j\}$ are linearly independent the multiplicity of eigenvalues of $\tilde{P}(0,\omega,0,0)$ are less than m if $\omega \neq 0$. We then fix $\omega \neq 0$ and proceed exactly as the same way in Sect. 4.5. Take an orthogonal T_0 so that $T_0^{-1} L(\omega) T_0 = \oplus_{i=1}^{p} \lambda_i I_{s_i}$. Then we have

$$\begin{aligned} Q(r,\theta,y,a) &= r^{-1} T_0^{-1} S(a)^{-1} P(r(\omega+\theta), y) S(a) T_0 \\ &= \tilde{L}(\omega) + \tilde{L}(\theta) + \tilde{P}(\omega; r, y, a) + O(|(r,\theta,y,a)|^2) \end{aligned}$$

where $\tilde{L}(\omega) = \oplus \lambda_i I_{s_i}$ and $\tilde{L}(\theta) = T_0^{-1} L(\theta) T_0 = (\tilde{L}_{ij}(\theta))_{1 \le i,j \le p}$. Let

$$\tilde{L}_{ii}(\theta) = \sum_{j=2}^{q} \tilde{F}_{ii}^j \theta_j$$

then we get

Lemma 4.23. $\{I_{s_i}, \tilde{F}_{ii}^j\}$ *span* $M_{s_i}^s(\mathbb{R})$.

Proof. Let $\tilde{\mathscr{L}}(x) = T_0^{-1} \mathscr{L}(x) T_0$, $x = (x_1, x_2, \dots, x_q)$. Since $(x_1, x_2, \dots, x_q) = (-\lambda_i, \omega)$ is a characteristic of $\tilde{\mathscr{L}}(x)$ of order less than m and hence nondegenerate by assumption. It is clear that the localization of $\tilde{\mathscr{L}}(x)$ at $(-\lambda_i, \omega)$ is

$$\tilde{\mathscr{L}}_{(-\lambda_i,\omega)}(x) = x_1 I_{s_i} + \sum_{j=2}^{q} \tilde{F}_{ii}^j x_j$$

because $\tilde{\mathscr{L}}(-\lambda_i, \omega)$ is diagonal. Noting that the non degeneracy of characteristics is invariant under changes of basis for $\mathbb{C}^m$, the matrices $\{I_{s_i}, \tilde{F}_{ii}^j\}$ span a subspace of dimension $s_i(s_i+1)/2$. Since $\tilde{F}_{ii}^j$ are symmetric this proves the assertion. □

Completion of the Proof of Theorem 4.9. In view of Remark at the end of Sect. 4.5, the rest of the proof of Theorem 4.9 goes exactly as the same way in Sect. 4.5. □

Taking into account the invariance of non degeneracy of characteristics under change of basis we have

Corollary 4.2. *Assume that every characteristic of $\mathscr{L}(x)$ of order less than m is nondegenerate and there is an orthogonal matrix $T \in O(m)$ such that* $\det S_{T^{-1}\mathscr{L}T}(x)$ *verifies (4.103). Then the same conclusion as in Theorem 4.9 holds.*

Remark. The condition (4.103) is not invariant under orthogonal changes of basis for $\mathbb{C}^m$. Indeed let

$$\mathscr{L}(x) = x_1 I_2 + \begin{bmatrix} 0 & x_2 \\ x_2 & 0 \end{bmatrix}$$

then it is obvious that $S_{\mathscr{L}}(x) = O$. On the other hand it is easy to see that there is an orthogonal $T \in O(2)$ such that $S_{T^{-1}\mathscr{L}T}(x)$ verifies (4.103).

Example 4.12. Let us take

$$L_1(x) = \begin{bmatrix} x_2 + x_5 & x_5 & x_5 \\ x_5 & x_3 + x_5 & x_5 \\ x_5 & x_5 & x_4 + x_5 \end{bmatrix}, \quad L_2(x) = \begin{bmatrix} x_2 & x_4 & x_5 \\ x_4 & x_3 & -x_5 \\ x_5 & -x_5 & x_4 \end{bmatrix}$$

for which constant hyperbolic perturbation must be trivial (see Definition 4.7 in the next section and Theorems 3.5 and 3.6 in [22]). Applying Theorem 4.9 we show that not only constant hyperbolic perturbations but also more general hyperbolic perturbation is trivial.

Note that it is easy to see that

$$\begin{aligned}
\det S_{L_1}(x) &= x_2^2 x_3 + x_3^2 x_4 + x_4^2 x_2 - x_2 x_3^2 - x_3 x_4^2 - x_4 x_2^2 \\
&= -(x_2 - x_3)(x_3 - x_4)(x_4 - x_2), \\
\det S_{L_2}(x) &= x_2^2 x_3 + x_3^2 x_4 + x_5^2 x_2 - x_2 x_3^2 - x_4 x_2^2 - x_3 x_5^2 \\
&= (x_2 - x_3)(x_2 x_3 - x_2 x_4 - x_3 x_4 + x_5^2).
\end{aligned}$$

Let $\Theta_1 = (1,1,1,0)$ and $\Theta_2 = (2,2,1,0)$. It is obvious that $L_i(\Theta_i)$ is positive definite. Let us set

$$\tilde{L}_i(x) = L_i(\Theta_i)^{-1/2} L_i(x) L_i(\Theta_i)^{-1/2}.$$

It follows from Theorems 3.5, 3.6 in [22] and Lemma 4.3 that

Lemma 4.24. *Every characteristic of $\tilde{L}_i(x)$, $i = 1,2$ of order less than 3 is nondegenerate.*

To apply Theorem 4.9 to $\tilde{L}_i(x)$ we examine that

Lemma 4.25. $\det S_{\tilde{L}_i}(x)$, $i = 1,2$ *verifies (4.103).*

Proof. We first note that $\det S_{L_i}(x)$ verifies (4.103). The assertion for $S_{\tilde{L}_1}(x)$ is clear because $L_1(\Theta_1) = I$. To prove the assertion for $S_{\tilde{L}_2}(x)$ we note that

$$C = L_2(\Theta_2)^{-1/2} = \begin{bmatrix} \alpha & \beta & 0 \\ \beta & \alpha & 0 \\ 0 & 0 & 1 \end{bmatrix}, \quad \tilde{L}_2(x) = C L_2(x) C$$

with $\alpha > 0, \beta > 0$ and $\gamma = \alpha^2 - \beta^2 > 0$. Let x be so that $\det S_{L_2}(x) = 0$. Then there is a $H \in M_3^s(\mathbb{R})$, $H \neq O$ with zero diagonal entries such that $[L_2(x), H] = O$. Setting $\tilde{H} = C^{-1}HC$ it follows that

$$\tilde{L}_2(x)\tilde{H} - \tilde{H}\tilde{L}_2(x) = O.$$

Hence we have $[\tilde{L}_2(x), \tilde{H}^s] = O$ where $\tilde{H}^s$ is the symmetric part of $\tilde{H}$. It is easy to check that the diagonal entries of $\tilde{H}$ and hence those of $\tilde{H}^s$ are zero. Thus we conclude that $\det S_{\tilde{L}_2}(x) = 0$. Since $\det S_{L_2}(x)$ verifies (4.103) by Remark at the end of Sect. 4.5 we get $\det S_{\tilde{L}_2}(x) = c \det S_{L_2}(x)$ with a constant $c \neq 0$ and hence the assertion. □

4.9 Stability of Symmetric Systems Under Hyperbolic Perturbations

We start with

Definition 4.7. Let $R(x)$ be a hyperbolic perturbation to $\mathscr{L}(x)$ near $x = 0$. We say that the perturbation is trivial if there exist real analytic $A(x)$, $B(x)$ defined near the origin with $A(0)B(0) = I$ such that $A(x)\mathscr{P}(x)B(x)$ becomes symmetric.

In this section we prove that *generically* every hyperbolic perturbation of symmetric system $\mathscr{L}$

$$\mathscr{L}(x) = x_1 I + \sum_{j=2}^{n} F^j x_j, \quad F^j \in M_m^s(\mathbb{R})$$

is trivial if $\dim \mathscr{L}$ is enough large. As in Sect. 4.7 we identify $\mathscr{L}(x)$ with the subspace $\mathscr{L} = \{\mathscr{L}(x) \mid x \in \mathbb{R}^n\}$.

Theorem 4.10. *Assume $d_m - m + 3 \leq l \leq d_m$. Then in the $(d_m - l)(l - 1)$ dimensional Grassmannian of l dimensional subspaces of $M_m^s(\mathbb{R})$ containing the identity, the subset for which every hyperbolic perturbation is trivial is an open and dense subset.*

As in Sect. 4.5 we study $S_{\mathscr{L}}(x)$ for symmetric $\mathscr{L}(x)$ when $\dim \mathscr{L} = d_m - \nu$ where $1 \leq \nu \leq m - 3$. We first examine the representation matrix $S_{\mathscr{L}}(x)$. Let

$$V_m = \{H = (h_{ij}) \in M_m^s(\mathbb{R}) | h_{ii} = 0\}$$

and recall that $S_{\mathscr{L}}(x)$ is defined as the linear map between two d_{m-1} dimensional linear subspaces V_m and $W_m = M_m^{as}(\mathbb{R})$

$$V_m \ni H \mapsto [\mathscr{L}(x), H] = K \in W_m$$

where $M_m^{as}(\mathbb{R})$ denotes the set of all real $m \times m$ anti-symmetric matrices. Let us write

$$\mathscr{L}(x) = (\phi_j^i(x))_{1 \le i,j \le m}, \quad \phi_j^i(x) = \phi_i^j(x). \tag{4.105}$$

For $H \in V_m$ we write $\check{H} = {}^t(h_{12}, h_{13}, h_{23}, h_{14}, h_{24}, h_{34}, \ldots, h_{m-1m}) \in \mathbb{R}^{d_{m-1}}$. Then the equation $[\mathscr{L}(x), H] = K$ can be written as

$$S_{\mathscr{L}}(x)\check{H} = \check{K}$$

where $S_{\mathscr{L}}(x)$ is a $d_{m-1} \times d_{m-1}$ matrix. For instance when $m = 3$ we have

$$S_{\mathscr{L}}(x) = \begin{bmatrix} \phi_1^1(x) - \phi_2^2(x) & -\phi_3^2(x') & \phi_3^1(x') \\ -\phi_3^2(x') & \phi_1^1(x) - \phi_3^3(x) & \phi_2^1(x') \\ -\phi_3^1(x') & \phi_2^1(x') & \phi_2^2(x) - \phi_3^3(x) \end{bmatrix}. \tag{4.106}$$

We turn to the case $\mathscr{L}(x)$ is a $m \times m$ matrix. Let

$$\mathscr{L}(x) = \begin{bmatrix} L(x) & l(x') \\ {}^t l(x') & \phi_m^m(x) \end{bmatrix}$$

where $l(x') = {}^t(\phi_m^1(x'), \ldots, \phi_m^{m-1}(x'))$ and $L(x)$ stands for $\mathscr{L}(x)$ in (4.105) with $m - 1$. For $H \in V_m$ and $K \in W_m$ we write

$$H = \begin{bmatrix} H_1 & h \\ {}^t h & 0 \end{bmatrix}, \quad K = \begin{bmatrix} K_1 & k \\ {}^t k & 0 \end{bmatrix}$$

with $H_1 \in V_{m-1}$, $K_1 \in W_{m-1}$ and $h = {}^t(h_{1m}, \ldots, h_{m-1m})$. Then it is easy to see that the equation $[\mathscr{L}(x), H] = K$ is written as

$$\begin{bmatrix} S_L(x) & c(l) \\ c'(l) & L(x) - \phi_m^m I \end{bmatrix} \begin{bmatrix} \check{H}_1 \\ h \end{bmatrix} = \begin{bmatrix} \check{K}_1 \\ k \end{bmatrix} = \check{K}$$

and hence we get

$$S_{\mathscr{L}}(x) = \begin{bmatrix} S_L(x) & c(l) \\ c'(l) & L(x) - \phi_m^m I \end{bmatrix}. \tag{4.107}$$

Our aim in this section is to prove

Proposition 4.11. *Assume that $1 \le \nu \le m - 3$. Then in the Grassmannian $G_{d_m, I}^{d_m - \nu}$, the subset of $\mathscr{L}$ for which the condition (4.103) is fulfilled for $T^{-1}\mathscr{L}T$ with some $T \in O(m)$ is an open and dense subset.*

Here we use a parametrization of the Grassmannian $G^l_{d_m,I}$ used in Sect. 4.7. We set $\Delta = \{(i,i) | 1 \le i \le m\}$ and let $1 \le k \le m-1$. We first remark that

Lemma 4.26. *Assume that* $1 \le k \le m-1$. *Then one can find finitely many* $S_1, \dots, S_N \in O(m)$ *such that for any* $\mathscr{L} \in G^{d_m - k}_{d_m,I}$ *there is* $S_i \in \{S_1, \dots, S_N\}$ *so that* $S_i^{-1} \mathscr{L} S_i \in \Omega_\sigma$ *with some* σ *verifying* $\sigma(\{1, \dots, k\}) \cap \Delta = \emptyset$.

Proof. In this proof we denote $|C| = \max_{i,j} |c_{ij}|$ for a matrix $C = (c_{ij})$. Let $T_{pq}(\epsilon)$ be the orthogonal matrix obtained replacing p-th and q-th, $p < q$, rows of the identity matrix by

$$(0, \dots, 0, f(\epsilon), 0, \dots, 0, \epsilon, 0, \dots, 0), \quad (0, \dots, 0, -\epsilon, 0, \dots, 0, f(\epsilon), 0, \dots, 0)$$

where $\epsilon^2 + f(\epsilon)^2 = 1$. We show that it is enough to take $\{S_i\}$ as the set of all

$$K_1 K_2 \cdots K_m$$

where

$$K_j \in \{I, T_{pq}(\epsilon_i) \mid \epsilon_i = (C_i m^{2^{i-1}})^{-1}, i = 1, \dots, m, 1 \le p < q \le m\}$$

and $C_1 < C_2 < \cdots < C_m$ will be chosen suitably. Let $\mathscr{L} \in G^{d_m-k}_{d_m,I}$ and let $A_1, \dots, A_k$ define $\mathscr{L}$ so that $\mathscr{L}$ consists of all $X \in M^s_m(\mathbb{R})$ such that $\mathrm{Tr}\,(A_j X) = 0$, $1 \le j \le k$ where A_j are linearly independent and $\mathrm{Tr}\, A_j = 0$. We first note that we may assume

$(H)_\mu$: there is an injective $\tau : \{1, \dots, \mu\} \to \{(i,j) | 1 \le i < j \le m\}$ such that $\tau(i)$-th entry of A_j is zero unless $i = j$ and $\tau(j)$-th entry of A_j is 1, $|A_j| \le a_\mu m^{2^{\mu-1}}$ for $1 \le j \le \mu$ where $a_1 = 1$, $a_{\mu+1} = B a_\mu C_\mu$ with a fixed large B and $A_{\mu+1}, \dots, A_k$ are diagonal matrices.

In fact if some A_j has a non-zero off diagonal entry we may assume that the off diagonal $\tau(1)$-th entry of A_1 is 1 and $|A_1| \le 1$. Replacing A_j by $A_j - \alpha_j A_1$, $j \ne 1$, with suitable α_j one can assume that $\tau(1)$-th entry of A_j is zero if $j \ne 1$. A repetition of this argument gives the assertion. If $\mu = k$ then $\tau(\{1, \dots, k\}) \cap \Delta = \emptyset$ and there is nothing to prove. Then we may assume that $\mu \le k-1$. Let $A_{\mu+1} = \mathrm{diag}\,(\lambda_1, \dots, \lambda_m)$. Since $\mathrm{Tr}\, A_{\mu+1} = 0$ it is easy to see that there are at least $m-1$ pairs $(i,j), i < j$ such that

$$3|\lambda_i - \lambda_j| \ge |\lambda_r|, \quad r = 1, \dots, m.$$

Since $\mu \le m-2$ there exists such a (p,q) with $(p,q) \notin \tau(\{1, \dots, \mu\})$. Let us set

$$A_j(\epsilon_\mu) = T_{pq}(\epsilon_\mu)^{-1} A_j T_{pq}(\epsilon_\mu), \quad 1 \le j \le k$$

and note that $|A_j(\epsilon_\mu) - A_j| \le B_1 a_\mu C_\mu^{-1}$, $1 \le j \le \mu$. Choose C_μ so that $a_\mu C_\mu^{-1}$ is small enough then taking $\tilde{A}_j(\epsilon_\mu) = \sum_{i=1}^{\mu} c_{ji} A_i(\epsilon_\mu)$, $1 \le j \le \mu$, with a non singular $C = (c_{ji})$ we may suppose that $\tau(i)$-th entry of $\tilde{A}_j(\epsilon_\mu)$ is zero unless $i = j$ and $\tau(j)$-th entry of $\tilde{A}_j(\epsilon_\mu)$ is 1 and $|\tilde{A}_j(\epsilon_\mu)| \le 2|A_j|$. Note that the off diagonal entries of $A_{\mu+1}(\epsilon_\mu)$ are zero except for (p, q), (q, p)-th entries which are $\epsilon_\mu f(\epsilon_\mu)(\lambda_q - \lambda_p)$. Set

$$\tilde{A}_{\mu+1}(\epsilon_\mu) = \{\epsilon_\mu f(\epsilon_\mu)(\lambda_q - \lambda_p)\}^{-1} A_{\mu+1}(\epsilon_\mu)$$

and hence $|\tilde{A}_{\mu+1}(\epsilon_\mu)| \le B_2 C_\mu m^{2^{\mu-1}}$. Replacing $\tilde{A}_j(\epsilon_\mu)$ by $\tilde{A}_j(\epsilon_\mu) - \alpha_j \tilde{A}_{\mu+1}(\epsilon_\mu)$ with suitable α_j we can conclude that $\tau(\mu+1) = (p, q)$-th entry of $\tilde{A}_j(\epsilon_\mu)$ is zero for $1 \le j \le \mu$ and $|\tilde{A}_j(\epsilon_\mu)| \le a_{\mu+1} m^{2^\mu}$, $1 \le j \le \mu + 1$. By subtraction again we may suppose that $A_j(\epsilon_\mu)$, $j \ge \mu + 2$ are diagonal matrices and then we get to $(H)_{\mu+1}$. The rest of the proof is clear. □

Proof of Proposition 4.11. We first assume that $\mathscr{L} \in \Omega_\tau$ with $\tau(\{1, \dots, \nu\}) \cap \Delta = \emptyset$ and let $A = (A_1, \dots, A_\nu) \in U_\tau$ be the coordinate of $\mathscr{L}$. Let us denote

$$\mathscr{L}(x) = \sum_{j=1}^{n} K_j x_j = (\phi^i_j(x))$$

where $\{K_j\}$, $1 \le j \le n = d_m - \nu$ is a basis for $\mathscr{L}$ and set $g(x) = \det S_{\mathscr{L}}(x)$. Let $J_\tau = \{(i, j) | 1 \le i \le j \le m\} \setminus \tau(\{1, \dots, \nu\})$ and note that $\phi^i_j(x)$, $(i, j) \in J_\tau$ are linearly independent and $\Delta \subset J_\tau$. With $A_k = (a^{(k)}_{ij})$ it is clear that the equations $\phi^i_j(x) = 0$, $(i, j) \in J_\tau \setminus \Delta$ and $\mathrm{Tr}\,(A_k \mathscr{L}(x)) = 0$ define a plane

$$\sum_{j=1}^{m} a^{(k)}_{jj} \phi^j_j(x) = \sum_{j=1}^{m-1} a^{(k)}_{jj} (\phi^j_j(x) - \phi^m_m(x)) = 0, \quad 1 \le k \le \nu \tag{4.108}$$

and $S_{\mathscr{L}}(x)$ is diagonal matrix on the plane with the determinant

$$g(x) = \prod_{1 \le i < j \le m} (\phi^i_i(x) - \phi^j_j(x)). \tag{4.109}$$

We show that there is a polynomial $\pi(A)$ in $a^{(k)}_{jj}$, $1 \le k \le \nu$, $1 \le j \le m - 1$ such that if $\pi(A) \ne 0$ then no two $\phi^i_i(x) - \phi^j_j(x)$, $i < j$ are proportional on the plane (4.108). To simplify notations we write y_i for $\phi^i_i(x) - \phi^m_m(x)$ so that

$$g(y) = \prod_{1 \le i < j \le m-1} (y_i - y_j) y_1 \cdots y_{m-1}$$

provided that $y\tilde{A} = 0$ where $y = (y_1, \dots, y_{m-1})$ and $\tilde{A} = (a^{(k)}_{jj})$ which is a $(m - 1) \times \nu$ matrix. Suppose that some two $y_i - y_j$ are proportional on the plane $y\tilde{A} = 0$ and hence $yb = 0$ with some $b \in \mathbb{R}^{m-1}$ for every y with $y\tilde{A} = 0$.

Then it is clear that rank $(\tilde{A}, b)$ = rank $\tilde{A}$. Note that at most two components of b are the constant of the proportionality c and the other components are either 0 or 1 (at most two 1 appear). Take a $(\nu + 1) \times (\nu + 1)$ submatrix of $(\tilde{A}, b)$ and expand the determinant with respect to the last column. Equating the determinant to zero we get a linear relation of ν-minors of $\tilde{A}$ with coefficients which are either 1 or the proportional constant c. Since $\nu + 1 \le m - 2$ we have at least $m - 1$ such linear relations. Elimination of c gives a quadratic equation in ν-minors of $\tilde{A}$. Denote this equation by $\pi(A) = 0$. Then we conclude that the rank of the matrix $(\tilde{A}, b)$ is $\nu + 1$ if $\pi(A) \neq 0$. This shows that no two $y_i - y_j$ are proportional if $\pi(A) \neq 0$.

Let $g(x) = \prod g_j(x)^{r_j}$ be the irreducible factorization in $\mathbb{R}[x]$. Without restrictions we may assume that the plane $y\tilde{A} = 0$ is given by $y_b = f(y_a)$, after a linear change of coordinates y if necessary, where $y = (y_a, y_b)$ is a partition of the coordinates y. Then we have

$$\prod g_j(y_a, f(y_a))^{r_j} = \prod p_i(y_a)$$

where $p_i(y_a)$ are linear in y_a and no two $p_i(y_a)$ are proportional if $\pi(A) \neq 0$. Then it follows that $r_j = 1$ and $g_j(y_a, f(y_a))$ is a product of some $p_i(y_a)$'s;

$$g_j(y_a, f(y_a)) = \prod_{i \in I_j} p_i(y_a).$$

From this it is obvious that $\{g_j(y_a, f(y_a)) = 0\}$ contains a regular point. Then it follows that $\{g_j(x) = 0\}$ contains a regular point. This shows that, in U_τ, the set of A such that $S_{\mathscr{L}}(x)$ does not verify (4.103) is contained in an algebraic set. We now study $\mathscr{L} \in \Omega_\sigma$ with $\sigma(\{1, \dots, \nu\}) \cap \Delta \neq \emptyset$. By Lemma 4.26 there is $S_i \in O(m)$ such that $S_i^{-1} \mathscr{L} S_i \in \Omega_\tau$ with some τ verifying $\tau(\{1, \dots, \nu\}) \cap \Delta = \emptyset$. Since $\{S_i\}$ is a finite set the proof is clear. □

Proof of Theorem 4.10. Let $d_m - m + 3 \le l \le d_m$. Then Theorem 4.10 follows immediately from Propositions 4.10, 4.11 and Corollary 4.2. □

4.10 Some Special Cases

In the case $m = 3$ one can improve Theorem 4.10.

Theorem 4.11. *Assume that $m = 3$ and $4 \le l \le 6 = d_3$. Then in the $(6-l)(l-1)$ dimensional Grassmannian of l dimensional subspaces of $M_3^s(\mathbb{R})$ containing the identity, the subset for which every hyperbolic perturbation is trivial is an open and dense subset.*

We assume $m = 3$ throughout the section. Let $\mathscr{L} \in G_{6,I}^l$ for $l = 4$ or 5. Taking a basis $\{K_j\}$ for $\mathscr{L}$, $\mathscr{L}$ is the image of

$$\mathscr{L}(x) = \sum_{j=1}^{n} K_j x_j.$$

We first study the case $l = 5$.

Lemma 4.27. *In the Grassmannian $G^5_{6,l}$, the subset of $\mathscr{L}$ for which the condition (4.103) is fulfilled for $T^{-1}\mathscr{L}T$ with some $T \in O(m)$ is an open and dense subset.*

Proof. Let $A = A_1 \in U_\sigma$ be the coordinate of $\mathscr{L}$ and assume that $\sigma(1) \cap \Delta = \emptyset$ so that the diagonal entries of $\mathscr{L}(x)$ are linearly independent. Considering $T^{-1}\mathscr{L}(x)T$ with suitable permutation matrix T, if necessary, we may assume that $\sigma(1) = (1, 2)$ so that with $\mathscr{L}(x) = (\phi^i_j(x))$ we have from $\mathrm{Tr}\,(A\mathscr{L}(x)) = 0$ that

$$-2\phi^1_2(x) = a_{11}(\phi^1_1 - \phi^3_3) + a_{22}(\phi^2_2 - \phi^3_3) + 2a_{13}\phi^1_3 + 2a_{23}\phi^2_3.$$

From (4.106), simplifying notations, it is enough to study

$$S(x,y) = \begin{bmatrix} x_1 - x_2 & -y_1 & y_2 \\ -y_1 & x_1 & \phi(x,y) \\ -y_2 & \phi(x,y) & x_2 \end{bmatrix}$$

where $\phi(x,y) = a_1x_1 + a_2x_2 + b_1y_1 + b_2y_2$. We show that if $a_1 + a_2 \neq 1$ and $4a_1a_2 - 1 \neq 0$ then the condition (4.103) is fulfilled. We first assume that $x_1x_2 - \phi(x,0)^2$ is irreducible. Note that $g(x,y) = \det S(x,y)$ is then irreducible. Indeed if $g(x,y)$ were reducible so that $g(x,y) = h(x,y)k(x,y)$ then from $g(x,0) = (x_1 - x_2)\psi(x)$ with $\psi(x) = x_1x_2 - \phi(x,0)^2$ we may suppose that

$$h(x,y) = \psi(x) + p(x,y), \quad k(x,y) = x_1 - x_2 + q(y)$$

where $p(x,0) = 0$, $q(y) = \alpha y_1 + \beta y_2$. Equating the coefficients of y_j in both sides of $g(x,y) = h(x,y)k(x,y)$ we see that $\alpha\psi(x)$, $\beta\psi(x)$ have a factor $x_1 - x_2$ which implies that $q = 0$. This gives $g(x,y) = h(x,y)(x_1 - x_2)$ which is a contradiction. Thus g is irreducible. It is clear that $\{g(x,0) = 0\}$ has a regular point and hence so does $\{g(x,y) = 0\}$. This proves the assertion.

Assume now that $\psi(x) = x_1x_2 - \phi(x,0)^2$ is reducible. From the assumption $4a_1a_2 - 1 \neq 0$ it follows that $\psi(x)$ has no multiple factor. Note that $a_1 + a_2 \neq \pm 1$ implies that $\psi(x)$ and $x_1 - x_2$ are relatively prime. The rest of the proof is a repetition of the last part of the proof of Proposition 4.11. □

We turn to the case $l = 4$. We show that

Lemma 4.28. *Assume that $l = 4$ and every double characteristic of $\mathscr{L}(x)$ is nondegenerate. Then the condition (4.103) is fulfilled for $T^{-1}\mathscr{L}(x)T$ with a suitable $T \in O(3)$.*

Proof. Following the proof of Theorems 3.5 and 3.6 in [22] we choose a specific basis for $\tilde{\mathscr{L}} = T^{-1}\mathscr{L}T$ with suitably chosen $T \in O(3)$ and show that (4.103) is fulfilled for $\tilde{\mathscr{L}}$ using this basis. From the proof of Theorem 3.3 in [22], if every double characteristic of $\mathscr{L}$ is nondegenerate, then only two cases occur, that is $\mathscr{L}$ has either four nondegenerate double characteristics or two nondegenerate double characteristics.

We first treat the case that $\mathscr{L}$ has four nondegenerate characteristics. Choosing a suitable $T \in O(3)$ we see from [22] that $A^{\pm} = \alpha_{\pm} \otimes \alpha_{\pm}$ and $B^{\pm} = \beta_{\pm} \otimes \beta_{\pm}$ is a basis for $\tilde{\mathscr{L}} = T^{-1}\mathscr{L}T$ where $\alpha_{\pm} = (a, \pm a, 1)$, $\beta_{\pm} = (b, \pm b, 1)$ and $a \neq b$, $ab \neq 0$. Now we can write

$$\tilde{\mathscr{L}}(x) = A^{+}x_1 + A^{-}x_2 + B^{+}x_3 + B^{-}x_4.$$

With $X = x_1 + x_2$, $Y = x_1 - x_2$, $Z = x_3 + x_4$, $W = x_3 - x_4$ we have

$$\tilde{\mathscr{L}} = \begin{bmatrix} a^2X + b^2Z & a^2Y + b^2W & aX + bZ \\ a^2Y + b^2W & a^2X + b^2Z & aY + bW \\ aX + bZ & aY + bW & X + Z \end{bmatrix}. \tag{4.110}$$

Therefore it follows from (4.106) and (4.110) that

$$S_{\tilde{\mathscr{L}}} = \begin{bmatrix} 0 & -aY - bW & aX + bZ \\ -aY - bW & cX + dZ & a^2Y + b^2W \\ -aX - bZ & a^2Y + b^2W & cX + dZ \end{bmatrix}$$

where $c = a^2 - 1$, $d = b^2 - 1$. Let $\tilde{g} = \det S_{\tilde{\mathscr{L}}}$. On the plane $a^2Y + b^2W = 0$, that is, if $W = -a^2Y/b^2 = eY$ we get

$$\tilde{g} = (cX + dZ)(aX + bZ + (a + be)Y)(aX + bZ - (a + be)Y).$$

Note that $a + be \neq 0$ because $a \neq b$ and no two factors in the right-hand side are proportional. Now, as the end of the proof of Proposition 4.11, it is easy to conclude that $\tilde{g}$ satisfies (4.103).

We next study the case $\mathscr{L}$ has two nondegenerate double characteristics. With a suitable $T \in O(3)$ we see that $\tilde{\mathscr{L}} = T^{-1}\mathscr{L}T$ contains $K^{\pm} = \alpha_{\pm} \otimes \alpha_{\pm}$ with $\alpha_{\pm} = (a, \pm a, 1)$, $a \neq 0$, which are intersections with $M_3^s(2; \mathbb{R})$. Since $\tilde{\mathscr{L}}$ contains the identity, as a member of basis for $\tilde{\mathscr{L}}$, one can take K_3

$$K_3 = \begin{bmatrix} 0 & 0 & -2a \\ 0 & 0 & 0 \\ -2a & 0 & 2(a^2 - 1) \end{bmatrix}$$

because $K^+ + K^- + K_3 = 2a^2 I$. The last member of basis for $\tilde{\mathscr{L}}$ can then be chosen of the form

$$K_4 = \begin{bmatrix} 0 & 0 & 0 \\ 0 & \lambda & \mu \\ 0 & \mu & \nu \end{bmatrix}.$$

Thus with $X = x_1 + x_2$, $Y = x_1 - x_2$, $Z = x_3$, $W = x_4$ and $c = a^2 - 1$ the matrix $K^+ x_1 + K^- x_2 + K_3 x_3 + K_4 x_4$ can be written

$$\tilde{\mathscr{L}} = \begin{bmatrix} a^2 X & a^2 Y & aX - 2aZ \\ a^2 Y & a^2 X + \lambda W & aY + \mu W \\ aX - 2aZ & aY + \mu W & X + 2cZ + \nu W \end{bmatrix}. \quad (4.111)$$

We examine if there are other double characteristics, that is, if $\tilde{\mathscr{L}}$ is of rank 1 for some (X, Y, Z, W) with $Z^2 + W^2 \neq 0$. It is not difficult to see that six 2-minors of (4.111) vanish for such (X, Y, Z, W) if and only if the equation

$$4a^2 Z^2 + 2(a^2 + 1)\lambda Z W + (\lambda\nu - \mu^2) W^2 = 0$$

has a real solution $(Z, W) \neq (0, 0)$. Thus in order that $\tilde{\mathscr{L}}$ has two nondegenerate double characteristics it is necessary and sufficient that

$$4a^2 \lambda\nu > 4a^2 \mu^2 + (a^2 + 1)^2 \lambda^2. \quad (4.112)$$

In particular λ and ν have the same sign. From (4.111) and (4.106) it follows that

$$S_{\tilde{\mathscr{L}}} = \begin{bmatrix} -\lambda W & -aY - \mu W & aX - 2aZ \\ -aY - \mu W & cX - 2cZ - \nu W & a^2 Y \\ -aX + 2aZ & a^2 Y & cX - 2cZ + (\lambda - \nu) W \end{bmatrix}.$$

If $c \neq 0$ then we consider $\tilde{g} = \det S_{\tilde{\mathscr{L}}}$ on $W = 0$ so that

$$\tilde{g} = (cX - 2cZ)(aX - 2aZ + aY)(aX - 2aZ - aY).$$

The same argument as before proves that (4.103) is verified for $\tilde{g}$. If $c = 0$ and hence $a^2 = 1$ then

$$\begin{aligned} \tilde{g} &= W(-\nu(aX - 2aZ)^2 + \lambda(\nu^2 - \mu^2)\alpha^{-1} Y^2 + (\lambda - \nu)\alpha(W - a\mu\alpha^{-1} Y)^2) \\ &= W h(X, Y, Z, W) \end{aligned}$$

where $\alpha = \lambda\nu - \mu^2$. From (4.112) it follows that $\alpha > 0$ and $\nu^2 - \mu^2 > 0$ because $\nu^2 + \lambda^2 \geq \lambda\nu > \mu^2 + \lambda^2$. Then the quadratic form h is indefinite and hence $\{h = 0\}$ contains a regular point. This proves the assertion. □

Proof of Theorem 4.11. If $l = 6$ then the assertion follows from Theorem 4.2 in [53]. If $l = 5$, combining Proposition 4.10 and Lemma 4.27 we get the result by Corollary 4.2. Let $l = 4$. Then by virtue of Proposition 4.10 and Lemma 4.28 one can apply Corollary 4.2 to get the assertion. □

4.11 Concluding Remarks

In [25], F. John discovered mysterious phenomena on the characteristics of hyperbolic systems. He considered the system P of 3 second order equations in 4 independent variables, which is the system discussed in Example 4.4. He showed that any system $\tilde{P}$ near P is hyperbolic if and only if $\tilde{P}$ has 4 double characteristics near the double characteristics of P. In [26] he showed that P is strongly hyperbolic. In [22], L. Hörmander studied hyperbolic systems with nondegenerate double characteristics. In particular, it was proved there that nondegenerate double characteristics are stable, that is we can not remove nondegenerate double characteristics by hyperbolic perturbations which shows a complexity of hyperbolic systems compared with the scalar case (see [58]).

For first order systems the notion of nondegenerate characteristics of any order is introduced in [53, 54]. We adapt this definition for higher order systems through the associated first order system in [57]. According to this definition, simple characteristics are nondegenerate characteristics of order 1 and nondegenerate double characteristics coincide with those studied in [4, 17, 22, 25, 26, 48].

Theorem 4.3 (in the real case) was proved for analytic first order systems in [53] and for systems with nondegenerate double characteristics in [22]. The results about hyperbolic perturbations of symmetric systems with constant coefficients are found in [54].

Problem. Generalize Theorem 4.4 to C^∞ $m \times m$ matrix valued $\mathscr{P}(x)$.

Problem. Determine the minimal l such that Theorem 4.10 holds.

Problem. Determine the minimal $\dim_{\mathbb{R}} \{L(x) \mid x \in \mathbb{R}^n\}$ such that Lemma 4.9 holds. In the real valued case it is known that 5 is optimal when $m = 3$ (see [59]).

References

1. L.V. Ahlfors, *Complex Analysis* (McGraw-Hill, New York, 1966)
2. M.F. Atiyah, R. Bott, L. Gårding, Lacunas for hyperbolic differential operators with constant coefficients, I. Acta Math. **124**, 109–189 (1970)
3. V.I. Arnold, Matrices depending on parameters. Uspehi Math. Nauk **26**, 101–114 (1971)
4. E. Bernardi, T. Nishitani, Remarks on symmetrization of 2 × 2 systems and the characteristic manifolds. Osaka J. Math. **29**, 129–134 (1992)
5. J. Chazarain, Opérateurs hyperboliques à caractéristiques de multiplicité constante. Ann. Inst. Fourier **24**, 173–202 (1974)
6. F. Colombini, T. Nishitani, Two by two strongly hyperbolic systems and the Gevrey classes. Ann. Univ. Ferrara Sci. Mat. Suppl. **XLV**, 291–312 (1999)
7. P. D'Ancona, T. Kinoshita, S. Spagnolo, On the 2 by 2 weakly hyperbolic systems. Osaka J. Math. **45**, 921–939 (2008)
8. P. D'Ancona, S. Spagnolo, On pseudosymmetric hyperbolic systems. Ann. Sc. Norm. Super. Pisa **25**, 397–417 (1997)
9. P. D'Ancona, S. Spagnolo, A remark on uniformly symmetrizable systems. Adv. Math. **158**, 18–25 (2001)
10. Y. Demay, Paramétrix pour des systèmes hyperboliques du premier ordre à multiplicité constante. J. Math. Pures Appl. **56**, 393–422 (1977)
11. N. Dencker, Preparation theorems for matrix valued functions. Ann. Inst. Fourier **43**, 865–892 (1993)
12. H. Flaschka, G. Strang, The correctness of the Cauchy problem. Adv. Math. **6**, 347–379 (1971)
13. S. Friedland, J. Robbin, J. Sylvester, On the crossing rule. Commun. Pure Appl. Math. **37**, 19–37 (1984)
14. K.O. Friedrichs, Symmetric hyperbolic differential equations. Commun. Pure Appl. Math. **7**, 345–392 (1954)
15. K.O. Friedrichs, P.D. Lax, On symmetrizable differential operators, in *Proceedings of the Symposium on Pure Mathematics*, ed. by A.P. Calderon. Singular Integrals, vol. 10 (American Mathematical Society, Providence, 1967), pp. 128–137
16. L. Gårding, Linear hyperbolic partial differential equations with constant coefficients. Acta Math. **85**, 1–62 (1951)
17. D.C. Hernquist, Smoothly symmetrizable hyperbolic systems of partial differential equations. Math. Scand. **61**, 262–275 (1987)
18. L. Hörmander, The Cauchy problem for differential equations with double characteristics. J. Anal. Math. **32**, 118–196 (1977)
19. L. Hörmander, *The Analysis of Linear Partial Differential Operators II* (Springer, Berlin, 1983)
20. L. Hörmander, *The Analysis of Linear Partial Differential Operators I* (Springer, Berlin, 1983)

T. Nishitani, *Hyperbolic Systems with Analytic Coefficients*, Lecture Notes in Mathematics 2097, DOI 10.1007/978-3-319-02273-4,

21. L. Hörmander, *The Analysis of Linear Partial Differential Operators III* (Springer, Berlin, 1985)
22. L. Hörmander, Hyperbolic systems with double characteristics. Commun. Pure Appl. Math. **46**, 261–301 (1993)
23. V.Ja. Ivrii, V.M. Petkov, Necessary conditions for the Cauchy problem for non strictly hyperbolic equations to be well posed. Uspehi Mat. Nauk **29**, 3–70 (1974)
24. V.Ja. Ivrii, *Linear Hyperbolic Equations in Partial Differential Equations IV* (Springer, Berlin, 1988), pp. 149–235
25. F. John, Algebraic conditions for hyperbolicity of systems of partial differential equations. Commun. Pure Appl. Math. **31**, 89–106 (1978)
26. F. John, Addendum to: algebraic conditions for hyperbolicity of systems of partial differential equations. Commun. Pure Appl. Math. **31**, 787–793 (1978)
27. K. Kajitani, Strongly hyperbolic systems with variable coefficients. Publ. RIMS Kyoto Univ. **9**, 597–612 (1974)
28. K. Kajitani, Cauchy problem for non-strictly hyperbolic systems. Publ. RIMS Kyoto Univ. **15**, 519–550 (1979)
29. K. Kajitani, T. Nishitani, *The Hyperbolic Cauchy Problem*. Lecture Notes in Mathematics, vol. 1505 (Springer, Berlin, 1991)
30. K. Kasahara, M. Yamaguti, Strongly hyperbolic systems of linear partial differential equations with constant coefficients. Mem. Coll. Sci. Univ. Kyoto Ser. A **33**, 1–23 (1960)
31. T. Kusaba, *Topics in Matrix Theory (in Japanese)* (Shokabo, Tokyo, 1979)
32. N.D. Kutev, V.M. Petkov, First order regularly hyperbolic systems. Annu. Univ. Sofia Fac. Math. Méc. **67**, 375–389 (1976)
33. P.D. Lax, Asymptotic solutions of oscillatory initial value problems. Duke Math. J. **24**, 627–646 (1957)
34. P.D. Lax, The multiplicity of eigenvalues. Bull. Am. Math. Soc. **6**, 213–214 (1982)
35. J. Leray, *Hyperbolic Differential Equations* (Institute for Advanced Study, Princeton, 1953)
36. W. Matsumoto, On the conditions for the hyperbolicity of systems with double characteristic roots I. J. Math. Kyoto Univ. **21**, 47–84 (1981)
37. W. Matsumoto, On the conditions for the hyperbolicity of systems with double characteristic roots II. J. Math. Kyoto Univ. **21**, 251–271 (1981)
38. W. Matsumoto, H. Yamahara, Necessary conditions for strong hyperbolicity of first order systems. Journées "Équations aux Dérivées Partielles" (Saint Jean de Monts, 1989), Exp. No. VIII, 16 pp., École Polytech., Palaiseau, 1989
39. W. Matsumoto, Levi condition for general systems, in *Physics on Manifolds* (Kluwer, Dordrecht, 1992), pp. 303–307
40. W. Matsumoto, Normal form of systems of partial differential and pseudo differential operators in formal symbol classes. J. Math. Kyoto Univ. **34**, 15–40 (1994)
41. J. Milnor, *Singular Points of Complex Hypersurfaces* (Princeton University Press, Princeton, 1968)
42. L. Mencherini, S. Spagnolo, Well-posedness of 2×2 system with C^∞ coefficients, in *Hyperbolic Problems and Related Topics* (International Press, Somerville, 2003), pp. 235–241
43. S. Mizohata, Some remarks on the Cauchy problem. J. Math. Kyoto Univ. **1**, 109–127 (1961)
44. S. Mizohata, *The Theory of Partial Differential Equations* (Cambridge University Press, Cambridge, 1973)
45. S. Mizohata, *On Hyperbolic Matrices*. Frontiers in Pure and Applied Mathematics (North-Holland, Amsterdam, 1991), pp. 247–265
46. T. Nishitani, The Cauchy problem for weakly hyperbolic equations of second order. Commun. Partial Differ. Equ. **5**, 1273–1296 (1980)
47. T. Nishitani, A necessary and sufficient condition for the hyperbolicity of second order equations with two independent variables. J. Math. Kyoto Univ. **24**, 91–104 (1984)
48. T. Nishitani, On strong hyperbolicity of systems, in *Hyperbolic Equations*. Research Notes in Mathematics, vol. 158 (Longman, New York, 1987), pp. 102–114

49. T. Nishitani, Une condition nécessaire pour systèmes hyperboliques. Osaka J. Math. **26**, 71–88 (1989)
50. T. Nishitani, Necessary conditions for strong hyperbolicity of first order systems. J. Anal. Math. **61**, 181–229 (1993)
51. T. Nishitani, On localizations of a class of strongly hyperbolic systems. Osaka J. Math. **32**, 41–69 (1995)
52. T. Nishitani, Symmetrization of hyperbolic systems with real constant coefficients. Sc. Norm. Super. Pisa **21**, 97–130 (1994)
53. T. Nishitani, Symmetrization of hyperbolic systems with non-degenerate characteristics. J. Funct. Anal. **132**, 251–272 (1995)
54. T. Nishitani, Stability of symmetric systems under hyperbolic perturbations. Hokkaido Math. J. **26**, 509–527 (1997)
55. T. Nishitani, Hyperbolicity of two by two systems with two independent variables. Commun. Partial Differ. Equ. **23**, 1061–1110 (1998)
56. T. Nishitani, Strongly hyperbolic systems of maximal rank. Publ. Res. Inst. Math. Sci. **33**, 765–773 (1997)
57. T. Nishitani, Hyperbolic systems with nondegenerate characteristics, in *Hyperbolic Differential Operators and Related Problems*. Lecture Notes in Pure and Applied Mathematics, vol. 233 (Marcel Dekker, New York, 2003), pp. 7–29
58. W. Nuij, A note on hyperbolic polynomials. Math. Scand. **23**, 69–72 (1968)
59. Y. Oshime, Canonical forms of 3×3 strongly hyperbolic systems with real constant coefficients. J. Math. Kyoto Univ. **31**, 937–982 (1991)
60. V.M. Petkov, Necessary conditions for the correctness of the Cauchy problem for nonsymmetrizable hyperbolic systems. Trudy Sem. Petrovsk. **1**, 211–236 (1975)
61. V.M. Petkov, The parametrix of the Cauchy problem for nonsymmetrizable hyperbolic systems with characteristics of constant multiplicity. Trudy Moskov. Mat. Obshch. **37**, 3–47 (1978)
62. V.M. Petkov, Microlocal forms for hyperbolic systems. Math. Nachr. **93**, 117–131 (1979)
63. J. Rauch, *Partial Differential Equations, Graduate Texts in Mathematics*, vol. 128 (Springer, Berlin, 1991)
64. A. Shadi Tahvildar-Zadeh, Relativistic and nonrelativistic elastodynamics with small shear strains. Ann.de l'I.H.P. Phys. Théorique **69**, 275–307 (1998)
65. S. Tarama, Une note sur les systèmes hyperboliques uniformément diagonalisable. Mem. Fac. Eng. Kyoto Univ. **56**, 9–18 (1993)
66. J. Vaillant, Symétrisabilité des matrices localisées d'une matrice fortement hyperbolique. Ann. Sc. Norm. Super. Pisa **5**, 405–427 (1978)
67. J. Vaillant, Sysèmes hyperboliques à multiplicité constante et dont le rang puet varier, in *Recent Developments in Hyperbolic Equations*. Research Notes in Mathematics, vol. 183 (Longman, New York, 1988), pp. 340–366
68. J. Vaillant, Diagonalizable complex systems, reduced dimension and hermitian systems, I, in *Hyperbolic Problems and Related Topics*. Graduate Series in Analysis (International Press, Somerville, 2003), pp. 409–421
69. J. Vaillant, Diagonalizable complex systems, reduced dimension and hermitian systems, II. Pliska Stud. Math. Bulgar. **15**,131–148 (2003)
70. S. Wakabayashi, On the Cauchy problem for hyperbolic operators of second order whose coefficients depend only on the time variable. J. Math. Soc. Jpn. **62**, 95–133 (2010)
71. W. Wasow, *Asymptotic Expansions for Ordinary Differential Equations* (Interscience, New York, 1965)
72. W. Wasow, On holomorphically similar matrices. J. Math. Anal. Appl.**4**, 202–206 (1962)
73. H. Yamahara, On the strongly hyperbolic systems II -A reduction of hyperbolic matrices. J. Math. Kyoto Univ. **29**, 529–550 (1989)

Index

T. Nishitani, *Hyperbolic Systems with Analytic Coefficients*, Lecture Notes in Mathematics 2097, DOI 10.1007/978-3-319-02273-4,

LECTURE NOTES IN MATHEMATICS

Springer

Edited by J.-M. Morel, B. Teissier; P.K. Maini

Editorial Policy (for the publication of monographs)

1. Lecture Notes aim to report new developments in all areas of mathematics and their applications - quickly, informally and at a high level. Mathematical texts analysing new developments in modelling and numerical simulation are welcome.
   Monograph manuscripts should be reasonably self-contained and rounded off. Thus they may, and often will, present not only results of the author but also related work by other people. They may be based on specialised lecture courses. Furthermore, the manuscripts should provide sufficient motivation, examples and applications. This clearly distinguishes Lecture Notes from journal articles or technical reports which normally are very concise. Articles intended for a journal but too long to be accepted by most journals, usually do not have this "lecture notes" character. For similar reasons it is unusual for doctoral theses to be accepted for the Lecture Notes series, though habilitation theses may be appropriate.

2. Manuscripts should be submitted either online at www.editorialmanager.com/lnm to Springer's mathematics editorial in Heidelberg, or to one of the series editors. In general, manuscripts will be sent out to 2 external referees for evaluation. If a decision cannot yet be reached on the basis of the first 2 reports, further referees may be contacted: The author will be informed of this. A final decision to publish can be made only on the basis of the complete manuscript, however a refereeing process leading to a preliminary decision can be based on a pre-final or incomplete manuscript. The strict minimum amount of material that will be considered should include a detailed outline describing the planned contents of each chapter, a bibliography and several sample chapters.
   Authors should be aware that incomplete or insufficiently close to final manuscripts almost always result in longer refereeing times and nevertheless unclear referees' recommendations, making further refereeing of a final draft necessary.
   Authors should also be aware that parallel submission of their manuscript to another publisher while under consideration for LNM will in general lead to immediate rejection.

3. Manuscripts should in general be submitted in English. Final manuscripts should contain at least 100 pages of mathematical text and should always include
   - a table of contents;
   - an informative introduction, with adequate motivation and perhaps some historical remarks: it should be accessible to a reader not intimately familiar with the topic treated;
   - a subject index: as a rule this is genuinely helpful for the reader.

   For evaluation purposes, manuscripts may be submitted in print or electronic form (print form is still preferred by most referees), in the latter case preferably as pdf- or zipped ps-files. Lecture Notes volumes are, as a rule, printed digitally from the authors' files. To ensure best results, authors are asked to use the LaTeX2e style files available from Springer's web-server at:

   ftp://ftp.springer.de/pub/tex/latex/svmonot1/ (for monographs) and
   ftp://ftp.springer.de/pub/tex/latex/svmultt1/ (for summer schools/tutorials).

Additional technical instructions, if necessary, are available on request from lnm@springer.com.

4. Careful preparation of the manuscripts will help keep production time short besides ensuring satisfactory appearance of the finished book in print and online. After acceptance of the manuscript authors will be asked to prepare the final LaTeX source files and also the corresponding dvi-, pdf- or zipped ps-file. The LaTeX source files are essential for producing the full-text online version of the book (see http://www.springerlink.com/openurl.asp?genre=journal&issn=0075-8434 for the existing online volumes of LNM). The actual production of a Lecture Notes volume takes approximately 12 weeks.
5. Authors receive a total of 50 free copies of their volume, but no royalties. They are entitled to a discount of 33.3 % on the price of Springer books purchased for their personal use, if ordering directly from Springer.
6. Commitment to publish is made by letter of intent rather than by signing a formal contract. Springer-Verlag secures the copyright for each volume. Authors are free to reuse material contained in their LNM volumes in later publications: a brief written (or e-mail) request for formal permission is sufficient.

Addresses:

Professor J.-M. Morel, CMLA,
École Normale Supérieure de Cachan,
61 Avenue du Président Wilson, 94235 Cachan Cedex, France
E-mail: morel@cmla.ens-cachan.fr

Professor B. Teissier, Institut Mathématique de Jussieu,
UMR 7586 du CNRS, Équipe "Géométrie et Dynamique",
175 rue du Chevaleret
75013 Paris, France
E-mail: teissier@math.jussieu.fr

For the "Mathematical Biosciences Subseries" of LNM:

Professor P. K. Maini, Center for Mathematical Biology,
Mathematical Institute, 24-29 St Giles,
Oxford OX1 3LP, UK
E-mail: maini@maths.ox.ac.uk

Springer, Mathematics Editorial, Tiergartenstr. 17,
69121 Heidelberg, Germany,
Tel.: +49 (6221) 4876-8259

Fax: +49 (6221) 4876-8259
E-mail: lnm@springer.com